Progress in Bioengineering

Progress in Bioengineering

Proceedings of an international seminar
held on the occasion of the
25th anniversary
of the
Strathclyde Bioengineering Unit, Glasgow
in September 1988

Edited by

J P Paul, J C Barbenel, J M Courtney and R M Kenedi

Editorial Associate

E M Smith

Bioengineering Unit
Wolfson Centre
University of Strathclyde

Adam Hilger, Bristol and New York

British Library Cataloguing in Publication Data

Progress in Bioengineering
   1. Bioengineering
   I. Paul, J. P. (John P)   II. Series
   610′.28

   ISBN 0-85274-085-9

Library of Congress Cataloging-in-Publication Data are available

Published under the Adam Hilger imprint by IOP Publishing Ltd
Techno House, Redcliffe Way, Bristol BS1 6NX, England
335 East 45th Street, New York, NY 10017-3438, USA

Printed in Great Britain by Galliard (Printers) Ltd, Great Yarmouth

Session Chairmen and Chairmen Associate

Session 1   Artificial Organs and Biomaterials: presented in Chapters 2 to 8

> *Chairman*   A C Kennedy
> *Chairman Associate*   J C Barbenel

Session 2   Rehabilitation: presented in Chapters 10 to 15

> *Chairman*   G Murdoch
> *Chairman Associate*   R M Kenedi

Session 3   Biomechanics: presented in Chapters 17 to 32

> *Chairman*   D Hamblen
> *Chairman Associate*   J P Paul

Session 4   Prosthetics and Orthotics: presented in Chapters 34 to 40

> *Chairman*   D C Simpson
> *Chairman Associate*   J Hughes

Session 5   Technological Advances: presented in Chapters 42 to 49

> *Chairman*   W H Reid
> *Chairman Associate*   J S Orr

Contents

x   *Contents*

Preface

This, the seventh volume in the series of the Strathclyde Bioengineering Seminars, somewhat like the seventh son of a seventh son is rather special. As its forerunners it surveys the 'state of the art' on the application of science and technology to problems in health care in selected fields. However, commemorating as it does the quarter centenary of the Bioengineering Unit at Strathclyde, the fields selected for overview were those which the Unit made very much its own. Further, the participating presenters were by invitation either distinguished alumni of the Unit or similarly distinguished close associates who through the past twenty-five years have helped to shape and orientate the activities of the Unit.

This 'family' atmosphere was further emphasised by the Unit's Reunion which perorated the proceedings. In this the Adam Thomson lecture became a genuine 'family' affair being given by its original instigator, the undersigned and in the presence of Emeritus Professor Adam Thomson whose wholehearted support for, and generous encouragement of, the Unit it commemorates.

The importance of the 25th Jubilee of the Unit was further emphasised by Strathclyde's Vice-Chancellor Sir Graham Hills consenting to confer the honorary degree of DSc on Professor Horst Klinkmann of the Wilhelm-Pieck University, Rostock as part of the reunion proceedings. This again was a 'family' celebration by its formal recognition of the long and fruitful collaborative association between Professor Klinkmann's department and the Bioengineering Unit.

As on many occasions before it is a particular pleasure and privilege to acknowledge the Bioengineering Unit's sense of indebtedness to the bodies listed on page xii as Associated Organisations for their generous and unstinting sponsorship of the social events and of the seminar as a whole, and particularly for the support of the University.

Listed last but regarded as foremost, grateful thanks are offered to present and past staff, students and associates for their contribution to the Unit in general and to this seminar in particular. Events, activities, even Units are ephemeral—in the ultimate only people matter.

R M Kenedi
J P Paul
Glasgow, 1989

Associated Organisations

Akzo, Wuppertal

Bio-Flo Ltd, Glasgow

City of Glasgow

Convatec Wound Healing Research Institute, Clwyd

Devro Ltd, Moodiesburn

Ethicon Ltd, Edinburgh

Gambro AB, Lund

Greater Glasgow Health Board

J E Hanger & Co. Ltd, London

Howmedica (UK) Ltd, London

Hugh Steeper Ltd, London

Hydro Polymers Ltd, Newton Aycliffe

Johnson & Johnson Orthopaedics Ltd, New Milton

Neurotech Ltd, London

Norit (UK) Ltd, Glasgow

Otto Bock Orthopaedics (UK), Egham

Penny & Giles Blackwood Ltd, Gwent

Scottish Development Agency, Glasgow

Vascutek Ltd, Paisley

Zimmer–Deloro, Swindon

1

BIOENGINEERING - THE FUTURE

J P Paul

The remit of this chapter was that the author should predict the future of Bioengineering in his own institution, the Bioengineering Unit in the University of Strathclyde. One way of predicting in advance is to look, for instance as in weather forecasting, if it rained yesterday, if it is raining today, it will probably rain tomorrow also. This may be a local motto in Glasgow but it is one of the principles of this chapter. An extrapolation from the past to the present and hence to the future is fraught with a very big error between the straightforward progression, the pessimistic and the optimistic as shown in figure 1. The Unit's activities cover a wide range of disciplines and two are selected for close attention.

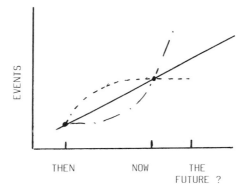

Fig. 1   Possible extrapolations to the future.

The research topic which was the instigation of the Unit was the collaboration between Professor R M Kenedi and Professor T Gibson in respect of the mechanical properties of skin and the application of these principles to clinical problems. Early findings showed that the familiar engineering materials had characteristics which unfortunately the biologically based materials did not have. Where the engineers were used to load/deformation relationships which were linear from zero load biological materials were found to be extremely non-linear. Materials familiar to engineers are mainly isotropic whereas skin from the abdomen is markedly anisotropic. Additionally this non-linear relationship was also found to be time dependent and dependent on a large range of plastic physiological parameters. One of the immediate clinical applications was in respect of surgery where the problem is to cover, without grafting, an area where a lesion has been removed. The surgeons were already using Z plasties and W plasties and analysis of the maximum skin deformability allowed the prediction of the optimum angles between the limbs of the Z and the W. This surprisingly subsequently led to studies in heart valves because in one generation of prosthetic heart valves the material utilised is in fact of animal origin and displays such anisotropy. Since the valve is a structure which is required to bend largely in one direction and not very much in another it is appropriate that it be constructed with the material characteristics oriented in a suitable direction. Professor Barbenel and clinical collaborators in cardiac surgery have been noteworthy in introducing this concept to the construction of cardiac valves. We look forward to applications of these principles in future but unfortunately it is difficult to predict where the next occasion for the application will be.

Another descendant from the original work in tissue is the study of the deformability of red cells and the study of the rheology and flow characteristics of blood. Having always the clinical application in mind the relationship between the size of a cerebral infarct and the haematocrit of the blood is obviously of great clinical relevance as an indicator to the method of treatment and to the prognosis for the patient. The study of the localised and general flow characteristics of the patient is strongly related to the detailed laboratory investigation of the blood characteristics. Again surprisingly this work in tissue mechanics led to a further studies in surgery using modern laser technology. In a number of units people found a laser of some kind and said let's see how we can use it in surgery. In the Bioengineering Unit it was realised that effective use of this tool required study of the physics of laser energy in respect of wavelength, power and pulse width with particular reference to the absorption characteristics of different biological materials. As a consequence particular lasers and particular modes of energy delivery were selected in order to accomplish surgically useful end results. Highly successful results have been obtained in the removal of tattoos and a small company has been set up to exploit this. A further system of treatment relates to the removal of port wine stain haemangiomas where systems of treatment are being developed and evaluated.

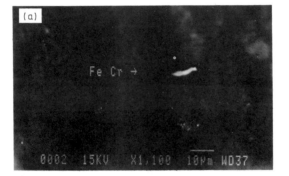

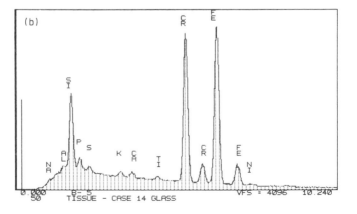

Fig. 2   a)   Particles showing on a specimen of tissue taken from near a hip joint replacement on the occasion of a revision procedure. These are identified as Iron and Chromium by back scattered X-ray analysis as shown in figure 2b.

          b)   The spectrum for the particles in figure 2a.

Further application of the lasers is in respect of joining of vessels in cardiac surgery and there is always this diversification which makes it difficult to predict where the studies in the Unit are going to go next.

The handmaiden of the studies of the biological materials is knowledge of the details of their structure: microscopy has been one of the Unit's basic facilities supporting work both in the mechanics of tissue and in the biocompatibility studies relating to artificial organs. The Unit is fortunate indeed that the UK's Medical Research Council recently funded a new instrument which as well as giving an image of a surface allows chemical analysis of the materials seen. Thus with a picture such as figure 2a, one can identify for instance a specific element by scanning for a particular energy level and wavelength as shown in figure 2b. Thus microscopy as well as being the servant of many of the other disciplines in the Unit is a research tool in its own right.

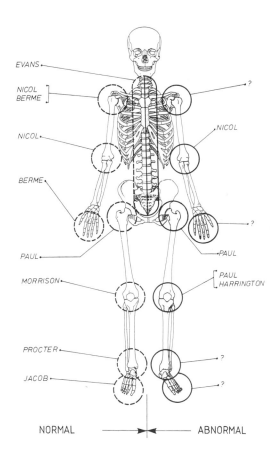

Fig. 3 Joints of the human body analysed by Unit members - normal joints to the left of the diagram - pathological or artificial joints on the right.

Early on in the history of the Unit we looked at orthopaedic mechanics. The author was asked to give a little time to an orthopaedic surgeon who wished to re-design the nail he was using for the fixation of fractured neck of femur and agreed to give maybe three to four weeks of time towards the exercise. This exercise transpires still to have 'spin-offs' some 25 years later and indeed the related investigations are of interest. If we take the mechanics of the human body there is a long list of members of the Bioengineering Unit who have looked at the load transmission in normal and abnormal individuals. Figure 3 indicates the studies on normals on the left and on joint replacements or pathological joints on the right. We have looked at the lumbar spine (Nachemson and Evans, 1968; Evans, 1989), the shoulder (Dvir and Berme, 1978), the elbow (Nicol *et al.*, 1977), the fingers (Berme *et al.*, 1977; Toft and Berme, 1980), the hip joint (Paul, 1976; Brown *et al.*, 1984), the knee joint (Morrison, 1968; Harrington, 1974; Paul, 1988), the ankle joint (Procter and Paul, 1982) and work is currently under way on the great toe (Jacobs, 1989).

In orthopaedic mechanics it was necessary to progress from primitive measuring instruments such as tape and stopwatch, which we have used initially and still are using in relevant situations, to develop our complex sophisticated gait analysis laboratory using television, force platforms, transducers in the load bearing structures or artificial limbs or orthoses (Jarret *et al.*, 1976). We have adapted force platforms and we are looking at different kinds of force transducers to be portable with the patient. We looked at goniometers and we have developed our own, marketed commercially for a one angle measurement and now becoming available for two dimensional angular measurement with plans afoot for the third dimension. The displacement measurement system based on television has gone from development from the work of Ing. Furnée from the Netherlands (Furnée, 1967), to our three dimensional system which is now marketed commercially with user friendly software (VICON). We have analysed the loads transmitted at the hip joint in joint replacements; we have collaborated in the design and development of the Souter/Strathclyde elbow joint replacement which is now in some 1,000 patients and 20 of them still have it 11 years after surgery. In artificial limbs we have measured the loads transmitted (Zahedi *et al.*, 1986) but the big problem we see at present is the manufacture of sockets and our colleagues in the National Centre for Training and Education in Prosthetics and Orthotics are actively working in this field using computer imaging techniques for the refinement of socket form leading to the forming of the end product mould on which the socket is produced.

The interacting of physical phenoma with living tissue is an area of considerable excitement at the present time. Since the body works by electrical signals it is definitely possible to produce effects in living systems by external stimulation. We have not yet developed, as our friends in Bulgaria, a system using lasers for acupuncture but we are working on the application of electromagnetic energy to cells in culture to see if the growth rate can be enhanced. If the rate of repair or the rate of production of cells cultured from humans can be enhanced cultured cells may be used as replacements or adjuncts to treatment of injury and appropriate stimulus parameters may assist regeneration of injured nerves.

Functional Electrical Stimulation is another exciting field where one applies the power of new technology in computing in association with classical physiology to a very simple and coarse technique of Faradic stimulation to produce contractile force in paralysed muscle. A major physiological difficulty exists since with electrical stimulation on a continuous basis the force in the muscle very quickly diminishes and therefore the spinal cord injured patient may have excessive hopes of effective use of the muscle only to be disappointed when the force produced by the stimulus diminishes. Our colleagues in Ljubljana have studied different 'cocktails' whereby the signal can be administered in active and rest periods to prolong the active function of the muscle and reduce the fatiguing effect (Kralj, 1986). We are exploring the application of this philosophy to the components of the quadriceps muscle in turn to give effectively continuous control of knee extension.

While it is a major and worthwhile objective to enable a patient to use the strength in his own legs ot move from sitting to standing by pressing a button on his stimulator, a more sophisticated system due to Brian Andrews is to enhance the effective time over which the stimulation can be employed by utilising a hybrid system. In this, mechanical support is combined with functional stimulation with a control system to apply stimulation only when stability is threatened. It is good that the patient can walk while tied to the computer. It is even better if he can walk freely in the general environment and this is one of the objectives still to be effectively realised.

Unit activities in the field of instrumentation are exemplified by gait analysis, prosthetics, stimulation and lasers. Other examples are instrumentation for the analysis of red cell size and deformability as Tony Fisher is doing with us at the present time or monitoring equipment as Jim MacGregor did many years ago. His early system of tape recorder captured data over a 24 period relating to cardiac frequency, acceleration signals and the posture of the patient.

Another area which was nowhere in our minds originally at the start of the studies with our clinical colleagues was the field of artificial organs exemplified by extracorporeal blood treatment. This requires appropriate material to transmit the blood and interface material through which the composition of the blood can be modified. Biocompatibility assessment involves a range of factors *in vitro* and *in vivo* and comparison of the characteristics of different materials in the presence of different anticoagulants. What seems to be a simple exercise in material selection in fact transpires to be complicated by all kinds of biomechanical, cellular, mechanical and physiological factors.

When the Unit started to explore the mechanical aspects of bioengineering there were those already in the field of electronics and instrumentation who looked somewhat down their noses at these simple mechanical people looking at the application of engineering techniques to the human body. It is foreseen that with the world's population having a continuously increasing elderly fraction, studies of instrumentation systems and engineering in medicine in general might be fostered by appropriate training courses (ACARD, 1986).

The Unit moved into teaching early on because research departments are greatly helped by extra pairs of hands and eyes in the form of research students. It became obvious that students coming from their individual distinctive disciplines needed training in areas not previously studied so we set up training courses to which in 1988 we recruited 45 students in comparison with 4 in 1964. Physical scientists and engineers, medics, paramedics and life scientists are the disciplines represented in the student body at the present time.

In respect of research, prediction is more difficult. Extrapolation allows the making of realistic predictions of the future from the past and the present but cannot cater for bright ideas, development of newly emerging technologies, use of new materials or even acquisition of more information on human body function. It is safe to predict however that twenty five years from now a Bioengineering Unit will still occupy the Wolfson Centre in this University existing solely by virtue of the strong collaboration with clinical colleagues in the city of Glasgow and elsewhere.

REFERENCES

ACARD (1986). Report on Medical Equipment, HMSO, London.
Berme N, Paul J P and Purves W K (1977). A biomechanical analysis of the metacarpophalangeal joint. *J. Biomech.anics*, **10**, 409-412.
Brown T R M, Nicol A C and Paul J P (1984). Comparison of loads transmitted by Charnley and C A D Muller total hip arthroplasties. *Proc. Conf. Engineering and Clinical Aspects of Endoprosthetic Fixation*, Inst. Mech. Eng., London, 63-68.
Dvir Z and Berme N (1978). The shoulder complex in elevation of the arm: a mechanistic approach. *J. Biomechanics*, **11** (5), 219-226.
Evans J H (1989). Biomechanics of interbody fusion of the spine, in P M Lin (ed.) *Lumbar Interbody Fusion*, Aspen Publications.
Furnée E H (1967). Hybrid instrumentation in prosthesis research. *Proc. 7th int. Conf. Medical and Biological Engineering*, Stockholm, 446.
Harrington I J (1974). The effect of congenital and pathological conditions on the load actions transmitted at the knee joint. *Total Knee Replacement*. Inst. Mech. Eng., London, pp 1-7.
Jacobs H A C (1989). Biomechanics of the forefoot. Ph.D. Thesis, University of Strathclyde, Glasgow.
Jarret M O, Andrews B J and Paul J P (1976). A television/computer system for the analysis of human locomotion. *Inst. Elec. Radio Eng. The Application of Electronics in Medicine*, 357-370.
Kralj A (1986). Electrical stimulation of lower extremities in spinal cord injury, in D N Ghista and H L Frankel (eds.) *Spinal Cord Injury Medical Engineering.* C C Thomas III, 439-509.
Morrison J B (1968). Bioengineering analysis of force actions transmitted by the knee joint. *Biomed. Engng.*, **3**, 164-170.
Nachemson A L and Evans J H (1968). Some mechanical properties of the third human interlamina ligament. *J. Biomech.anics.*, **1** (3), 211-220.

Paul

Nicol A C, Berme N and Paul J P (1977). A biomechanical analysis of elbow joint function. *Joint Replacement in the Upper Limb*, Inst. Mech. Eng., London, pp 45-51.

Paul J P (1976). Forces transmitted by joints in the human body. *Proc. Roy. Soc.*, **192**, 163-172.

Paul J P (1988). Mechanics of the knee joint and certain joint replacements, in S Niwa, J P Paul and S Yamamoto (eds.) *Total Knee Replacement*, Springer Verlag, Tokyo, pp 25-35.

Procter P and Paul J P (1982). Ankle joint biomechanics. *J. Biomech.anics*, **15** (9), 627-634.

Toft R and Berme N (1980). A biomechanical analysis of the joints of the thumb. *J. Biomech anics*, **13** (4), 353-360.

Zahedi M S, Spence W D, Solomonidis S E and Paul J P (1986). Alignment of lower limb prostheses. *J. Rehab. Res. Dev.*, **23** (2), 2-19.

2

PROGRESS IN ARTIFICIAL ORGANS

H Klinkmann

INTRODUCTION

It is convenient to begin this examination of progress in artificial organs by contrasting the present situation with that which existed 20 years ago. In the past, problems relating to material performance, durability and control greatly exceeded the problem of clinical application. However, during the last 20 years there has been considerable development in material technology and current emphasis is on questions of clinical application. This examination will focus on the artificial kidney and the artificial heart, which are the most applied and perhaps the most advanced artificial organs.

ARTIFICIAL KIDNEY

Biocompatibility

One of the difficulties in the progress in artificial organs is the establishment of a suitable definition of biocompatibility. A generally accepted definition is the "no definition", (Klinkmann et al., 1984, 1987), which implies that a biocompatible material must not induce any thrombogenic, toxic, allergic or inflammatory reactions and must cause no deformation of cellular elements, no changes in plasma proteins or enzymes, no immunological reaction, no carcinogenic effect and no deterioration in adjacent tissue. Of course, such a material does not exist. Biocompatibility parameters currently receiving attention range from the fall in the white blood cell level to anaphylactic reactions and one purpose of this paper is to examine which of the parameters being investigated are of real clinical relevance.

Biocompatibility and Haemocompatibility

I am concerned with the misuse of the terms "biocompatibility" and "haemocompatibility", in particular interchanging haemocompatibility, which relates to blood-material interactions, with biocompatibility, which denotes the complete response of the biological organism in the human body to the artificial device. Formerly, the emphasis was on the testing of a material, with the quality of devices linked to the biocompatibility of the material. However, our studies have shown (Klinkmann et al., 1987) that in terms of the clinical relevance of biocompatibility, the material cannot be divorced from the device.

The Device Cannot Be Separated From the Procedure

It is also true that the device cannot be separated from the procedure used. This can be demonstrated by considering the generation rate of the complement component C3a in two different therapeutic approaches, haemodialysis and haemofiltration. C3a is produced during complement activation (figure 1) and is a widely used parameter for demonstrating activation of the complement system by a device (Gordon and Hostetter, 1986).
    As shown in figure 2, the generation rate of C3a is obtained from the mass balance of substances and factors passing across a membrane (Falkenhagen et al., 1986). Differentiation can be made between the effective generation rate and the total generation rate, which includes all contact areas and permeation processes of the device.

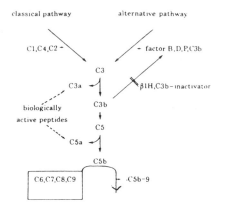

Fig. 1   Complement acticiation - classical and alternative pathways.

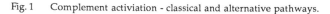

$$G_t = (Q_{B_0} \times C_0 + Q_F \times C_F) - Q_{B_I} \times C_I$$

effective generation rate

$$G_e = (Q_{B_0} \times C_0) - (Q_{B_I} \times C_I)$$

$$Q_{B_I} . C_I \rangle \quad \boxed{\qquad} \quad \longrightarrow Q_{B_0}.C_0$$

$$\downarrow Q_F.C_F$$

Fig. 2   The difference between total generation rate (G_t) and effective generation rate (G_e).
C_I, C_O, C_F - CONCENTRATION OF INFLOW, OUTFLOW, DIALYSATE/FILTRATE.
Q_{BI}, Q_{BO}, Q_F - FLOW RATES.

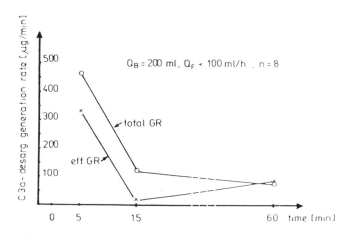

Fig. 3   Total and effective generation rates of C3a desarg in haemodialysis with the F60 membrane.

It is the generation rate which is of interest to the clinician, since this describes the amount entering the patient and therefore influencing the patient's condition. We measured the generation rates for a highly permeable polysulphone membrane (Fresenius, FRG) in both diffusive and convective modes. In the diffusive mode, as in dialysis, there is little difference between the total and effective generation rates, indicating that the patient receives almost the full dose of the so-called bioincompatibility, which is expressed in figure 3 as the C3a desarg concentration. With the same membrane in the convective mode, as in haemofiltration, the difference between the total and effective generation rates is considerably higher than in haemodialysis, clearly showing that the patient receives almost none of the C3a dose when the same membrane is used in a convective mode of treatment (figure 4).

Therefore, describing a material by itself and not taking into account the mode of application has resulted in incorrect interpretation, which has been commonly misused by industry for promoting products.

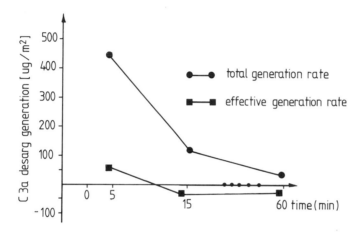

Fig. 4   Total and effective generation rates of C3a desarg in haemofiltration with the F60 membrane.

The Procedure Cannot Be Separated From The Patient

In recent years, it has emerged from the literature that beta-2-microglobulin is perhaps one of the factors leading to long term complications, being responsible for causing amyloidosis in chronic haemodialysis patients (Gejyo *et al.*, 1986). Therefore, the removal of this substance is under strong discussion as an effective index of the quality of a device. We measured beta-2-microglobulin concentration as part of a joint clinical study (the International Cooperative Biocompatibility Study) in which five centres in Japan, USA and Europe compared seven different haemodialysers. We found a marked increase in beta-2-microglobulin concentration after dialysis, in comparison to the concentration before dialysis, using dialysers with regenerated cellulose membranes. However, prior to results published in the Lancet (Bergström and Wehle, 1987), no account was taken of the influence of the patient's response on the distribution of the extracellular and intracellular compartments. On correcting for this, a completely different picture is obtained of beta-2-microglobulin as it appears in the patient. This is an example of the fact that in a consideration of the procedure, the patient must not be forgotten.

In considering biocompatibility parameters with possible clinical relevance, attention should be given to the release phenomenon (figure 5), including interleukin 1, thromboxane, tumour necrosis factor and prostaglandins from different cells, which probably play a much more important role than the cellular response. It is my belief that the white blood cell drop could be discarded as a biocompatibility parameter but that interest in the coagulation cascade may increase.

The biocompatibility of a dialyser does not relate only to the material.   The individual patient response, which depends on sterilisation, flow geometry and membrane characteristics, must be taken into consideration.

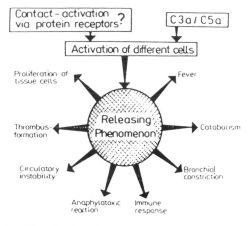

Fig. 5   Possible effects of the releasing phenomenon.

Diffusion Versus Convection

Since starting dialysis treatment in patients, we have been looking for the optimal mode in the treatment of end stage renal failure.   The objective remains the highest removal of toxins in the shortest possible time compatible with vascular stability and connected with the maximal physical well-being of the patient.

The patient himself is influenced by his intake of toxins, which depends clearly on his nutrition.   The concentrations of toxins in the body fluids and their removal are the final determinants of the detoxification process.   There is ongoing discussion about the improvement of patient well-being and the treatment of complications.

Reports from the European Dialysis and Transplant Association (Wing *et al.*, 1983) indicate the increase in the survival rate of dialysis patients attributable to the introduction of the currently available conventional dialysers and vascular access procedures.   However, fatigue remains in patients and in comparison to 10 years ago, has only changed in degree.   Consequently, discussion continues on whether convection or diffusion is better for the patient.

Convection is the transport mechanism in the membrane separation process of the human kidney and is used in haemofiltration.   Therefore, the elimination of solutes from the blood in haemofiltration depends mainly on the filtration flow and the cut off characteristics of the membrane used.   The basic transport mechanism in haemodialysis is diffusion and the main factors for diffusion are the concentration difference between blood and dialysate, and the thickness, surface area and pore size distribution of the membrane used.   Haemodiafiltration describes the use of diffusion and convection for solute removal.

A clinical question is whether or not to reduce the time of treatment.   This reduction can be achieved by the effectiveness of haemofiltration and haemodiafiltration and these procedures, utilising convection, resemble more closely the natural kidney than haemodialysis.   However, clinical results for the reduction of treatment time are disturbing.   Short time haemodialysis has been reported to be associated with most frequent causes of pulmonary congestion, hypertension, hypotension and hyperkalaemia, whereas a switch to long term haemodialysis has been reported to cause a considerable decrease in clinical complications (Laurent *et al.*, 1983).   In contrast, a reduction of treatment time using high flux dialysis has been reported to be beneficial (von Albertini *et al.*, 1984).   If the advantages of convection and diffusion are examined, the improved biocompatibility, arising from better sodium and fluid balance and higher removal of biologically active molecules such as C3a and interleukin 1, supports the view that convection should be used to a greater extent in the future. However, the competitiveness of haemodialysis as a therapy has been enhanced by the application of new membranes, such as polysulphone, polyamide and acrylonitrile copolymers, in combination with new technical approaches. The new trends in haemodialysis have overcome many of the problems previously in existence and added to the economic advantages of this therapy have led to the question of diffusion or convection remaining unsettled.

The resolution of the value of diffusion or convection requires consideration of therapy, membrane design, the optimal combination of both transport mechanisms, the introduction of erythropoietin and new anticoagulants such as low molecular weight heparin and hirudin, water quality, back filtration, beta-2-microglobulin removal and patient response.

At the present time, I believe that a favoured approach for technical development is the biocompatible high flux dialyser, combined with controlled ultrafiltration and an adjustable sodium concentration in a dialysate containing bicarbonate as buffer.

ARTIFICIAL HEART

About 20 years ago, the artificial heart development obtained a breakthrough with the survival for periods greater than 100 days of calves maintained by total artificial hearts. Further development was climaxed when the group from Salt Lake City, with which we were closely collaborating, implanted the first artificial heart into a patient.

At present, there are three types of artificial heart, the diaphragm type, the sac type and the pusher plate heart. The three types require materials which are as nonthrombogenic as possible. Different polyurethanes, such as Pellethane and Avcothane, are used in association with artificial heart valves and vascular grafts. I believe that biomaterials utilised in the development of artificial hearts face the same problems as in 1972 with respect to thrombus formation. The evidence of our calf experiments demonstrates that thrombus formation in the outflow valve of the left ventricle causes the deaths of most long term survival animals. The likelihood of this valve stenosis appearing in the clinical application of the total artificial heart brings us to a completely new field, termed biostability, which I believe will play an important role in future biomedical engineering.

Biomaterial-Related Infection

Biodegradation is well known to those involved with long term implantation of biomaterials. We must now address a new problem, that of biomaterial-related infection. I have tried to obtain from the literature the manner by which biomaterial-related infections arise and how they can influence the material and the organism. I believe that our group in the Department of Microbiology, Wilhelm Pieck University, Rostock, was the first to demonstrate that staphylococci epidermis, considered to be apathogenic, can in long term contact with a biomaterial grow little filaments which attach them to the surface. This process reflects a selection of those types of staphylococci having this attachment capacity due to the presence of filaments. Because of multiplication of these attaching germs, the staphylococci turn into pathogenic bacteria. Putting a biomaterial into the human body may cause a tremendous change in the pathogenesis of those bacteria long considered to be apathogenic and the biomaterial may cause a tremendous infection in the patient.

SUMMARY

In conclusion, I would like to summarise the features which influence device failure in artificial organs. These are the material, the material processing, device design, device fabrication, the medical application and the patient's condition. Artificial organs medicine, which was considered in 1982 by the NIH as a therapeutic approach in end stage disease, is now shifting towards elective restoration of chronically damaged structures. A statement that artificial organs may soon be considered for preventive maintenance in early stage disease may be a sign of looking a little too far ahead. However, with cooperation between medicine and bioengineering, cooperation between universities and colleges of medicine and cooperation regardless of boundaries, this role for artificial organs medicine can be achieved.

REFERENCES

Bergström J and Wehle B (1987).   No change in correct ß$_2$ - microglobulin concentration after cuprophane haemodialysis. *Lancet*, **1**, 628-629.

Falkenhagen D, Brown G S, Thomaneck U and Klinkmann H (1986). Biocompatibility, in Ad van Berlo (ed.), *New Trends in Blood Purification*, Eindhoven University of Technology, pp 37-52.

Gejyo F, Odani S, Yamada T, Honma N, Saito H, Suzuki Y, Nakagawa Y, Kobayashi H, Maruyama Y, Hirasawa Y, Suzuki M and Arakawa M (1986). ß$_2$- microglobulin: a new form of amyloid protein associated with chronic hemodialysis. *Kidney int.*, **30**, 390-395.

Gordon D L and Hostetter M K (1986).   Complement and host defence against microorganisms. *Pathology*, **18**, 365-375.

Klinkmann H, Wolf H and Schmitt E (1984). Definition of biocompatibility. *Contr. Nephrol.*, **37**, 70-77.

Klinkmann H, Falkenhagen D and Courtney J M (1987). Clinical relevance of biocompatibility - the material cannot be divorced from the device, in H J Gurland (ed.), *Uremia Therapy*, Springer-Verlag, Berlin, pp 125-140.

Laurent G, Calemand E and Charra Ḅ (1983). Long dialysis: a review of fifteen years experience in one centre 1968-1983. *Proc. Eur. Dialysis Transplant Assoc.*, **20**, 122-139.

Von Albertini B, Miller J H, Gardner P W and Shinaberger J H (1984). High-flux hemodiafiltration: under six hours/week treatment. *Trans. Am. Soc. artif. intern. Organs.*, **30**, 127-131.

Wing A J, Broyer M, Brunner F P, Brynger H, Challah S, Donckerwolke R A, Gretz N, Jacobs C, Kramer P and Selwood N H (1983).   Combined report on regular dialysis and transplantation in Europe, XIII, 1982. *Proc. Eur. Dialysis Transplant Assoc.*, **20**, 2-71.

3

THROMBUS FORMATION IN ARTIFICIAL ORGANS

C D Forbes, J M Courtney, A R Saniabadi
and L M A Morrice

INTRODUCTION

The practice of modern technological medicine could not exist if there were no implantable devices for the support of vital life functions. Every foreign material implanted into the body provokes a reaction, the scope and extent of which vary with the material implanted and the site of the implant. Nowhere in the body is there a greater reaction than in the circulation. Every foreign body in the vascular tree (i.e. anything which is not normal human endothelium) provokes a reaction which results in the deposition of fibrin thrombus on its surface and which limits its function and reduces its operating efficiency. Alternatively, thrombus may obstruct the circulation or act as a source of embolism. The mechanism by which this process occurs has been recently reviewed (Forbes and Courtney, 1987; Salzman and Merrill, 1988) and is shown in figure 1. In addition blood proteins are actively adsorbed and this may lead to activation or depletion of plasma levels. Also white cells may be adsorbed and activated.

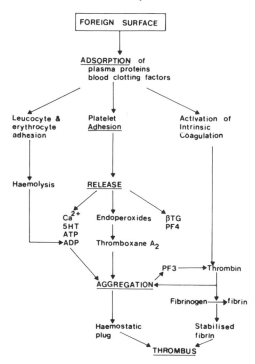

Fig. 1   Sequence of events following artificial surface exposure.

These in turn may activate components of the complement system and this may account for some of the adverse effects noted clinically (Clagett, 1988).

This chapter will cover only the alterations in blood and its components, especially in coagulation proteins, platelets and red and white cells following exposure to artificial surfaces.

COAGULATION FACTOR ADSORPTION

Protein adsorption is promoted by the amphipathic (polar/non-polar) character of protein molecules and their limited solubility and is inevitable during exposure to artificial surfaces (Brash, 1983). Exposure results in rapid adsorption of proteins from simple solutions or plasma and whole blood (Vroman *et al.*, 1977; Gendreau *et al.*, 1981). Subsequent reactions are strongly influenced by the composition of the adsorbed protein layer e.g. adsorption of fibrinogen and γ-globulins actively enhances adhesion of blood platelets and adsorption of albumin tends to diminish platelet adhesion (Jenkins *et al.*, 1973; Absolom *et al.*, 1983). This property of albumin has been used to enhance the biocompatibility of artificial membranes for human use (Fougnot *et al.*, 1984). In addition to enhancing platelet adhesion, adsorbed fibrinogen and γ-globulin may also encourage white cell adhesion, aggregation and activation thus inducing clinical sequelae in the patient (Szycher, 1983; Vroman, 1983). Fibronectin may also play a significant role in leucocyte adhesion as it is actively adsorbed onto most surfaces and probably has a receptor site for white cells (Vroman, 1983).

A range of blood coagulation proteins of the intrinsic thromboplastin system is also adsorbed and is activated to a greater or lesser extent. These include the so-called "contact" factors; factors XII (Hageman factor), XI (plasma thromboplastin antecedent), high molecular weight kininogen (HMWK) and prekallikrein (Griffin and Cochrane 1979). Interaction of these four factors initiates the coagulation mechanism. The mechanism is thought to start with a cellular protease activating prekallikrein to kallikrein and with HMWK acting as a cofactor this converts factor XII to its active form (XIIa) and this in turn activates factor XI to its active form XIa. These active products are also known to initiate the activation of the fibrinolytic pathway (via plasminogen) and also of complement, the renin-angiotensin system and generate kinins (Griffins and Bouma, 1987).

The end result of activation of the intrinsic coagulation system is fibrin generation which is then laid down on the surface in a manner which is dependent on the velocity of flow of the blood as well as the physical characteristics of the surfaces.

PLATELET ADHESION AND AGGREGATION

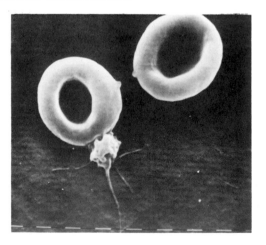

Fig. 2   Scanning EM of a Cuprophan membrane exposed for one minute to whole blood to show platelet adhesion with production of pseudopodia.

Exposure of an artificial surface to blood inevitably leads to platelet adhesion and aggregation (Mason *et al.*, 1976). Platelet adhesion is enhanced by prior adsorption of γ-globulin and fibrinogen but reduced by albumin. It is thought that the mechanism of adsorption is via incomplete heterosaccharides of these proteins and glycosyl transferases on the platelet membrane (Kim *et al.*, 1974). The inhibitory effect of albumin may be due to the absence of such saccharide chains. Adherence of platelets to a surface leads to a rapid change from the circulating platelet form, which is globular and about 2-3 µm in diameter, to a form in which pseudopodia are thrown out (figure 2). This is usually followed rapidly by platelet release in which a variety of active constituents are liberated into the surrounding plasma, these include adenosine diphosphate (ADP), serotonin (5-hydroxytryptamine, 5-HT), platelet factor 4 and β-thromboglobulin (ß-TG).

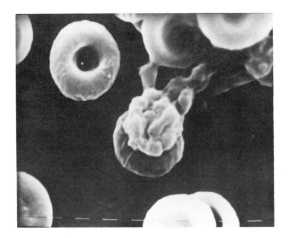

Fig. 3   Aggregation of platelets on a Cuprophan membrane after 4 minutes exposure to whole blood. The platelets have undergone adhesion and release of their constituents and in this example are seen adhering to red cells as well as the membrane. As yet no fibrin has formed.

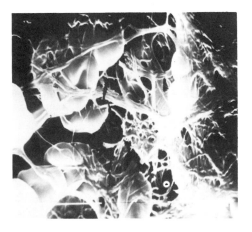

Fig. 4   Fibrin deposition on a Cuprophan membrane with entrapment of red cells and white cells. This growing mass of thrombus significantly interferes with the function of the artificial surface.

In addition there is activation of the platelet membrane phospholipase and this triggers the generation of platelet thromboxane A$_2$ (TXA$_2$) (Holmsen *et al.*, 1969). The presence of ADP, 5-HT and TXA$_2$ leads to rapid irreversible platelet aggregation at the site of the original platelet adhesion and on artificial surfaces these are deposited in a flow dependent fashion (Baumgartner *et al.*, 1976) (figure 3). In addition 5-HT and TXA$_2$ have powerful vasoconstrictor actions and this may be manifested by pulmonary vasoconstriction on return of the blood to the patient. Platelet factor 4 is also important as an activator of the coagulation cascade mechanism and its phospholipid rich micelles adsorb both coagulation factors VIII and V and accelerate the speed of the coagulation reaction many fold. The end result is that masses of platelets are deposited on the exposed surface and in this position may become surrounded by fibrin clot which further entraps red cells and white cells (figure 4).

ACTIVATION OF BLOOD COAGULATION

As stated before, the four blood contact sensitive factors are rapidly adsorbed and activated by foreign surface contact. Activated factor XI (XIa) then triggers off a sequential series of enzymatic steps in which inactive plasma precursors are activated. The active form then triggers the next step (figure 5). This has been called a waterfall or cascade sequence (Davie and Ratnoff, 1964). Understanding of these complex coagulation reactions has led to the possibility of therapeutic intervention as we now more fully understand the actions of therapeutic heparin, low molecular weight heparins, dermatan sulphate and oral anticoagulants. These currently act as the prime methods in clinical use for the reduction of fibrin formation on surfaces.

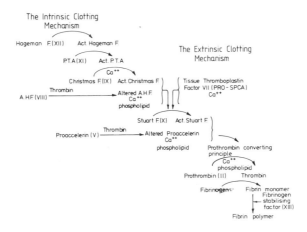

Fig. 5   Waterfall or cascade theory of blood coagulation in which there is activation of a series of inert contact factors and these in turn result in sequential activation of the next part of the system. The end result is production of a meshwork of fibrin on the artificial surface. This then entraps red and white cells and interferes with function of the material. Redrawn from Davie and Ratnoff.

RED AND WHITE CELLS

In *in vitro* testing it can easily be shown that isolated red cells adhere to the protein layer on an artificial surface. Additionally it may be that in flowing blood the presence of fibrin enhances cell adhesion. In some conditions the red cell is haemolysed and thus liberates large amounts of ADP which in turn accelerates platelet aggregation. Also the red cell "ghost" which is rich in phospholipid, acts as a tissue thromboplastin and further activates the coagulation mechanism (Stormorken, 1971). Entrapped red cells may coat the surface and alter its function, capacity and flow characteristics (Bruck, 1980). It is probable that leucocytes actively adhere to membranes unlike the passive role of the red cell (Szycher, 1983).

Leucocyte adhesion to artificial membranes is well recognised and there is preferential adsorption of polymorphonuclear leucocytes rather than lymphocytes (Wright *et al*, 1978). Leucocytes may contribute to platelet aggregation and fibrin formation by enzymatic release and may also participate in fibrinolysis, the enzymatic process by which deposited fibrin is removed (Cumming, 1980; Lindsay *et al.*, 1980). It is probable that leucocyte adhesion is intimately linked with complement activation which mediates chemotactic, adhesion and phagocytic reactions involving white cells and fractions $C3_a$, $C4_a$ and $C5_a$ are of particular relevance (Herzlinger and Cumming, 1980; Farrell, 1984).

The final result of these processes is that proteins, platelets, fibrin, red cells and white cells are deposited to a greater or lesser extent on all artificial surfaces. The site, extent and composition however vary greatly and are dependent on a multiplicity of physical factors such as the nature of the surface exposed, the thickness and supporting structure of the surface, the flow characteristics, the composition of the blood and the levels of the various blood proteins and formed elements. The other dimension of maximum importance is time, as the related amount of the various components adsorbed from the blood is likely to alter greatly according to duration of exposure, as shown in table 1.

Within minutes of exposure of artificial surface to blood:	Water and ion interactions
	Protein adsorption
	Cell adhesion
	Local fibrin deposition
	Fibrinolysis
	Complement activation
	Inflammation
	Embolisation
Within days of exposure:	Changes in proteins
	Continued cell adhesion
	Thrombus formation
	Fibrinolysis
	Chronic inflammation
	Embolisation
Within months:	Embolisation
	Calcification
	Changes in materials fatigue

Table 1.   Sequence of events with artificial surfaces. The effects of time.

When the process of contact is short, as in dialysis, haemoperfusion or catheter insertion it seems likely that all the above reactions occur rapidly. Proteins may be denatured or activated on the membranes and may accumulate and be subsequently strippped off by the blood flow and return to the patient as emboli.

It is unusual in a short exposure situation to observe depletion of levels of any protein. Platelet aggregates in association with fibrin may accumulate and may be stripped off to form emboli which return to the patient. This is probably of little clinical relevance usually, but it has been suggested that it may be one of the factors in adult respiratory distress syndrome. In addition while complement factors, fibrinolysis and inflammation are activated there are usually few clinical sequelae because of the short duration of exposure. This however is not the case if exposure is over a period of days in which the process of coagulation and platelet activation persists and activated products of complement and white cells have swamped the normal protective inhibitory proteins. This is the situation which is present in artificial left ventricular assist devices and artificial hearts to support the failing circulation.

A different set of problems arises in long term vascular implants such as heart valves and vascular grafts. Here the clinical problems are mainly platelet-fibrin emboli and local occlusion due to thrombus formation or calcification due to the body's reaction to the implant.

METHODS OF MONITORING THROMBUS FORMATION

Compatibility of surfaces should be measured *in vitro, ex vivo, in vivo* in animals and finally *in vivo* in man. The methods for this have already been extensively reviewed (Philp, 1981). It must be remembered that short *in vitro* experiments have to be interpreted with caution as the blood movement is not representative of the clinical use of the material and will produce different results. Also the use of animals may lead to misinterpretation as the processes of animal coagulation, platelet function and the other triggered enzyme systems are quite different from human. However such testing can have value when correctly interpreted.

Coagulation assays	Prothrombin time
	Activated partial thromboplastin time
	Fibrinogen
	Factor VIII
	Fibrinopeptide A
Platelet function	Platelet count
	Aggregation/retention tests
	Circulating platelet aggregates
	Platelet release products (ßTG, PF4)
	Thromboxane B_2
General	Haematology - Haemoglobin
	White cell count
	Plasma haemoglobin
	Haptoglobin
	White cell aggregation
	Scanning electron microscopy
Complement activation	C3a, C5a
Fibrinolysis	Fibrinogen/fibrin degradation products

Table 2.   Tests for monitoring blood-surface interactions.

			Result
1)	*Proteins*	Albumin	Adsorption
		γ-globulin	Adsorption
		Factors XII/XI/	Conformation changes
		prekallikrein/HMWK)	and activation
2)	*Lipids*		Adsorption
3)	*Formed elements*	(a) Platelets	Adhesion/release/aggregation
			Platelet thrombi
			Fall in circulating count
			Release of ßTG/PF4
		(b) Red cells	Adhesion/haemolysis
			Release of ADP
			Enhancement of clotting
		(c) White cells	Adhesion/loss of function
			Activation of complement

Table 3.   Changes in blood components after blood-surface interactions.

In the human situation there is no single 'best test' which tells all about the compatibility of a surface. It is therefore usual to perform a profile of tests of coagulation, platelet function and complement and white cells (table 2). These tests have been extensively used for comparison of materials. Because of cost it is not possible to perform all tests and in most situations selected tests are used to illustrate a particular characteristic. Table 3 shows the most common changes which are found after blood surface interaction.

CONCLUSION

To date no biomaterial has been produced which exactly matches the functional characteristics of normal human endothelium. Attempts have of course been made to enhance the seeding of endothelium over the implanted organ but these have, as yet, not been successful. However by investigation of the factors which initiate thrombus formation it has been possible to substantially enhance the viability of implanted devices. This has been done by alteration of the chemical composition and surface physical characteristics of the material in association with the use of newer therapeutic agents to alter the function of blood coagulation, platelet activation, white cell adhesion and release and complement activation.

REFERENCES

Absolom D R, Zingg W, Policova Z and Neumann A W (1983). Determination of the surface tension of protein coated materials by means of the advancing solidification front technique. *Trans.Am.Soc.artif.intern.organs*, 29, 146-151.

Baumgartner H R, Muggli R, Tschopp T B and Turitto V J (1976). Platelet adhesion, release and aggregation in flowing blood: effects of surface properties and platelet function. *Thromb. Haemostas*, 35, 124-138.

Brash J L (1983). Protein adsorption and blood interactions in M. Szycher (ed.), *Biocompatible Polymers, Metals, and Composites*, Technomic, Lancaster, Pennsylvania, pp 35-52.

Bruck S D (1980). *Properties of Biomaterials in the Physiological Environment*, CRC Press, Boca Raton.

Clagett G P (1988). Artificial devices in clinical practice in R W Colman, J Hirsh, V J Marder and E W Salzman (eds.), *Hemostasis and Thrombosis - Basic Principles and Clinical Practice*, Lippincott, Philadelphia, Ch 89, pp 1348-1365.

Cumming R D (1980). Important factors affecting initial blood-material interactions. *Trans. Am. Soc. artif. inter. Organs*, 26, 304-308.

Davie E W and Ratnoff O D (1964). Waterfall sequence for intrinsic blood clotting. *Science*, 145, 1310-1315.

Farrell P C (1984). Biocompatibility aspects of extracorporeal circulation in J P Paul, J D S Gaylor, J M Courtney and T Gilchrist (eds.), *Biomaterials and Artificial Organs*, MacMillan, London, pp 342-350.

Forbes C D and Courtney J M (1987). Thrombosis and artificial surfaces, in A L Bloom and D P Thomas (eds.), *Haemostasis and Thrombosis*, Churchill Livingstone, Edinburgh, 2nd edition, pp 902-921.

Fougnot C, Labarre D, Jozefonwicz J and Jozefowicz M (1984). Modifications to polymer surfaces to improve blood compatibility in G W Hastings and P Ducheyne (eds.) *Macromolecular Bilmaterials*, CRC Press, Boca Raton, pp 215-238.

Gendreau R M, Winters S, Leininger R I, Fink D, Hassler C R and Jakobsen R J (1981). Fourier transform infrared spectroscopy of protein adsorption from whole blood: ex-vivo dog studies. *Appl. Spectroscopy*, 35, 353-357.

Griffin J H and Bouma B M (1987). The contact phase of blood coagulation, in A L Bloom and D P Thomas (eds.), *Haemostasis and Thrombosis*, Churchill Livingstone, Edinburgh, pp 101-115.

Griffin J H and Cochrane C G (1979). Recent advances in the understanding of contact activation reactions. *Seminars Thromb. Hemostas.* 5, 254-273.

Herzlinger G A and Cumming R D (1980). Role of complement activation in cell adhesion to polymer blood contact surfaces. *Trans. Am. Soc. artif. intern. Organs*, 26, 165-170.

Holmsen H, Day H J and Stormorken J (1969). The blood platelet release reaction. *Scand. J. Haematol.*, (Suppl) 8, 1-26.

Jenkins C S P, Packham M A, Guccione M A and Mustard J F (1973). Modification of platelet adherence to protein-coated surfaces. *J. Lab. clin. Med.*, 81, 280-290.

Kim S W, Lee R G, Oster H, Coleman D, Andrade J D, Lentz D J and Olsen D (1974). Platelet adhesion to polymer surfaces. *Trans. Am. Soc. artif. Intern. Organs*, 20, 449-455.

Lindsay R M, Mason R G, Kim S W, Andrade J D and Hakim R M (1980). Blood surface interactions. *Trans. Am. Soc. artif. intern. Organs*, **26**, 603-610.

Mason R G, Mohammad S F, Chuang H Y K and Richardson P D (1976). The adhesion of platelets to subendothelium, collagen and artificial surfaces. *Seminar Thromb. Hemostas.*, **3**, 98-116.

Philp R B (1981). *Methods of Testing. Proposed Antithrombotic Drugs.* CRC Press, Boca Raton, Florida.

Salzman E W and Merrill E W (1988). Interaction of blood with artificial surfaces in R W Colman, J Hirsh, V J Marder and E W Salzman (eds.), *Hemostasis and Thrombosis - Basic Principles and Clinical Practice*, Lippincott, Philadelphia, Ch 88, pp 1335-1347.

Stormorken H (1971). Platelets, thrombosis and hemolysis. *Fed. Proc.*, **30**, 1551-1556.

Szycher M (1983). Thrombosis, hemostasis, and thrombolysis at prosthetic interfaces in M Szycher (ed.), *Biocompatible Polymers, Metals and Composites*. Technomic, Lancaster, Pennsylvania, pp 1-33.

Vroman L, Adams A L, Klings M, Fischer G C, Munoz P C and Solensky R P (1977). Reactions of formed elements of blood with plasma proteins at interfaces. *Ann NY Acad. Sci.*, **283**, 65-76.

Vroman L (1983). Protein/surface interaction in M Szycher (ed.), *Biocompatible Polymers, Metals, and Composites*, Technomic, Lancaster, Pennsylvania, pp 81-88.

Wright D G, Kauffman J C, Terpstra G K, Graw R G, Deisseroth A B and Gallin J I (1978). Mobilization and exocytosis of specific (secondary) granules by human neutrophils during adherence to nylon wool in filtration leukapheresis (FL). *Blood*, **56**, 770-782.

4

BLOOD COMPATIBILITY OF BIOMATERIALS IN ARTIFICIAL ORGANS

J M Courtney, L M Robertson, C Jones, L Irvine, J T Douglas, M Travers,
C J Ryan and G D O Lowe

INTRODUCTION

Our involvement in artificial organs is with those based on extracorporeal blood purification and hence relevant biomaterials are membranes, sorbents and blood tubing. Membranes are fundamental to current extracorporeal blood purification, sorbents offer a means of extending or enhancing such purification and blood tubing is an essential component of extracorporeal blood purification systems. For biomaterials used in artificial organs, an acceptable level of compatibility is a basic requirement for clinical application and artificial organs would benefit from biomaterials with improved blood compatibility. In the case of polymeric biomaterials, improved blood compatibility can be obtained by exploiting the fact that the blood response is dependent on polymer features such as formulation and chemical structure. With the objective of acquiring information relevant to the development of improved biomaterials, we have utilised selected blood compatibility evaluation procedures in order to determine the influence on the blood response of particular polymer features. In this publication, the features considered are polymer formulation, polymer modification and polymer synthesis.

INVESTIGATION OF BLOOD-MATERIAL INTERACTIONS

Parameter Selection

Examination of the blood compatibility of biomaterials is linked to the study of blood-material interactions and thus linked to the topic of blood compatibility assessment. A general objective is the utilisation of relevant parameters in order to establish a structure-property relationship (Klinkmann, 1984; Klinkmann et al., 1984). Thus, the evaluation of biomaterials often attempts to correlate some characteristic of a material with alterations in blood constituents induced by blood-material contact. A basic problem is the selection of parameters representative of the blood response to an artificial surface.

Consideration of the formation of a thrombus (Szycher, 1983) draws attention to the importance of three mechanisms, platelet activities, the intrinsic pathway and the extrinsic pathway, and numerous parameters could be utilised in a study of these mechanisms. While it is possible to ignore the extrinsic pathway in dealing with biomaterials and thrombus formation on artificial surfaces, the situation with respect to parameter selection remains complicated. In fact, a focus on platelets and the intrinsic coagulation represents too narrow a perspective for the interactions of blood with artificial surfaces (Murabayashi and Nosé, 1986; Forbes and Courtney, 1987). Other aspects must be taken into account and for membranes this has led to an emphasis on interaction with the complement system (Chenoweth, 1984; Farrell, 1984; Klinkmann et al., 1987).

If importance is given to haemostatic and complement systems in blood compatibility evaluation, it is essential to have the capability of determining a range of parameters. However, in the formulation of a blood compatibility evaluation programme, the advantages of multiparameter assessment must be balanced against practical limitations and the need to achieve consistency with a selected parameter.

Blood Compatibility Evaluation Procedures

The blood compatibility evaluation programme we have undertaken is outlined in table 1. The programme is based on a combination of *in vitro* and *ex vivo* procedures (Courtney *et al.*, 1986).

	Feature	Parameters
In vitro	assessment in the absence or presence of antithrombotic agents	platelet adhesion platelet aggregate formation platelet release complement activation
Ex vivo	rat extracorporeal circuits	erythrocytes leucocytes platelets

Table 1.   Selected blood compatibility evaluation for biomaterials in artificial organs.

In Vitro Assessment

The *in vitro* procedures examine platelet adhesion, platelet aggregate formation, platelet release and complement activation. Tests can be made in the absence or presence of antithrombotic agents.

The influence of a biomaterial on platelet adhesion and platelet aggregate formation is determined by a modification to the Wu and Hoak method for measuring circulating platelet aggregates (Bowry *et al.*, 1985; Courtney *et al.*, 1987). The procedure is based on the ability of ethylene diamine tetraacetic acid (EDTA) to disperse platelet aggregates and that of an EDTA - formalin mixture to fix any aggregates present. Platelet counts in EDTA and EDTA - formalin are made for blood before and after material contact. The procedure determines platelets lost to platelet aggregate formation and platelets lost to adhesion in the absence of platelets lost to platelet aggregate formation.

The relationship between a biomaterial and the platelet release reaction is monitored by measuring the concentration of the platelet-specific protein beta thromboglobulin (BTG) (Bowry *et al.*, 1984; Travers *et al.*, 1986).

The effect on the complement system of blood-material contact is evaluated by measuring the concentration of the complement component C3a (Travers *et al.*, 1989).

For the *in vitro* assessment of flat sheet biomaterials (Bowry *et al.*, 1984; Courtney *et al.*, 1986), a test cell is used in which blood is retained within two sheets of the material under evaluation. The test cell consists of two Perspex sections, each containing a groove. An elastomeric O-ring, which fits into the grooves on clamping, acts as a blood port. The test cell is oscillated during the evaluation period.

Oscillation is also used for the *in vitro* assessment of tubular biomaterials (Jones *et al.*, 1989). The tubing is formed into a loop by closure with a latex rubber cuff.

Ex Vivo Assessment

The *ex vivo* procedures measure levels of erythrocytes, leucocytes and platelets in rat extracorporeal circuits. The animal evaluation procedures are derived from a haemoperfusion circuit (Ryan *et al.*, 1979; Ryan *et al.*, 1980). Relevant features are the selection of a mass of material appropriate to the blood volume of the animal, a priming volume sufficiently small to render priming with donor blood unnecessary, elimination of the need for anaesthesia during treatment and the institution of a minimum 21h period between cannulation and extracorporeal circulation. The haemoperfusion circuit is suitable for the evaluation of granular biomaterials such as sorbents. The examination of flat sheet biomaterials in a rat extracorporeal circuit utilises a test cell designed as a miniature dialyser (Robertson, 1988). Blood is circulated withing two sheets of the material under evaluation. The contact area is 35 cm^2, corresponding to 1 m^2 for a human.

POLYMER FORMULATION

An example of utilising polymer formulation in order to alter the blood compatibility of a biomaterial is the selection of a plasticiser for poly(vinylchloride) (PVC) blood tubing.

Plasticisers for PVC blood tubing have been investigated in collaboration with Hydro Polymers Ltd. The plasticisers are di-2-ethyl-hexyl phthalate (DEHP), trioctyl trimellitate (TOTM) and polymeric adipate (PA). DEHP, the most widely used plasticiser for PVC, is extracted in contact with biological fluids and concern has been expressed over potential toxicological hazards.

As a solution to the extraction problem, Hydro Polymers have produced PVC plasticised by the higher molecular weight substances TOTM and PA. Our objective is to assess if the improved resistance to extraction is matched by an improvement in blood compatibility. Polymers plasticised by DEHP, TOTM and PA have been converted into tubing for evaluation.

In the first phase of the examination of the blood response to plasticised PVC, an *in vitro* study has been made of the influence of the different polymers on platelet adhesion and platelet aggregate formation. Figure 1 compares the results obtained for TOTM and PA relative to those for DEHP. The results determined in the absence of an anticoagulant indicate that platelet adhesion is clearly influenced by the plasticiser, with TOTM causing an improvement and PA inducing a sharp rise between 6 and 9 min. The results for platelet aggregate formation also indicate an improvement for TOTM, with the decline for PA corresponding to the sharp rise in platelets lost to adhesion.

The results suggest that the blood response to PVC is influenced by the nature of the plasticiser and support additional *in vitro* and *ex vivo* evaluation.

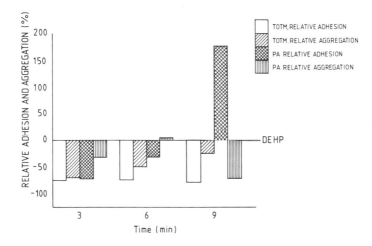

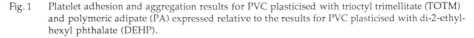

Fig. 1   Platelet adhesion and aggregation results for PVC plasticised with trioctyl trimellitate (TOTM) and polymeric adipate (PA) expressed relative to the results for PVC plasticised with di-2-ethyl-hexyl phthalate (DEHP).

POLYMER MODIFICATION

Chemical modification of a polymer structure is an important option for improving the blood compatibility of certain biomaterials. It is particularly relevant for cellulose haemodialysis membranes, since there is the possibility of improving membrane compatibility without adversely altering membrane compatibility (von Sengbusch *et al.*, 1987).

Our interest has been a comparison of the regenerated cellulose membrane Cuprophan with the modified cellulose membrane Hemophan, in which modification is achieved by the introduction, to a low degree of substitution, of diethylaminoethyl groups. Both membranes were supplied in flat sheet form by Akzo (Enka AG).

Values of C3a and BTG, obtained 6 min blood-membrane contact in the absence of an anticoagulant, are listed in table 2 for both Cuprophan and Hemophan. The results support the clinical evidence (Falkenhagen *et al.*, 1985; Bosch *et al.*, 1986) that complement activation, as determined by the measurement of C3a, can be strongly reduced by cellulose modification.

Membrane	C3a (ng/ml)	BTG (ng/ml)
Cuprophan	1273 ± 74	57 ± 7.8
Hemophan	475 ± 78	40 ± 12.9

Table 2.   C3a and BTG levels following 6 min. blood-membrane contact. No anticoagulant. Mean ± standard deviation (n = 5).

Results from the investigation of the platelet release reaction, as represented by the measurement of BTG, also indicate the superiority of Hemophan.

Platelet counts measured over a 4h period in the rat extracorporeal circuit are listed for both membranes in table 3. The results suggest that the advantage gained by cellulose modification with respect to complement activation may be offset by a higher platelet adhesion.

On balance, our study supports cellulose modification as an appropriate route for the preparation of novel membranes.

Membrane	Platelet Count ($\times 10^9$/litre) Contact Period (min)			
	0	15	60	120
Cuprophan	631 ± 126	587 ± 149	560 ± 168	494 ± 154
Hemophan	639 ± 105	569 ± 135	415 ± 128	408 ± 150

Table 3.   Platelet levels in the normal rat following contact with haemodialysis membranes. Mean ± standard deviation (n = 5).

POLYMER SYNTHESIS

To study the influence of polymer synthesis, an investigation has been made of hydrogel polymers (Robertson, 1988). These hydrogels have been developed at the University of Aston for artificial liver support and were supplied in bead form. Hydrogels offer a means of obtaining sorbents different markedly from conventional materials, a point illustrated in figures 2 and 3, which show platelet results obtained in the rat haemoperfusion circuit.

Figure 2 compares platelet loss for a hydrogel with losses for two uncoated granular charcoals, Norit RBX1 extruded and Asahi spherical. The pattern is not unexpected in that there is a high platelet loss with Norit RBX1, a smaller loss with the spherical charcoal and the smallest loss with the hydrogel. However, these values for before and after a 4h perfusion do not represent the complete picture. Figure 3, which is based on counts during haemoperfusion, demonstrates that the sharp fall shown by the two charcoals contrasts strongly with the sharp fall and rapid rise shown by the hydrogel.

The overall investigation of hydrogels (Robertson, 1988) has shown that the selection of both the polymerisation method and comonomers provides the possibility of strongly influencing the cellular response in terms of platelets and leucocytes, while erythrocytes are unaffected. Therefore, the synthesis of hydrogel polymers has important implications for the development of biomaterials with improved properties.

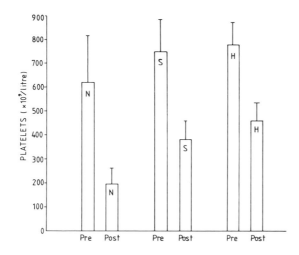

Fig. 2   Platelet counts pre and post 4h rat haemoperfusion over Norit RBX1 charcoal (N), Asahi spherical charcoal (S) and hydrogel beads (H).

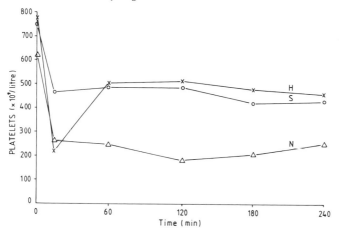

Fig. 3   Platelet counts during 4h rat haemoperfusion over Norit RBX1 charcoal (N), Asahi spherical charcoal (S) and hydrogel beads (H).

DISCUSSION

The evaluation of the blood compatibility of biomaterials in artificial organs has demonstrated that, in terms of the parameters utilised, compatibility improvement may be achieved by polymer formulation, polymer modification and polymer synthesis. However, the development of better biomaterials depends not only on improvements to polymers but also on improvements to evaluation procedures. There is a need to study additional parameters, which would provide a more complete representation of the influence of biomaterials on blood, particularly in the clinical situation. In this respect, attention can be drawn to our clinical investigation of the release of granulocyte elastase during haemodialysis with cellulose and synthetic membranes (Irvine *et al.*, 1989). The investigation has shown that this parameter is membrane-dependent but the gradual increase in granulocyte elastase during haemodialysis with both cellulose and polysulphone membranes contrasts sharply with the pattern for changes in C3a, where there is marked transient increase with the cellulose membrane and little effect with the polysulphone membrane.

The study of granulocyte elastase acts as a reminder that parameters determined in clinical application may not always be interpreted so readily as in the case of the relationship between C3a generation and the structures of cellulose and non-cellulose haemodialysis membranes (Chenoweth, 1984). Consideration of granulocyte elastase also encourages interest in the contact activation proteins, since elastase release has not only a possible role in clinical disorders but may interact with high molecular weight kininogen, factor XII and kallikrein to promote the contact activation phase of blood coagulation (Gustafson and Colman, 1987). Estimation of the influence of biomaterials on contact activation would be a beneficial addition to blood compatibility evaluation procedures.

In conclusion, progress in a complex area such as the blood compatibility of biomaterials in artificial organs requires a proficiency in parameters relevant to the blood response and an awareness of what can be accomplished by polymer formulation, polymer modification and polymer synthesis.

REFERENCES

Bosch T, Schmidt B, Samtlebem W and Gurland H (1986). Biocompatibility and clinical performance of a new modified cellulose membrane. *Clin. Nephrol.*, **26**, (Suppl), 22-29.

Bowry S K, Courtney J M, Prentice C R M and Douglas J T (1984). Utilization of the platelet release reaction in the blood compatibility assessment of polymers. *Biomaterials*, **5**, 289-292.

Bowry S K, Prentice C R M and Courtney J M (1985). A modification of the Wu and Hoak method for the determination of platelet aggregates and platelet adhesion. *Thromb. Haemostas.*, **53**, 381-385.

Chenoweth D E (1984). Complement activation during hemodialysis: Clinical observations, proposed mechanisms, and theoretical implications. *Artif. Organs*, **8**, 281-287.

Courtney J M, Travers M, Douglas J T, Lowe G D O, Forbes C D, Aslam M and Ryan C J (1986). Preclinical investigation of membranes and sorbents, in S Dawids and A Bantjes (eds.), *Blood Compatible Materials and their Testing*, Martinus Nijhoff, Hingham M A, pp 137-144.

Courtney J M, Travers M, Bowry S K, Prentice C R M, Lowe G D O and Forbes C D (1987). Measurement of platelet loss in the blood compatibility assessment of biomaterials. *Biomaterials*, **8**, 231-233.

Falkenhagen D, Zinner D, Falkenhagen U, Ahrenholz P, Holtz M, Behm E and Klinkmann H (1985). A modified cellulose membrane (MC) with reduced complement activiation. *Kidney int.*, **28**, 331.

Farrell P C (1984). Biocompatibility aspects of extracorporeal circulation, in J P Paul, J D S Gaylor, J M Courtney and T Gilchrist (eds.), *Biomaterials in Artificial Organs*, Macmillan, London, pp 342-350.

Forbes C D and Courtney J M (1987). Thrombosis and artificial surfaces, in A L Bloom and D P Thomas (eds.), *Haemostasis and Thrombosis*, Churchill Livingston, Edinburgh, pp 902-921.

Gustafson E J and Colman R W (1987). Interaction of polymorphonuclear cells with contact activation factors. *Seminars Thromb. Hemostas.*, **13**, 95-105.

Irvine L, Travers M, Simpson K, Lowe G D O and Courtney J M (1989). Influence of haemodialysis membranes on the release of granulocyte elastase. *Int. J. artif. Organs*, 12, 502-504.

Jones C, Courtney J M, Robertson L M, Biggs M S and Lowe G D O (1989). Influence of plasticised poly(vinyl chloride) on platelet adhesion and platelet aggregates. *Int J. artif. Organs.*, 12, 466-470.

Klinkmann H (1984). The role of biomaterials in the application of artificial organs, in J P Paul, J D S Gaylor, J M Courtney and T Gilchrist (eds.), *Biomaterials in Artificial Organs*, Macmillan, London, pp 1-8.

Klinkmann H, Wolf H and Schmitt E (1984). Definition of biocompatibility. *Contr. Nephrol.*, **37**, 70-77.

Klinkmann H, Falkenhagen D and Courtney J M (1987). Biomaterials and biocompatibility in hemodialysis. *Contr. Nephrol.*, **55**, 231-249.

Murabayashi S and Nosé Y (1986). Biocompatibility: Bioengineering aspects. *Artif. Organs*, **10**, 114-121.

Robertson L M (1988). Blood compatibility of modified biomaterials: Application of selected in vitro and ex vivo procedures. PhD Thesis, University of Strathclyde, Glascow.

Ryan C J, Courtney J M, Wood C B, Hood R G and Blumgart L H (1979). Activated charcoal haemoperfusion via an extracorporeal circuit in the unrestrained and unanaesthetised rat. *Br. J. Expl. Path.*, **60**, 400-410.

Ryan C J and Courtney J M (1980). Hemoperfusion in the rat: Application to adsorbent evaluation, in S Sideman and T M S Chang (eds.), *Hemoperfusion: Kidney and Liver Support and Detoxification*, Hemisphere, Washington, pp 27-36.

von Sengbusch G, Lemke H D and Vienken J (1987). Evolution of membrane technology: Possibilities and consequences, in H J Gurland (ed.) *Uremia Therapy*, Springer-Verlag, Berlin, pp 111-122.

Szycher M (1983). Thrombosis, hemostasis, and thrombolysis at prosthetic interfaces, in M Szycher (ed.) *Biocompatible Polymers, Metals, and Composites*, Technomic, Lancaster, pp 1-33.

Travers M, Courtney J M, Douglas J T, Forbes C D, Lowe G D O, Falkenhagen D and Klinkmann H (1986). Hemodialysis membranes and the platelet release reaction in Y Nosé, C Kjellstrand and P Ivanovich (eds.), *Progress in Artificial Organs* - 1985, ISAO Press, Cleveland, pp 1056-1058.

Travers M, Simpson K, Courtney J M, Bradley H E, Pollock J C and Forbes C D (1989). Biomaterials and the immune response, in J C Barbenel, A C Fisher, J D S Gaylor, W J Angerson and C D Sheldon (eds.), *Blood Flow in Artificial Organs and Cardiovascular Prostheses*, Clarendon, Oxford, pp 113-121.

5

MEMBRANE DEVELOPMENT FOR BLOOD PURIFICATION

H Göhl, R Buck and L Smeby

INTRODUCTION

After the first description of haemodialysis (Haas, 1928), it was more than thirty years before the method was applied clinically (Kolff and Berk, 1944), and another twenty years before haemodialysis treatment became available to patients with chronic end-stage renal disease.

The first membranes employed in dialysers were similar to sheets manufactured for the food packaging industry. The development of flat sheet and hollow fibre membranes for hemodialysis and optimisation of these allowed for more efficient removal of urea and low molecular weight uraemic toxins, but all membranes remained cellulose-based for many years.

At the same time as the NIH artificial kidney programme was introduced (in the 60's), development of new synthetic membranes such as polysulphone (Amicon) and polyacrylonitrile (Rhone Poulenc) started. These membranes had higher permeability (e.g. higher ultrafiltration and higher diffuse permeability for larger molecular weight substances), and therefore allowed for modification of the dialysis treatment creating new treatment modalities such as haemofiltration, haemodiafiltration, arteriovenous filtration and mixed therapies (see figure 1).

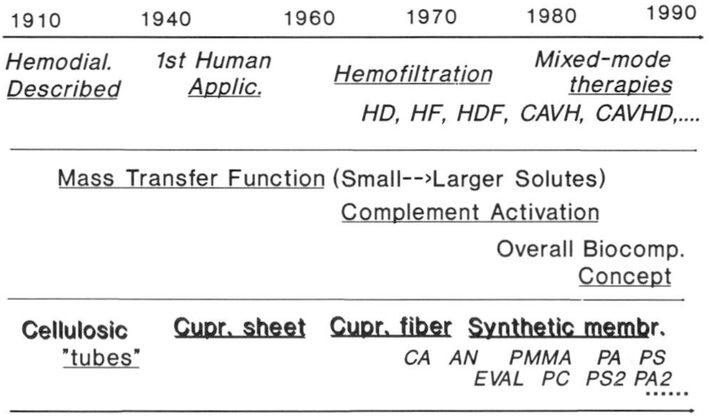

1910	1940	1960	1970	1980	1990
Hemodial. Described	*1st Human Applic.*		*Hemofiltration*	*Mixed-mode therapies*	
				HD, HF, HDF, CAVH, CAVHD,....	

Mass Transfer Function (Small-->Larger Solutes)

Complement Activation

Overall Biocomp. Concept

Cellulosic "tubes"	**Cupr. sheet**	**Cupr. fiber**	**Synthetic membr.**
			CA AN PMMA PA PS *EVAL PC PS2 PA2* ......

Fig. 1   Historical overview of dialysis development, topics of interest and appearance of different membranes used in dialysis.
More detail about membrane types are given in the legend to figure 2.

MEMBRANE CLASSIFICATION

Recently, new synthetic membranes with ultrafiltration characteristics similar to the cellulosic, but with improved permeability for larger molecules, have been developed. All these membranes can be used for standard haemodialysis without special equipment to control fluid removal and are therefore classified as haemodialysis (HD) membranes. It should be noted that there are variations in performance among membrane devices in this class (figure 2), but in general, diffusive permeability (Pm) for small solutes like urea is

28

high, while transport of larger molecules (vitamin B12, ß2-microglobulin) and water is relatively low.

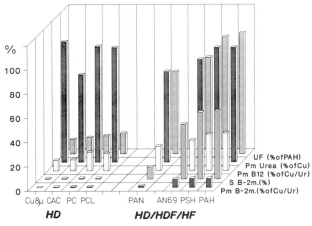

Fig. 2   The performance of different membranes used for dialysis (HD), haemodiafiltration (HDF) and haemofiltration (HF). Values are approximate and given as a percentage of urea diffusion (Pm Ur.) in 8µ Cuprophan.
Pm B12 refers to the relative diffusion of vitamin B_{12}, and Pm B-2m refers to diffusion of ß2 microglobulin. UF values give the maximum ultrafiltration for given conditions in relation to what can be obtained with the PAH membrane.

Cu8µ	=	8 micron Cuprophan R membrane (ENKA)
CaC	=	Cellulose acetate membrane (CD Medical)
PC	=	Polycarbonate membrane Gambrane R (Gambro)
PSL	=	Low flux Polysulphone membrane (Fresenius)
PAN	=	Polyacrylonitrile membrane (ASAHI)
AN69	=	Acrylonitrile membrane (Hospal)
PSH	=	High flux Polysulphone membrane (Fresenius)
PAH	=	High flux Polyamide membrane (Gambro)

The first synthetic membranes to be developed had increased permeability for larger molecular weight substances and higher ultrafiltration capacity than the cellulosic membranes. These led to new treatment modalities such as high flux dialysis and haemodiafiltration (HDF). The mass transfer properties of these membranes (polyacrylonitrile, polysulphone, cellulose triacetate, polyamide) are closer to the natural kidney permeability (figure 3) and can be described as combining diffusive and convective transport to achieve the total flux I_S of a solute.

$$I_S = PmA\Delta C_S \quad + \quad I_V(1-\sigma_S).C_S$$
$$\text{diffusive} \qquad\qquad \text{convective}$$

where the water flux I_V is given by

$$I_V = Lp\ (\Delta P - \sigma_S.\Delta\pi)$$

These parameters are mainly determined by membrane properties and in general it can be said that the solute permeability (Pm) is governed by membrane hydrophilicity and thickness. The reflection coefficient (σ_S) is controlled by the pore diameter, while the hydraulic permeability (Lp) is related to the pore density and pore size of the membrane.

Membrane area (A), hydrostatic pressure (ΔP) and osmotic pressure (π) are additional factors, which influence the total performance of a device. In modern synthetic HDF membranes, these parameters are optimised to obtain both high diffusive and convective transfer of given substances without losing essential blood components (see figure 2).

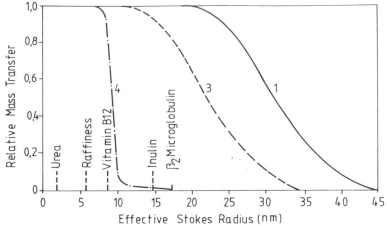

Fig. 3   Relative mass transfer vs. Stoke's Radius for kidney (1), HDF and HF (3), and HD
          membranes (4) (adapted from Ward *et al.*, 1984)

In the haemofiltration modality (HF), substances are removed by convective mass transfer. Pore density (for the Lp) and pore size (for σ) are, therefore, the most important parameters for haemofiltration membrane performance; they can, however, be influenced by protein interaction, as discussed later in this chapter under the heading Blood-Membrane Interactions.

MEMBRANE STRUCTURE

As mentioned above, the permeability of membranes is closely related to the membrane structure. Most haemodialysis membranes have a homogeneous gel-like composition. In the scanning electromicrograph (figure 4), distinctive pores cannot be seen. The haemofiltration and haemodiafiltration membranes are often built with heteroporous structures, most of them in asymmetric heteroporous configuration (figure 4), in which the more dense layer - usually defined as skin - determines the effective pore size and the remaining layer provides the mechanical support.

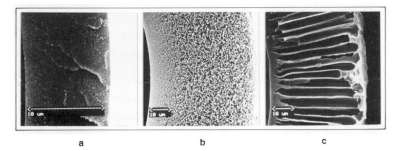

a   b   c

Fig. 4   Scanning electromicrographs cross sections of three typical structures of membranes used
          in blood purification;
          (a)   haemodialysis membrane with homogeneous structure, wall thickness 11 μm; (PC,
                     Gambrane R);
          (b)   asymmetric sponge like structure for HDF and HF, wall thickness 40 μm
                     (Polysulphone);
          (c)   asymmetric finger like structure, wall thickness 50 μm, (PAH, Polyamide).

This configuration offers the opportunity to vary convective and diffusive permeability independently; the convective transport of solutes by the pore size and pore density in the skin layer, and the diffusive permeability by the thickness and density of the support layer and the hydrophilicity of the membrane material.

The following example shows how different structures can be built with one polymer material. Polyamide membranes are manufactured - like all other membranes for dialysis - with the phase inversion process. A polymer is dissolved in a solvent and the solvent is exchanged during the membrane forming process by a nonsolvent. The dissolved polymer is thereby inverted to the solid stage (phase inversion), and the structure of the 'solid' is the formed membrane. It can be influenced by many parameters, like polymer concentration, polymer type and polymer mixtures and temperature (Strathmann, 1980; Kesting, 1985; Göhl and Konstantin, 1986).

Figure 5 shows membrane with typical heteroporous structures, formed by the same material - polyamide - but at different conditions: sponge (a) - macrovoid (c) and fingerlike (b) structure. The permeability properties are different for the different structures and this is the reason for making a "composite" membrane with three "layers", a skin (1 µm), a sponge layer (5 µm) and a finger structure as shown in 5d. A higher magnification (60 000 x) of the skin region for the same membrane (5d) shows that the skin looks similar to a sponge membrane, with increasing polymer density toward the inner surface giving the smallest effective pore size at the blood - membrane interface. This asymmetric structure can be important in avoiding "plugging" of pores during highly convective treatment modalities.

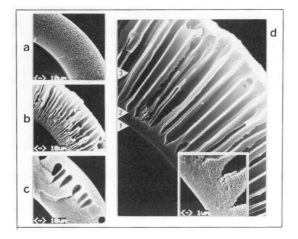

Fig. 5   Scanning electromicrograph of different heteroporous structures made from polyamide, a = sponge, b = finger, c = macrovoid, d = 3-layer structure with detail sketch.

BLOOD - MEMBRANE INTERACTIONS

Mass transfer characteristics of membranes are mainly determined by the membrane material and the surface structure but can also be influenced by the interaction with blood components. When blood comes in contact with a membrane, proteins are immediately adsorbed on the surface, thus changing both membrane performance and haemocompatibility (figure 6).

After the initial protein adsorption, the modified surface interacts with cells and blood components related to the complement and the coagulation system (figure 6). The mechanisms of the subsequent activations and stimulations are very complex and not yet fully understood but there are clear differences between different membrane materials. To measure overall response parameters on membranes and to characterise a membrane's "biofunction", is one of the most challenging tasks in the development of blood purification membranes. A better understanding of how and why protein adsorbs on membrane surfaces therefore appears to be essential in the process of improving the haemocompatibility of membranes. Blood proteins can change their configuration dependent on pH, ionic strength of the solution and many other factors. They can have different surface domains (figure 7) and can interact with membrane surfaces by hydrogen bonds, charge and hydrophilic/hydrophobic forces.

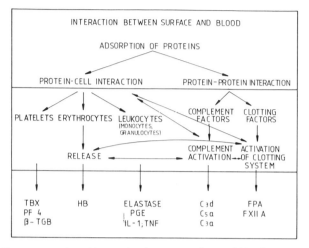

Fig. 6   Schematic representation of interaction between surface and blood components, indicating release of various substances: TBX_{A2} = thromboxane, PF4 = platelet factor, B-TGB = beta thromboglobin, HB = haemoglobin, elastase, PGE_2 = prostaglandin, C_{3D}, C_{5A}, C_{3A} = anaphylatoxin products of complement activation, FPA = fibrinopeptide activator, F XII A = factor XII A.

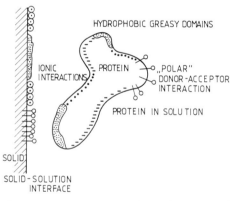

Fig. 7   Schematic view of protein interaction with a solid surface (i.e. membrane). The solid surface as well as the protein molecule can have a number of surface domains with hydrophobic, charged and polar characteristics (Andrade, 1985).

Membranes are usually characterised by studying each of these factors separately, i.e. hydrogen bonding by protein dissolving procedures (Notohamiprodjo *et al.*, 1986); charge can be studied by measuring zeta potential (Ikada, 1986), and hydrophilic/hydrophobic forces by contact angle measurements.

It has been shown (Ikada, 1986) that contact angles of membranes from different polymers vary over a wide range and that these values influence 'biofunction', with respect to haemocompatibility parameters, such as platelet adhesion (figure 8) and complement activation (Ikada, 1986).

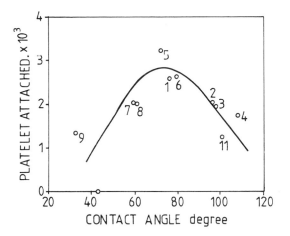

Fig. 8   Platelet deposition on materials with different contact angles. (Y Ikada, 5th Conference
PIMS, 1986). Polymer surfaces: 1. PE, 2. PP, 3. PTFE, 4. GF, 5. PET, 6. PA 66, 7. PMMA,
8. VAECO, 9. Cellulose, 10. PVA, 11. PDMS.

With regard to platelet adhesion, it is interesting to note that it is not linearly
related to hydrophilicity/hydrophobicity as shown in figure 8. Low platelet adsorption
can be obtained both with very hydrophilic and very hydrophobic membranes, and that
the "worst case" appears when the contact angle is about 90°.

As previously observed, the degree and type of protein adsorption to a membrane
also influence the mass transfer of membranes. Figure 9 shows sieving curves of two
membranes made from the same polymer but with different hydrophilicity. The contact
angle was 60° for the hydrophilic version and 80° for the more hydrophobic version. Their
permeabilities, expressed by the sieving coefficient S, as measured with saline, and with
plasma (*in vitro*), are clearly different. With the hydrophilic membrane, the sieving
coefficient for ß2-microglobulin is reduced from 0.80 in saline to 0.65 when plasma is used,
and from 0.80 to 0.30 for the more hydrophobic membrane. These *in vitro* measurements
with plasma are similar to actual values obtained *in vivo* with whole blood (Floege *et al.*,
1987).

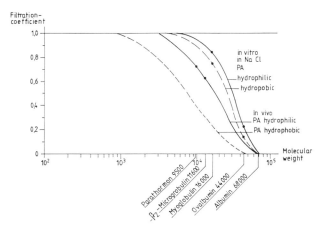

Fig. 9   Sieving coefficient *in vivo/in vitro* with NaCl-solution and plasma for hydrophilic and
hydrophobic polyamide membranes.

On the same two membranes, protein adsorption measurements were performed. The membranes were loaded with bovine plasma, rinsed thoroughly both from the fibre lumen and with backfiltration. After rinsing, the remaining proteins were eluted by three different dissolving fluids: urea to dissolve the hydrogen bound proteins, NaCl to dissolve the ionic bound and SDS to dissolve the hydrophilic/hydrophobic bound. In the electropherogram (silverstaining) a clear band can be found at albumin (MW 68 000) on the hydrophobic, but not on the hydrophilic type of membrane.

Another example related to the performance of hydrophilic versus hydrophobic membranes is illustrated in figure 10. Haemofiltration with three different polyamide filters (PA1, PA2, PA3) shows a wide range of total protein "leakage", different relative distribution between small (< 67 000) and large (> 67 000) ("size selectivity") proteins, and differences in the sieving coefficient for ß2-microglobulin during the treatment (Wizemann and Birk personal communication).

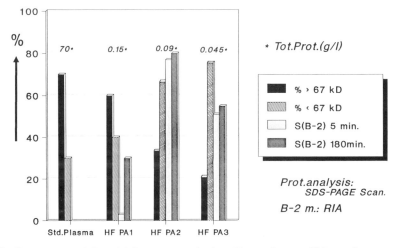

Fig. 10   Protein permeability of different types of polyamide membranes. Values refer to measured concentrations of total protein and the relative amount (%) of proteins greater or smaller than 67 000 dalton in plasma and in the filtrate using different filters PA1, PA2 and PA3. Sieving coefficients for ß2-microglobulin measured after 5 minutes of treatment (S(B-2) 5 min) and after 180 minutes of treatment (S(B-2) 180 min) are also expressed in % (i.e. 100% = 1.0).

PA1 was the most hydrophobic membrane and also had the highest "leakage" of total protein. This membrane showed a low sieving of ß2-microglobulin at the start (5 minutes), which increased to about 30% after 180 minutes of treatment, indicating an adsorption process changing with time.

PA2 was a more hydrophilic membrane with lower total protein "leakage" (0.09 vs 0.15), better "size selectivity" (66% of total protein < 67 kD vs 40% < 67 kD for PA1), a much higher sieving (S $\approx$ 80% vs S $\approx$ 5-30% for PA1) and virtually no ß2-microglobulin adsorption (S(B-2) 5 min. $\approx$ S(B-2) 180 min). By changing the "pore size" (PA3), but keeping a high degree of hydrophilicity, total protein leakage was further reduced (to 0.045 g/l), "size selectivity" was improved ($\approx$ 78% of total protein < 67kD) and the sieving for ß2-microglobulin was still much higher than for PA1.

Block copolymers having hydrophobic - hydrophilic microdomain structure are reported to have selective adsorption for proteins and lead to remarkable improvements of thrombogenicity (Grainger *et al.*, 1986).

In our own investigation, we varied block length and distribution of hydrophilic polyethyleneglycol blocks (PEG) in a polycarbonate-polyether block copolymer (figure 11). Fresh human blood was incubated and platelet release parameters were measured (Schultze *et al.*, 1986). With increasing PEG-block length, increased fraction (%) of PEG, and with even block distribution, the released platelet parameter (ß-thromboglobulin) was reduced, although the complement activation was unchanged by the modifications (figure 12).

Membrane	Polymerization Process	PEG Fraction Mol. Wt.
PCa	Chemical Separation	20%
☐☑☑☐	(Random block posit)	8000 MW
PCb	Interfacial Process	20%
☐☑☑☐	(Even block distr)	8000 MW
PCc (Even distr blocs)	_ _ " _ _	30% 8000 MW
PCd (Even distr blocs)	_ _ " _ _	20% 20000 MW

Fig. 11   Modification of PC-PE (PC, Gambrane R) block co-polymers with different fractions and different molecular weights of the polyether (PE) block.

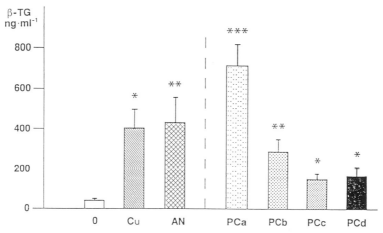

Fig. 12   Appearance of beta-thromboglobulin (ß-TG) in plasma after 10 minutes' incubation with different membranes. Cu = Cuprophan, AN = Acrylonitrile, PC(a) = Polycarbonate. Polyether block co-polymer (PC-PE, Gambrane R) with randomised block distribution, PC(b) = PC-PE with a statistically even block distribution, PC(c) = 30% PE fraction with mol wt. 8000 dalton, PC(d) = 20% PE fraction with mol wt. 20 000 dalton. The mol wt. and the PE fraction in both PC(a) and PC(b) is 8000 dalton and 20%. 0 refers to control sample from plasma without membrane incubation. (Schultze *et al.*, 1987).

Complement activation on synthetic membranes is generally reduced, when compared to regenerated cellulosic membranes. The OH-end groups at the surface of cellulosic membranes have been shown to be strongly related to complement activation (Chenoweth, 1984), and a modification of Cuprophan[R] (cellulose), where a small fraction of these end groups were substituted by DEAE (diethylaminaethyl) groups, reduced the complement activation by more than 50%. Other properties remained unchanged.

The early studies on complement activation initiated a series of studies related to interaction between membranes in presently available haemodialysers and blood. Today, these phenomena are accepted as only one of many aspects related to the biocompatibility of haemodialysis, but research into this area will undoubtedly provide answers which can be used to modify membranes for blood purification. An excellent summary of classical hypotheses and examples of membrane modifications to improve haemocompatibility is given by Andrade (1985).

CONCLUSION

Synthetic membrane materials offer a wide range of possibilities for improvement of haemocompatibility and performance due to:

a large choice of different basic polymer (PA, PC, PS, PAN,...)

the possibility of copolymer variation (block polymers, block sequences, block length,...)

membrane-forming technique for wide performance range (structure, porosity, pore size,...)

ease of surface modifications (hydrophilicity, charge,...)

Future developments of membranes to improve the overall biocompatibility of blood purification systems is mainly dependent on a better understanding of the biological processes in order to provide clear definitions of the physico-chemical membrane characteristics, giving optimal conditions at the blood-membrane interface, and optimal performance.

REFERENCES

Andrade J D (1985). In *Surface and Interfacial Aspects of Biomedical Polymers Vol. 2: Protein Adsorption,* Plenum Press, New York.

Chenoweth D E (1984). Complement Activation During Hemodialysis: Clinical Observations, Proposed Mechanisms, and Theoretical Implications. *Artif. Organs,* **8** (3), 281-287.

Flöge J, Granolleras C, Smeby L, Göhl H, Koch K M and Shaldon S (1987). Hydrophilic high flux polyamide membranes are more efficient for ß2-microglobulin removal than hydrophobic polyamide membranes. *Am. Soc. artif. intern. Organs,* **10** (3), 309-311.

Göhl H and Konstantin P (1986). Membranes and filters for hemofiltration, in L Henderson, E A Quellhorst, C A Baldamus and H J Lysaght (eds.), *Hemofiltration,* Springer-Verlag, Heidelberg, pp 41-82.

Grainger D, Okano T and Kim S W (1986). Multiblock PEO-PST copolymers - synthesis, characterisation and blood compatibility. *Proc. 5th int. Conf. Polymers in Medicine and Surgery,* Plastics and Rubber Inst., London, 8/1.

Haas G (1928). Uber Blutwaschung, *Klin. Wschr.,* **7**, 1356.

Ikada Y (1986). Development of a polymer surface with non-adherent platelet properties, *Proc. 5th. int. Conf. Polymers in Medicine and Surgery,* Plastic and Rubber, Inst., London 6/1.

Kesting R E (1985). *Synthetic Polymer Membranes,* John Wiley and Sons, New York.

Kolff W J and Berk T H (1944). The artificial kidney: a dialyser with great area. *Acta Med.,* **117**, 121-134.

Notohamiprodjo M, Andrassy K, Bommer J and Ritz E (1986). Dialysis membranes and coagulation systems, *Blood purification,* **4**, 130-141.

Schultze G, Mahiout A, Göhl H, Deppisch R and Molzahn M (1987). Reduced thrombogenicity of modified polycarbonate dialysis membranes. *Abstr. 24th Eur. Dialysis Transplant Ass.,* 136.

Strathmann H (1980). Development of new membranes. *Desalination,* **35**, 39.

Ward R A, Feldhoff P W and Klein E (1984). Therapeutic Applications of Membranes in Medicine Polymeric Materials Science and Engineering *Proc. 187th natn. Meet. Am. chem. Soc.,* **50**, 339-343.

6

THE DEVELOPMENT OF A PRESEALED DACRON

VASCULAR PROSTHESIS

R Maini

INTRODUCTION

The knitted Dacron vascular prosthesis has become the standard vascular graft for replacement of arterial vessels of 6mm and greater. However, while this has many features required by a surgeon - ease of handling, suturability and conformability, it has one major disadvantage, it is not blood-tight. The knitted structure, by its nature, is a porous structure, which is what is required for the rapid incorporation by tissue ingrowth from the host. At the time of surgery, the surgeon has to "preclot" the graft using some of the patient's own blood, which is taken prior to heparinisation; a time-consuming process, which may be difficult to carry out satisfactorily. This prevents its use when patients are heparinised, as in cardiopulmonary bypass, and in emergency aneurysmal surgery when preclotting is not possible.

CONCEPT

There was a demand from the medical profession for a graft that did not require preclotting but retained the advantages of the successful knitted prosthesis. The design criteria were - a comfortable graft with high suture retention, a graft with an initial porosity of zero to prevent blood loss but that would not prevent tissue ingrowth through the structure of the graft to provide good incorporation (Wesolowski et al., 1963).

The commercially successful Vascutek Triaxial fabric, which had been produced in the Ayr factory since 1981, was used as the base fabric to be sealed and tested in 1983 when the project was started. It was decided to use gelatin to seal the interstices of the graft as this had been shown to be effective elsewhere (Bascom, 1961), although these required careful handling (Jordan et al., 1963).

DESIGN

There are several ways of crosslinking gelatin; baking, radiation, glutaraldehyde and formaldehyde. Similarly, gelatins are available in different viscosities (Bloom strengths) and with chemical modifications. It was felt that the gelatin should seal the graft for at least 24 hours but be totally removed within 14 days. We have been able to regulate the rate of degradation of the sealant by using formaldehyde as the crosslinker and a mix of gelatins that have been chemically modified, so that a known proportion of the amide groups, which are potential crosslinking groups, has been substituted by succinyl groups. This renders them unavailable for crosslinking. This process, the control of degradation, has since been patented by Vascutek.

FABRICATION

The fabricated Triaxial graft is pressure impregnated with an aqueous solution of gelatin, which has been filtered through 0.2µm and depyrogenating filters.

37

The impregnated grafts are then immersed in a formaldehyde solution in order to effect crosslinking. Ethylene oxide is used to sterilise the graft. The resultant product is soft and pliable and ready for use by the surgeon.

IN VITRO PRE CLINICAL TESTING

The extent of crosslinking in the protein impregnation determines its rate of degradation. The assumption at this stage was that the coating would be removed primarily by hydrolysis at the flow surface. An *in vitro* test circuit was devised and constructed to produce a means of testing the longevity of the various coatings by continuous exposure to flowing saline at 37°C. The impregnation was considered to have hydrolysed when flow through a scavenging loop was equal to flow into the graft, i.e. when all the flow into the graft came through the graft wall. It was shown that the permeability was gradual (usually about 5 hours) rather than catastrophic. Minimal crosslinking resulted in a life of approximately 5 hours, while normal production grafts with full crosslinking lasted approximately 40 hours.

Fig.1   Platelet bland gelatin coated fibres, after exposure.

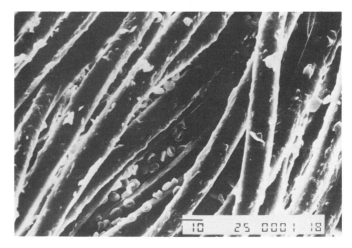

Fig.2   Thrombogenic Dacron fibres, after blood contact.

The coated grafts were also tested for thrombogenicity against a normal Dacron graft. It was known that gelatin had been used as a plasma expander without any major complications (Lundsgaard-Hansen and Tschiren 1981) and by the Cleveland Clinic to render their blood contacting left ventricle assist device (LVAD) surfaces more biocompatible (Szycher, 1983). It was hoped that the coated grafts would give a similar response.

Samples of treated and untreated Dacron were inserted into a flat plate blood contacting chamber, primed with Krebs-Ringer solution and then with oxygenated blood at 37^0C. The graft was exposed for 3 minutes at 104 ml/min. The samples were removed, rinsed in saline, fixed and examined under scanning electron microscopy (Figures 1 &2). Later, a simple 15-30 minute platelet dip was used to remove the problem of red cells. The results of these tests showed that the coating was relatively platelet bland, having a much lower thrombogenicity than bare Dacron.

ANIMAL EXPERIMENTS

Having completed these tests, the graft was considered ready for animal trials. The infrarenal aorta canine model was considered to be the most suitable. 30 mongrel dogs were used and 8 preclotted control and 22 coated grafts. The grafts were crimped, 50-60mm in length with a 6mm internal diameter. The operating procedure was standardised with 100 IU/kg IV heparin administered prior to clamping of the aorta. End to end anastomoses were made with continuous 5/0 monofilament polypropylene sutures. The heparin was not reversed.

The grafts were explanted at various times to give a picture of gelatin resorption and incorporation. After injection of 100 IU/kg IV, the grafts were removed, cut longitudinally and rinsed in saline. The explants were fixed in formalin and sectioned for histology. All the grafts were patent at sacrifice, with no evidence of toxicity or pyrogenicity observed. It was also confirmed that the gelatin was almost completely absorbed by two weeks, as predicted by the *in-vitro* work. The healing process and tissue reaction are similar to those of a normally preclotted Dacron prosthesis, with good tissue incorporation and a well attached pseudointima being found. These findings have been confirmed by further animal studies carried out at Harvard Medical School (Jonas *et al.*, 1988). The graft, now called Gelseal, was considered suitable for clinical trials.

CLINICAL TRIALS

These started at the Glasgow Royal Infirmary in August 1985 and by February 1987 more than 100 Gelseal aortic bifurcation grafts had been implanted. The surgeons were asked to complete a questionnaire about suturability, handling and blood loss. All patients received routine pre and post operative blood tests. A smaller study group was also tested for plasma C-reactive protein (CRP) and alpha 1 acid glycoprotein, which are significant indications of an inflammatory reaction.

In this study there was no operative mortality and no discernible graft-related problems. Performance of all grafts was similar to previously implanted prostheses. There was no measurable blood loss through the graft and it was important to note that 74% of patients did not require any blood transfusions during or after surgery (Drury *et al.*, 1987). The CRP study showed that there was no significant difference in the Gelseal group and no increased inflammatory reaction. There was also no difference in platelet profiles.

PRODUCTION PLANT

The initial pilot plant was based at the Coats Anchor Mills in Paisley where a small clean room was constructed to evaluate the production technique. The first clinical trial grafts were produced at Paisley and extra staff were recruited as the pilot capacity was increased, when more centres began using Gelseal in clinical trials. The equipment design and fabrication processes evolved in this period. The full production facility was installed when Vascutek's factory moved from the small Ayr plant to the custom-built facility at Inchinnan in 1986. Polypropylene tanks, which were designed by Vascutek, were custom built and installed in a separate clean room. This production plant is semi-automated using electronic controllers.

THE FUTURE

The present production plant is in full operation and running near capacity since the introduction of the second coated graft - Gelsoft. This has a lighter weight base fabric with different handling characteristics but still has the advantage of being sealed.

Currently, second generation tanks are being constructed, which will allow for future production demand.

REFERENCES

Bascom J U (1961). Gelatin sealing to prevent blood loss from knitted arterial grafts. *Surgery*, **50**, 504-512.

Drury J K, Ashton T R, Cunningham J D, Maini R and Pollock J G (1987). Experimental and clinical experience with a gelatin impregnated Dacron prosthesis. *Ann. Vasc. Surg.*, **5**, 542-547.

Jonas R A, Ziemen G, Schoen F J, Britton L and Castaneda A R (1988). A new sealant for knitted Dacron prostheses: Minimally cross-linked gelatin. *J. Vasc. Surg.*, **7** (3), 414-419.

Jordan G L, Stump M M, Allen J, De Bakey M E and Halpert B (1963). Gelatin-impregnated Dacron prosthesis implanted into porcine thoracic aorta. *Surgery*, **63**, 45-49.

Lundsgaard-Hansen P and Tschiren B (1981). Joint WHO/IABS Symposium on the Standardization of *Albumin, Plasma Substitutes and Plasmapheresis Geneva 1980, Develop. Biol. Standard*, **48**, 251-256.

Szycher M (1983). In M Szycher (ed.) *Biocompatible Polymers, Metals and Composites*, Technomic, Lancaster, Chapters 8 and 9, pp 179-212.

Wesolowski S A, Fries C C, Domingo R, Leibig W J and Sawyer P N (1963). The compound prosthetic vascular graft: A pathologic survey. *Surgery*, **53**, 20-44.

7

INVESTIGATION OF SOLVENT-CAST POLYURETHANE

FILM AS A MATERIAL FOR FLEXIBLE LEAFLET PROSTHETIC

HEART VALVES

J Fisher, A C Fisher, V M Evans and D J Wheatley

INTRODUCTION

Prosthetic heart valves have been implanted clinically for over 25 years (Black *et al.*, 1983). Approximately half the valves currently implanted have rigid mechanical occluders such as tilting discs, while the other half have flexible leaflets composed of chemically treated animal tissue. The main advantage of flexible leaflet bioprosthetic valves is that most patients receiving these valves do not require long-term anticoagulation therapy. The main disadvantage of bioprostheses is that long term durability is unpredictable with valve failure occurring due to tissue abrasion, wear and fatigue (Fisher *et al.*, 1987) or due to calcification of the valve leaflets (Goffin and Bartik, 1987). Two types of bioprostheses have been developed: valves made from porcine aortic valves (Reece *et al.*, 1985) are currently more popular than valves with leaflets fabricated with bovine pericardium (Fisher *et al.*, 1987). Improvements have been made to the design and function of both porcine (Reul *et al.*, 1986) and pericardial valves (Fisher and Wheatley, 1987) but long-term failures associated with calcification remain a major concern (Schoen, 1987). Considerable efforts are being made to develop biochemical treatments to reduce the incidence of calcification in tissue. It has been suggested that calcification is associated with glutaraldehyde preservation of the tissue (Schoen, 1987).

In an attempt to develop flexible leaflet heart valves which are more resistant to calcification and fatigue failure alternative materials have been considered. At present, most research work is focused on investigating the use of polyurethane materials (Gogolewski, 1987). These materials have been used clinically in artificial arteries (Fisher *et al.*, 1985, Annis, 1987) and in artificial blood pumps (Pierce *et al.*, 1981). There is considerable variation in the chemical composition of the currently available medical grade polyurethane which affects both the physical properties (Hayashi, 1987), and the blood compatibility and biostability (Coury *et al.*, 1987; Gilding and Reed, 1987). It is not clear which (if any) of these materials is suitable for long-term implantation in the body under repeated cyclic loading in flexible leaflet heart valves.

Leaflet geometry, thickness distribution and elastic modulus are important parameters in the design of flexible leaflet valves (Ghista *et al.*, 1976; Fisher *et al.* 1988) and the type of polyurethane used can affect both the design and method of manufacture. Vacuum forming techniques can be used to fabricate leaflets from thermoplastic polyurethanes (Jansen *et al.*, 1986), but with most solvent cast polyurethanes, a complex dip-moulding technique is used to manufacture leaflets of the correct geometry and thickness (Herold *et al.*, 1987). Valves manufactured by these techniques are currently being assessed in animal trials in the natural heart (Herold *et al.*, 1987) and in the artificial heart (Jansen *et al.*, 1986).

We have investigated the use of two grades of Biomer polyetherurethane (Ethicon Inc., New Jersey) as a material for flexible leaflet heart valves and have developed a method of leaflet manufacture from solvent cast film of uniform thickness. Valve function has been analysed in a pulsatile flow test apparatus and short-term performance assessed in accelerated durability tests.

METHODS

Material Preparation

Two grades of Biomer polyurethane (standard and low modulus) were supplied as a 24 percent solution in dimethylacetamide (DMAc). The stock solutions were diluted to 8 percent concentration for casting the films in a moulding tray. The films were cast in a dry air oven at 70°C for 24 hours to an average thickness of 130 μm with a thickness variability of less than 10 μm.

Material Testing

Material thickness was measured with a Mitutoya thickness gauge with 10 mm diameter contact pads allowing 5 seconds for material creep. The elastic modulus of the material, in the physiological loading range, was determined in uniaxial tensile tests on an Instron tensile test instrument at strain rates of 0.001 s^{-1} (Barbenel *et al.*, 1987). Sample sizes of 80 x 10 x 0.13 mm were used and the elastic load deformation curve determined over the range of 0 to 1.0 MPa.

Valve Development

Prototype valves were fabricated using the frame design concepts developed for pericardial valves (Fisher and Wheatley, 1987). Size 27 mm tissue annulus diameter mitral valves were used for this initial development work: the valve frames have an outer diameter of 25 mm and an internal diameter of 22.8 mm. The acetal homopolymer frame has an array of small radially-projecting pins on which the leaflets are mounted. Predetermined holes in the leaflets allow precise positioning of the leaflets on the frame. Valves were manufactured with uncoated frames and with frames coated with polyurethane. Repeated dip-coating in low modulus Biomer and drying at 20°C was carried out until a coat thickness of approximately 200 μm was produced along the edge of the valve frame.

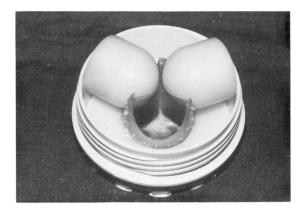

Fig. 1   A photograph of the jig and moulds used to form the leaflet geometry.

A spherical leaflet geometry was used for these initial prototype valves (Fisher *et al.*, 1988) which was defined by a hemispherical mould. At the free edge of the leaflets the geometry was defined by the area of contact between adjacent leaflets. The cut out in the valve frame was matched to the spherical geometry of the leaflets. Flat leaflets were cut out of the polyurethane film from templates and mounted on the valve frame using the predetermined holes. Leaflets were deformed using the hemispherical moulds (figure 1). The valve, moulds and constraining jig were immersed in the 20 percent DMAc solution for 2-3 hours at 70°C. At the end of the period the leaflets were permanently set in the specified geometry.

In the valves manufactured with coated-frames the leaflets adhered to the polyurethane coating on the outside of the frame and along the edge of the frame cut outs. To facilitate hydrodynamic testing, the valve assembly was completed with a thin acetal sleeve (figure 2). A number of prototype valves were manufactured to optimise the fabrication process and one valve manufactured from each material with and without coated frames were tested hydrodynamically (table 1).

Fig.2   The assembled valve including the thin sleeve prepared for hydrodynamic testing.

Valve Number	Material Modulus	Description
1	Low	Uncoated frame
2	Standard	Uncoated frame
3	Low	Coated frame
4	Standard	Coated frame

Table 1 Specifications of valves tested hydrodynamically.

Valve Testing

Valve function was assessed in our hydrodynamic flow simulator (Fisher *et al.*, 1986) under steady and pulsatile flows. The minimum steady flow and pressure difference required to open all three leaflets was measured along with the mean pressure difference, effective orifice area and closing regurgitant volume under pulsatile flows at a cardiac output of 4 l.min^{-1} and rate 70 beat.min^{-1}. Two valves with coated frames were also tested in accelerated durability tests.

RESULTS

The elastic modulus measured for the two types of Biomer used was 16 MPa for the standard material and 8.5 MPa for the low modulus material. Valves were manufactured with both types of material with and without coated-frames and the results of the hydrodynamic function tests are given in table 2.

Valve Number	1	2	3	4
Steady flow Minimal to open the leaflets ml.s^{-1}	120	145	133	125
Minimal pressure difference to open the leaflet mmHg (kPa)	2.8 (3.7)	3.2 (4.2)	2.5 (3.3)	3 (4.0)
Pulsatile flow Mean flow ml. s^{-1}	140	142	140	145
Mean pressure difference mmHg (kPa)	1.0 (1.3)	1.1 (1.5)	1.7 (2.2)	1.8 (2.4)
Effective orifice area (EOA) cm^2	2.5	2.3	2.1	1.9
Regurgitant volume during closure ml	6.6	5.7	4.6	3.9

Table 2 Hydrodynamic test results

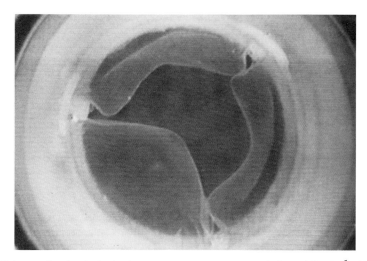

Fig. 3   Valve number 4 in the hydrodynamic test apparatus at a steady flow of 80 ml.s^{-1} with one leaflet in the closed position.

The minimal flow and pressure difference required to open all three leaflets were similar in all the valves. The small difference in elastic modulus did not have a major effect on the leaflet opening characteristics. Figure 3 shows valve number 4 and a low steady flow of 80 ml.s^{-1} in the pulse duplicator where one leaflet had failed to open. Under a pulsatile flow (Cardiac output 4 l.min^{-1}) all the leaflets of this valve opened up symmetrically (figure 4). Under these pulsatile flow conditions a peak flow rate of greater than 200 ml.s^{-1} passed through the valve to effect the leaflet opening. The effective orifice area of valves 1 and 2 with uncoated frames was greater than the valves with coated-frames (3 and 4), as in the coated frames the leaflets adhered to the frame along the edge of the scallop. This effectively moved the pivot point inwards, and reduced the portion of the leaflet that could flex to the open position. The closing regurgitant volumes on valves 3 and 4 were lower, due to the leaflets not opening as wide and the open position being less stable.

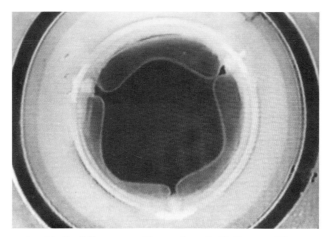

Fig. 4   Valve number 4 in the fully open position under pulsatile flow conditions at a cardiac output of 4 l.min^{-1} corresponding to a peak flow rate of 200 ml.s^{-1}.

Valves 3 and 4 were cycled in accelerated durability tests up to 35 million cycles when both valves had leaflet tears at the top of the post close to the pin attachment point. In both valves, the leaflets had separated from the coated-frame producing a high stress at the pin location point. There was no visual damage in the rest of the leaflets, nor any signs of fatigue cracks under low power optical microscopy.

DISCUSSION

Flexible leaflet heart valves manufactured from glutaraldehyde treated animal tissue have acceptable function and blood compatibility, but limited long-term durability due to microscopic fatigue failure of the collagen fibres and to intrinsic calcification. Synthetic materials such as medical-grade polyurethanes have the potential of providing similar function and biocompatibility to existing tissue valves but greater resistance to fatigue and calcification. The most recent work suggests that certain polyurethanes have acceptable function and durability *in vitro*, and in animal trials have showed surface calcification but to a lesser degree than pericardial bioprosthetic valves (Herold *et al.*, 1987). In this work the valve leaflets were manufactured from solvent-cast polyurethanes using a dip moulding technique which required a considerable level of technical expertise to produce leaflets of the correct thickness distribution and geometry. In this study we have investigated the used of two alternative Biomer polyurethanes and developed a novel means of leaflet fabrication with precisely controlled thickness and geometry from solvent-cast films.

A solution of 20 percent DMAc in water allowed precise moulding of the leaflets in the desired geometry, but the resulting bond, attaching the leaflet to the coated-frame, was not sufficiently strong to withstand repeated cycling in the durability tests.

A leaflet thickness of 130 µm was selected to compensate for the higher elastic modulus of the polyurethane compared to the chemically-treated animal tissue usually used in bioprosthetic heart valves. Bovine pericardium treated with glutaraldehyde can have an initial modulus of less than 1 MPa (Barbenel *et al.*, 1987) compared to 8.5 MPa and

16 MPa for the two materials used in this study. The pressure difference required to open a spherical leaflet is proportional to the product of the elastic modulus and the 4th power of the thickness and inversely proportional to the 4th power of the radius of curvature (Ghista, 1976). The reduction of leaflet thickness in the polyurethane valve increases the membrane stresses in the closed leaflets compared to the pericardial valve. In the four polyurethane valves tested, the minimal opening pressures (2.5 - 3.0 mmHg; 3.3 - 4.0 kPa) at steady flows of 120 - 145 ml.s^{-1} are acceptable for the aortic position, but are too high for the mitral position where the flows and pressure differences are much lower. However, only spherical leaflet geometries were used in these prototype valves and small changes in leaflet geometries in tissue valves have been shown to improve leaflet-opening significantly (Fisher *et al.*, 1988). Thus modifications to the leaflet geometries are expected to produce satisfactory leaflet openings at low flows, making them satisfactory for use in the mitral position. Under pulsatile flows, the valve functioned adequately and figure 4 shows the fully-open position of valve number 4.

The adhesion of the leaflets to the coated-frames (valves 3 and 4) did reduce the leaflet opening, but the pressure difference and effective orifice area of valves (3 and 4) were similar to the values expected in pericardial valves (Fisher and Wheatley, 1987). Valves 3 and 4 also showed reduced regurgitation during closure, because the leaflets did not open as widely and the open position was less stable than valves 1 and 2.

Valves 3 and 4 failed relatively early in the accelerated durability tests (after 35 x 10^6 cycles) due to failure of the adhesive bond between the leaflets and the coated-frame and, as a consequence, tears originated from the locating pin at the top of the post. Alternative methods of bonding the leaflets to the frame are now being considered.

In this initial study we have developed a new method of leaflet fabrication from solvent-cast polyurethane film which gives leaflets of uniform thickness and geometry. Further development of leaflet geometries are required to improve leaflet function at low flows, and improved methods of bonding leaflets to the frame are being investigated.

ACKNOWLEDGEMENT

Dr J Fisher was supported by the British Heart Foundation during this study. Mr V M Evans was supported by an SERC Advanced Course Studentship. Biomer polyurethane was supplied by Ethicon Inc., New Jersey.

REFERENCES

Annis D (1987). Polyether-urethane elastomers for small diameter arterial prostheses. *Life Support Systems*, 5, 47-52.

Barbenel J C, Zioupos P and Fisher J (1987). The mechanical properties of bovine pericardium in A Pizzoferrato, P G Marchetti, A Ravaglioli and A J C Lee (eds.), *Biomaterials and Clinical Applications*, Elsevier Science, Amsterdam, pp 421-426.

Black M M, Dury P J and Tindale W B (1983). Twenty five years of heart valve substitutes: a review. *J. Roy. Soc. Med.* 76, 667-680.

Coury A J, Stokes K B, Cahalan P T and Slaikeu P C (1987). Biostability considerations for implantable polyurethanes. *Life Support Systems*, 5, 25-39.

Fisher J, Jack G R and Wheatley D J (1986). Design of a function test apparatus for prosthetic heart valves. *Clin. Phys. Physio. Meas*, 7, 63-73.

Fisher J, Reece I J, Jack G R, Cathcart L and Wheatley D J (1987). Laboratory assessment of the design, function and durability of pericardial bioprostheses. *Engng. Med.*, 16, 105-109.

Fisher J and Wheatley D J (1987). An improved pericardial bioprosthetic heart valve. *Eur. J. Cardio Thorac. Surg.*, 1, 71-79.

Fisher J, Jack G R, Cathcart L and Wheatley D J (1988). Leaflet design and manufacture in pericardial heart valves in J C Barbenel A C Fisher, J D S Gaylor, W J Angerson and C D Sheldon (eds.), *Blood Flow in Artificial Organs and Cardiovascular Prostheses*, Clarendon Press, Oxford, pp 22-30.

Fisher A C, How T V, de Cossart L and Annis D (1985). The longer term patency of a compliant small diameter arterial prosthesis. *Trans. Am. Soc. artif. inter. Organs*, 31, 324-328.

Ghista D N (1976). Towards an optimum prosthetic trileaflet aortic valve design. *Med. Biol. Engng.*, 122-128.

Gilding D K and Reed A M (1987). Significant variables that must be controlled in the design and synthesis of polyurethanes for use in medicine. *Life Support Systems*, **5**, 19-24.

Goffin Y A and Bartik M A (1987). Porcine aortic versus pericardial valve: a comparative study of unimplanted and from patient explanted valves. *Life Support Systems*, **5**, 127-143.

Gogolewski S (1987). Implantable segmented polyurethanes: Controversies and Uncertainties. *Life Support Systems*, **5**, 41-46.

Hayashi D (1987). Tensile and fatigue properties of segmented polyether polyurethanes in H Planck, I Eyre, M Dauner and G Egbers (eds.), *Polyurethanes in Biomedical Engineering II*, Elsevier Science, Amsterdam, pp 129-149.

Herod M, Lo H B, Reul H, Muckter H, Taguchi K, Giersiepen M, Birkle G, Rau G and Messmer B J (1987). The Helmholtz Institute tri-leaflet polyurethane heart valve prosthesis in H Planck, I Eyre, M Dauner and G Egbers (eds.), *Polyurethanes in Biomedical Engineering II*, Elsevier Science, Amsterdam, pp 231-257.

Jansen J, Grevelink J M J, Kim S W, Kolf W J and Reul H (1986). New polyurethane trileaflet valves, performance and blood compatibility. *Life Support Systems*, **4** (2), 130-132.

Pierce W S, Parr G V S, Myers J L, Pae W E, Bull A P and Waldhausen J A (1981). Ventricular assist pumping in patients with cardiogenic shock after cardiac operations. *New Eng. J. Med.*, **305**, 1606-1610.

Reece I J, Anderson J P, Wain W H, Black M M and Wheatley D J (1985). A new porcine prosthesis, in vitro and in vivo evaluation. *Life Support Systems* , **3**, 207-227.

Reul H, Giersiepen M, Schindehutte H, Effert S and Rau G (1986). Comparative in vitro evaluation of porcine and pericardial bioprostheses. *Z. Kardiol.*, **75** (2), 223-231.

Schoen F J (1987). Cardiac valve prostheses: Review of clinical status and contemporary biomaterials issues, *J. Biomed. Mat. Res.*, **21**, 91-117.

8

AN ASSAY FOR THE INVESTIGATION OF LEUCOCYTE FUNCTION

T C Fisher, J J F Belch, J C Barbenel
and A C Fisher

INTRODUCTION

Within the microcirculation the formed elements in the blood pass through capillaries with a diameter of about 5-8µm. Red blood cells are highly deformable and rapidly traverse these vessels which are smaller than the greatest dimension of the cell. The passage of leucocytes, especially of polymorphonuclear leucocytes (PMNs), is more difficult because their size, their lower deformability, which is associated with increased time dependence, and their tendency to adhere to endothelial surfaces. In health, PMNs may be temporarily arrested in the entrance regions of small capillaries, but are released within a few seconds, and flow resumes (Branemark, 1976).

In myocardial infarction (MI), and perhaps many other diseases, it is known that permanent obstruction of capillaries can occur, resulting in tissue damage (Engler et al., 1983). Whether PMNs will actually flow through capillaries depends upon the balance between driving pressure and factors tending to promote adhesion, which may derive from either the cells themselves or the vessel wall. There will be areas in the ischaemic myocardium in which blood flow is borderline, promoting PMN entrapment, and following infarction, necrotic tissue will release PMN chemoattractants. However, there is also evidence to suggest that PMNs are abnormal before MI occurs. Several epidemiological studies have shown elevated white cell count (WCC) to be highly correlated with incidence, mortality and risk of recurrence in MI (Friedman et al., 1974; Lowe et al., 1985), and in stroke (Prentice et al., 1982).

To determine whether PMN activity rather than just cell count differs in these diseases, an assay is required. Various PMN functions may be studied: enzymes or other PMN granule contents may be measured by biochemical techniques, or the rate of migration of PMNs through filters may be assessed by microscopy.

A much more straightforward alternative is an aggregation assay, which measures the adhesion of PMNs to each other in an agitated suspension. The difficulty with previous aggregation techniques is that they need a pure population of PMNs suspended in buffer (Craddock et al., 1977). This requires the separation of the PMNs from whole blood, a procedure which is very time-consuming and may impair PMN function, defeating the object of the assay. In addition interaction between PMNs and other blood cell types is precluded.

To avoid these difficulties we have developed a whole-blood PMN aggregation assay which is more physiological than the previous techniques requiring cell separation, and also allows a much faster throughput of samples. This was evaluated with normal subjects and then applied to the investigation of PMN function in MI.

MATERIAL AND METHODS

Principles

Single PMNs and aggregates have different sizes and each can be counted by using an aperture impedance cell sizing device (Coulter counter with volume analyser). PMN aggregates are labile, and spontaneously disaggregate in the diluent for cell counting, particularly if EDTA is present. This may be overcome by the addition of formaldehyde, which stabilises the aggregates.

The main difficulty arising from the use of whole blood is the presence of the erythrocytes, which outnumber leucocytes by 800 to 1, and which must be removed before the cells are counted. Initial studies showed that proprietary erythrocyte-lysing agents, supplied for the purpose by cell counter manufacturers, were satisfactory if the leucocytes and aggregates had been previously fixed with a minimum concentration of formaldehyde. Too much fixation prevents adequate lysis, but a formaldehyde concentration of 0.2% was found to prevent disaggregation while still permitting lysis, providing that samples were processed within about 10-15 minutes. Leucocytes were also attacked by the lysing agent, although this was reduced by fixation. This has a differential effect on lymphocytes and PMNs, causing them to be resolved into two separate peaks on the volume distribution histogram - a very useful phenomenon.

The major chemotactic tripeptide produced by E Coli (Marasco *et al.*, 1984) N-formyl-methionyl-leucyl-phenylalanine (fMLP) was selected as the most appropriate aggregating stimulus because the response is reversible, allowing a check on the total number of cells. In addition fMLP has no direct effects upon platelets and hence does not produce large platelet aggregates which could be confused with leucocytes during volumetric classification.

Practice

Assays were performed using 2.5 ml of heparinised (10 U/ml) venous blood which was incubated at 37^0C in a 15 mm diameter polystyrene vessel, while continuously stirred with a 10 by 3 mm Teflon coated stir bar at 150 rev/min. After five minutes pre-warming a 100 µl aliquot was removed and dispersed in 20 ml of diluent (Isoton II; Coulter Electronics Ltd) to which 0.2% formaldehyde had previously been added. The aggregating stimulus was then added to the remaining blood, and further 100 µl aliquots were taken at appropriate time intervals. Three or four drops of the lysing agent (Ultra Lyse II; Clay-Adams Corp.) were then added to the diluted samples and the size distribution of the remaining leucocytes and aggregates recorded for each. Appropriate corrections were applied for the volume of the added stimulus.

Volume analysis was performed on a model ZM Coulter counter (Coulter Electronics Ltd) equipped with a volume analyser of our own design, the PWHA, described in detail elsewhere (Fisher and Fisher 1989). The main benefit of this system was that data could be transferred directly to a microcomputer for immediate analysis and that software procedures were developed for specific data characterisation, e.g. determination of the areas under each peak. This might present difficulties with other systems, which do not allow the user such flexibility in data transfer and handling. Subject to this reservation, any aperture impedance cell-sizing instrument could, in principle, be used.

The method was first evaluated using blood from normal young (21 29) male volunteers. Once validated, the assay was used to conduct a pilot study of PMN function in the very early period after MI. Samples were taken within 6 hours of onset, and MI was confirmed by ECG and enzyme changes. Twelve patients with MI were compared with an equal number of normal controls matched for age, sex and smoking habit.

RESULTS

Figure 1 shows three superimposed leucocyte volume distribution histograms obtained from a normal subject before stimulation and at two and five minutes after stimulation. The number of particles present in each population is given by the areas under each corresponding peak. The control trace shows only two peaks, corresponding to lymphocytes and single PMNs, with very few larger particles present. Two minutes after the addition of the stimulus, the number of single PMNs falls, and this is associated with the appearance of new peaks at double and triple the size of single PMNs produced by aggregates of two and three cells respectively. By five minutes, partial reversal of these changes is evident. The lymphocyte peak remains unchanged throughout, because lymphocytes are unaffected by the stimulus, and this forms a useful internal reference in the assay; changes in lymphocyte count indicate pipetting errors or blockage of the Coulter aperture.

Results from a group of eight normal volunteers showed maximal aggregation between one and two minutes in all cases. Disaggregation of duplicate samples with EDTA recovered over 95% of the PMNs at each time point, demonstrating that the fall in single PMN count was due to aggregation, with minimal adherence to the incubation vessel. The dose-response curve for serial dilutions of fMLP is shown in figure 2. The threshold for activation occurred at 10^{-9}mol/1, reaching a plateau at 10^{-7}mol/1, which is comparable with the previously published results for separated-cell assays (Craddock *et al.*, 1977; O'Flaherty *et al.*, 1977).

The myocardial infarction patients showed enhanced aggregability. In normal subjects, aggregate peaks of greater than three cells (triplets) were rarely observed, but following MI, when large numbers of aggregable cells were present, aggregates of up to 5 cells were clearly discernible (figure 3). The number of PMNs aggregating was also greater following MI; the cumulated results from the study are shown in figure 4. A highly significant increase ($p<0.005$, Mann-Whitney U-test) is seen in the aggregability of the PMNs in myocardial infarction. Total WCC was also much elevated, and the differential counts revealed that this increase is due mainly to a rise in the PMNs, as expected.

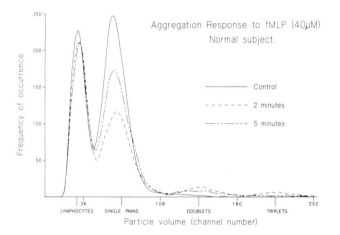

Fig. 1   Aggregation response to fMLP. The height of the peak produced by PMNs falls in response to aggregation, and there is an associated appearance of double and triple cell aggregates. Partial disaggregation has occurred by five minutes.

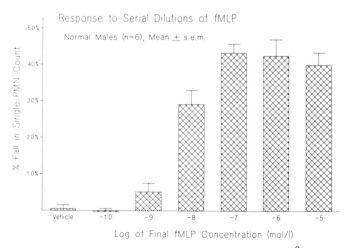

Fig. 2   Dose-response curve for fMLP. The activation threshold occurs at 10^{-9}mol/1.

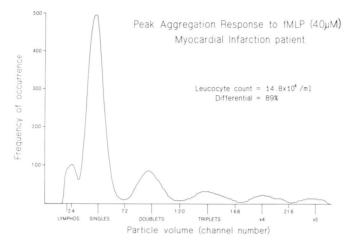

Fig. 3   Enhanced aggregability in myocardial infarction patients. Aggregates containing 5 PMNs are apparent.

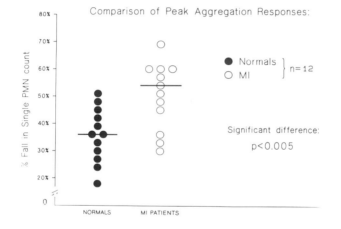

Fig. 4   Comparison of peak aggregation in normal subjects and myocardial infarction patients. The patients show significantly increased aggregability of PMNs.

DISCUSSION

The assay was found to be simple and reliable. The fall in single PMN count and appearance of aggregate peaks show clearly that aggregation occurs in response to the stimulus and that these changes are readily quantified. Reproducibility was very good, with duplicate samples giving distributions virtually identical at each time point,   and the intra-assay coefficient of variation based on nine replicates of the same sample was less than 5%.

The cell count did not appear to influence the assay. In normal subjects no correlation was observed, but the range of cell counts was small (4-7 x 10^6/ml). Cell counts below 2 x 10^6/ml do not produce satisfactory volume distributions with our system. The possibility that enhanced aggregation in the MI samples was merely a function of the high PMN counts was assessed by diluting samples with buffy-coat free blood to within normal values. No alteration of aggregation response was observed following this manoeuvre, confirming the presence of a more-aggregable population of cells.

The advantages of using the whole blood technique compared to separated cell assays are:

Speed of processing; samples can be taken from the donor and processed within ten minutes. Sample throughput is much enhanced, and time-dependent effects on PMN function are minimised.

Avoidance of the separation procedure which may modified PMN behaviour through contact with the various foreign surfaces and reagents used for density gradient centrifugation.

Aggregation occurs in a near natural milieu in the presence of all blood cell types and plasma constituents, allowing them to exert any modifying influences on PMN function which might occur *in vivo*.

This latter point also suggests one minor potential problem - that the other cell types may be affected by the stimulus and interfere with the volume distribution. However, the lymphocytes do not respond to fMLP or any other chemotaxin, and although monocytes do respond to fMLP, no noticeable effect has been observed to date presumably because too few are present. Large platelet aggregates are formed spontaneously on vigorous agitation of whole blood, which are counted in the volume distributions. To avoid this problem, samples were mixed by very gentle inversion after venepuncture, and the speed of stirring during the assay kept to an absolute minimum.

The MI study revealed that very soon after MI an increased number of more aggregable PMNs is found in the peripheral blood. This may be entirely a secondary response to trauma but it might also reflect, to some extent, the pre-morbid condition, which would fit with the results of the epidemiological studies. Further studies in stable and unstable angina are planned to clarify this point. Whatever the case, as the treatment of acute MI is orientated increasingly towards achieving early reperfusion, the potential for these activated PMNs to immediately occlude capillaries in the reperfused area would suggest that concurrent anti-PMN therapy might be beneficial.

REFERENCES

Branemark P I (1976). Microvascular function at reduced flow rates, in R M Kenedi, J M Cowden and J T Scales, (eds.), *Bedsore Biomechanics*, Macmillan Press, London, pp 63-68.

Craddock R P, Hammerschmidt D E, White J G, Dalmasso A P and Jacob H S (1977). Complement (C5a)-induced granulocyte aggregation in vitro: a possible mechanism of complement-mediated leukostasis and leukopaenia. *J. clin. Invest.*, 60, 260-264.

Engler R L, Schmid-Schonbein G W and Pavelec R S (1983). Leukocyte capillary plugging in myocardial ischaemia and reperfusion in the dog. *Am. J. Path.*, 111, 98-111.

Fisher T C and Fisher A C (1989). The PWHA, a signal analysis device for the accurate determination of cell volumes, in J C Barbenel, A C Fisher, J D S Gaylor, W Angerson and C Sheldon (eds.), *Blood Flow in Artificial Organs and Cardiovascular Prostheses*, Oxford University Press, Oxford, pp 219-226.

Friedman G D, Klatsky A L and Siegelamb A B (1974). The leukocyte count as a predictor of myocardial infarction. *New Engl. J. Med.*, 290, 1275-1278.

Lowe G D O, Machado S G, Krol W F, Barton B A and Forbes C D (1985). White blood cell count and haematocrit as predictors of coronary recurrence after myocardial infarction. *Thromb. Haemstas.*, 54, 700-703.

Marasco W A, Phan S H, Krutsch H, Showell H J, Feltner D E, Nairn R, Becker E L and Ward P A (1984). Purification and identification of N-formyl-methionyl-leucyl-phenylalanine as the major peptide chemotactic factor produced by Escherichia coli. *J. Biol. Chem.*, 259, 5430-5436.

O'Flaherty J T, Kreutzer D L and Ward P A (1977). Neutrophil aggregation and swelling induced by chemotactic agents. *J. Immun.*, 119, 232-239.

Prentice R L, Szatrowski T P, Kato H and Mason M W (1982). Leucocyte counts and cerebrovascular disease. *J. Chronic Dis.*, 35, 703-714.

9

DISCUSSION: ARTIFICIAL ORGANS AND BIOMATERIALS

Kennedy began the discussion by asking Klinkmann to comment on the present situation with respect to knowledge of uraemic toxins. In reply, *Klinkmann* stated that the definition of uraemic toxins remained a topic of great difficulty, although it appeared unlikely that there is a single uraemic toxin. Recalling the pioneering work of Kennedy with the disequilibrium syndrome, Klinkmann added that it was still necessary to consider the consequences of blood purification which was performed too quickly.

Courtney asked Klinkmann if he envisaged the use of the total artificial heart as an independent treatment or in conjunction with a heart transplant. *Klinkmann* replied that at the present stage of development, particularly with respect to the extracorporeal driving unit, he believed there was no place for the permanent replacement of heart function by an artificial heart. He was convinced that current artificial heart systems are only justified as bridging for heart transplants. Indeed, recent evidence suggested that implantation of an artificial heart prior to transplantation is beneficial to the recipient. Klinkmann saw an analogy between present artificial heart studies and the previous application of the artificial kidney, which was restricted to acute renal failure. In his view, the future of the artificial heart does not depend on the development of a pump but on the development of an implantable driving system, which would make the total device implantable and avoid infection. This was not likely for 20 years. *Malagodi* asked Klinkmann to comment on the status of the constant flow artificial heart versus the pulsatile flow. *Klinkmann* replied that this was the basis of ongoing discussion, although constant or non-pulsatile flow has regained favour. Klinkmann's presentation had dealt with what can be termed conventional designs for the artificial heart. New designs are being developed, including a constant flow pump by the group of Nosé at Cleveland Clinic. At present, clinical application of the artificial heart was based mainly on pulsatile flow, although non-pulsatile flow may find increased application in cardiac assist devices.

Forbes asked Maini two questions on the assessment of the coated or sealed vascular prosthesis. The first question related to the C-reactive protein (CRP) test and whether or not the CRP level was a consequence of the operation or was induced by the graft. *Maini* replied that the test compared the sealed prosthesis with an unsealed and preclotted Dacron graft as a control. The high peak at the beginning could be attributed to the trauma of the operation but evidence for the suitability of the sealed graft came from the fact that there was no difference between the two CRP levels post discharge. The second question was whether or not Maini believed that there is likely to be a graft failure in a small number of patients irrespective of the type of graft used. *Maini* replied that information was still being collated but he could make a statement on the basis of 200 patients at Glasgow Royal Infirmary for periods ranging from 3 months to 3 years. The expected patency for the aortal/bifermoral area was 95% but to date 100% had been achieved, although this level was unlikely to be maintained. *Gorham* asked Maini if labelling techniques or similar procedures had been performed to show how rapidly the implant impregnate is turned over *in vivo* or if there was evidence other than histological to demonstrate how quickly the gelatin disappears. *Maini* stated that labelling was an involved technique and, as degradation was by hydrolysis, he was satisfied that this process could be modelled in the laboratory. *Gorham* added that since the process was also mediated by lysosomal enzymes, variation among individuals was certain. *Maini* agreed but stated that he was primarily interested in the solubility of gelatin and its passage into the bloodstream rather than its phagocytosis.

Bowry raised a point in relation to the presentations of Forbes, Courtney and Smeby. With respect to blood compatibility assessment, numerous parameters and test systems had been described but it was important to bear in mind that contact with materials other than the test material is often unavoidable and care is required in the interpretation of the data. *Forbes* agreed with this point as it related to the problems of obtaining the blood and of the possible introduction of needles, cannulae and anticoagulants. *Courtney* stated that it was necessary to be clear as to the objective of *in vitro* tests. They could be utilised, as described by Smeby, in screening procedures for the evaluation of a large number of polymers or they could be used for the more difficult objective of relating *in vitro* performance to a characteristic property of a polymer. Courtney saw an analogy to the early development of toxicity tests in that it was possible to rank materials, particularly with procedures based on release reactions, such as the release of beta thromboglobulin. In these cases, it was possible to obtain a correlation between *in vitro* and clinical performances. *Smeby* agreed that *in vitro* tests can be good for screening but he believed that a final judgement should be based on the *in vivo* situation where clinical results could be seen.

Maini raised the topic of heparinised surfaces, in particular the use of albumin-heparin conjugates, and asked Courtney and Forbes whether they thought that leaching of heparin from the surface is essential for the retention of the antithrombotic capability. *Courtney* replied that one view is that binding of heparin to a surface can lead to a different protein adsorption pattern and improved compatibility in the absence of heparin release. With respect to albumin - heparin conjugates, albumin adsorbs readily to polymers and heparin can react with the albumin to produce a surface where contributions to improved blood compatibility come from both albumin and heparin moieties. *Forbes* added that the potential benefit of albumin - heparin conjugates had been demonstrated in the recent thesis of Robertson (1988).

Klinkmann urged caution over the terms haemocompatibility and biocompatibility and warned that misunderstanding can result from the use of these terms as synonyms.

Drawing attention to the general acceptance of the strong influence of protein adsorption on blood compatibility, *Klinkmann* asked Forbes if he believed that the nature of the adsorbed protein related more to the metabolism of the patient or to the composition of a biomaterial.

Forbes felt there was no absolute answer but that the question was a timely reminder that biomaterials are used with patients, who have abnormalities in their plasma proteins, particularly in the coagulation proteins. It seemed relevant that biomaterials should be assessed in the clinical situation, for example, in patients with renal failure or with thrombotic tendency due to chronic vascular disease, because it is known that fibrinogen and factor VIII will be high and other factors are in a prothrombotic situation. *Courtney* added that with protein adsorption it has to be borne in mind that *in vitro* tests can produce misleading results. An example is fibrinogen adsorption, which *in vitro* follows a classical adsorption pattern but, as shown by the thesis of McLaughlin (1988) behaves differently in the clinical situation.

Klinkmann asked Smeby to comment on the relevance of the influence of membrane thickness on platelet adhesion in comparison to the influence on complement activation. In reply, *Smeby* stated that there was a clear relationship between membrane thickness and complement activation, since activating components can be removed and less returned to the patient. However, the situation with respect to platelet activation and platelet deposition was less certain. The roughness of a membrane would have a definite influence but he was unaware of literature relating to thickness. *Klinkmann* felt it was important to have confirmation that platelet adhesion was only influenced by membrane composition in contrast to complement activation, which was influenced by both membrane composition and performance characteristics.

Forbes raised a general question on assessment as to which features among clotting factors, platelets, complement and permeability factors should be given preference. *Klinkmann* indicated that as broad a range as possible was preferable and *Kennedy* agreed.

Srivastava asked Courtney if he had found that the surface properties of a membrane could be influenced by the degree of modification. *Courtney* replied that for certain parameters there is not always a correlation with the degree of modification and he contrasted the influence on complement activation of the substitution of the hydroxyl groups of cellulose with acetate or diethylaminoethyl groups, where a small amount of diethylaminoethyl groups causes a marked reduction. This finding and evidence from the University of Aston on fibroblast attachment to modified hydrogels support the view that the nature of the substituting group may have greater importance than the degree of substitution.

REFERENCES

Robertson L (1988). Blood compatibility of modified biomaterials: Application of selected in vitro and ex vivo procedures. Ph.D. Thesis, University of Strathclyde, Glasgow.

McLaughlin K M (1988). The interaction of proteins with polymer surfaces. M. Phil Thesis, University of Strathclyde, Glasgow.

10

WHEELCHAIR BIOMECHANICS

C A McLaurin and C E Brubaker

INTRODUCTION

For many persons with physical disabilities, the wheelchair is the primary mode of transportation and in many cases mobility is dependent upon the occupant's arm motion. Thus wheelchair biomechanics is primarily a study of how a person utilises the arms to propel the wheelchair and how this may be optimised. Unlike walking each propulsion cycle can be intermittent, with a short period of rest between each stroke, depending on the coasting characteristics of the wheelchair upon the surface in question. This frequency of stroke will affect the power input and muscle activity and segmental arm motion all of which are used to describe the stroke activity. Other factors which affect the stroke are work rate, speed and seat position. The last named has considerable influence when using handrims and deserves careful attention.

All of these factors have considerable influence on propulsion efficiency which holds special interest for wheelchair propulsion since, unlike walking, the work done is easily measured on a treadmill. Efficiency is perhaps the single most important indicator of the effectiveness of the propulsion stroke for a specific condition. Several studies have shown that levers, cranks and gears can greatly improve propulsion efficiency, but simplicity, low cost and easy control offered by handrims have so far excluded the use of these alternate means of propulsion for general use.

POWER REQUIRED

The power required to propel a wheelchair is easily measured by placing the wheelchair suitably loaded on a treadmill and tethering it to a force gauge while the treadmill is run at various speeds. Since modern bearings offer very little resistance, the force to pull the wheelchair is dependent almost entirely upon the tyres and the total weight. High pressure pneumatic tyres offer the least resistance while clay filled solid rubber tyres may require four times the force. Modern synthetic solid or foam filled tyres approach the low drag of pneumatics and offer better wear characteristics with a harder ride. Large diameter wheels offer less drag than small diameter, therefore to reduce drag more weight should be placed on the large main wheels and less on the small caster wheels.

The drag force is essentially proportional to the weight and independent of speed. The power, of course, is directly proportional to speed. Treadmill tests have shown that alignment is very important. Two degrees toe-in or toe-out can double or triple the rolling resistance but up to 10^0 of camber has little effect.

The suface over which the wheelchair is rolling is more difficult to assess. If the surface is firm and smooth, friction will not play a part unless there is misalignment. Carpet has been measured, resulting in double the drag force for firm carpet and four times the drag for shag carpet.

At normal wheelchair speeds, wind resistance is not a factor, but a head wind can increase the drag force considerably. Studies in the NASA Langley (Coe, 1979) low speed wind tunnel indicated a drag coefficient of 1.56 with an occupant wearing loose clothing. This results in a drag force of 12 N with a head wind of 20 km/h. The drag force is proportional to the square of the wind velocity. The force required to propel a wheelchair up a ramp can be determined by multiplying the total weight by the sine of the angle of the ramp.

On a hard smooth surface, a modern wheelchair with a 75 kg occupant should have a drag force of only about 10 N but on an incline or with a head wind, this may be increased to 100 N or more. A force of this magnitude, while quite possible to overcome, will result in a much lower speed in order to limit the power required.

Other factors that affect the rolling resistance are side slopes and caster flutter. a 2^0 side slope may require twice the power input to overcome the tendency to turn downhill (with a front-castered wheelchair) while caster flutter can induce ten times the normal drag.

TYPICAL STROKE

Most of the biomechanical studies at the University of Virginia Rehabilitation Engineering Centre (UVa REC) have been conducted on a dynamometer (Figure 1) which simulates wheelchair propulsion. The dynamometer measures torque and speed with preset values of work output corresponding to level or ramp conditions. The seat position can be adjusted and either rims or levers can be used with left and right independence. Arm segment position is recorded by mechanical linkages to potentiometers. Because the subject is stationary, arm motion, emg., heart rate and oxygen consumption are relatively easy to monitor.

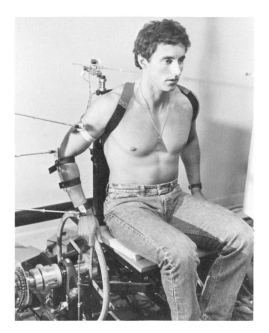

Fig.1   Dynamometer showing subject working at 4 km/h with a load of 15 W. The wands attached to the right arm monitor motion of the shoulder, elbow and wrist.

Studies with this apparatus have used subjects with normal and impaired lower extremities (paraplegics) with no observable difference in performance. A typical stroke (Figure 2) begins with the forward motion of the hand prior to contacting the handrim. At this point, the hand, still accelerating may not be up to rim speed and it is not uncommon to see a slight negative torque at initial contact. The torque input then begins and rises steeply to either a single or double peak. With a single peak a discontinuity usually occurs either on the up slope as shown in Figure 2 or the down slope of the torque curve. This is attributed to the change in muscle pattern from pulling to pushing as the hand moves over top centre. At the end of the

torque input, the hand releases the grip and begins the recovery stroke, swinging back to the
starting position. It should be noted that with a typical total stroke time of 1.2 s, torque input
occurs for only 0.3 s. The muscle function is indicated in Figure 3. Since surface electrodes were
used, discretion is needed when interpreting these results particularly with respect to muscle
effort. The muscle activity, arm position and torque curve, all vary considerably depending on
several factors described in the next section. Ideally muscle activity should be maximum during
the torque input phase and minimal during the return stroke. It should be noted in Figure 2 that
the arm motion cannot be considered to be acting in one plane only.

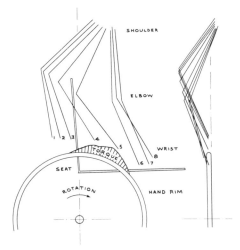

Fig. 2   Stick Diagram showing side and front views of a typical forward stroke for a 74 kg subject (Figure
          1) in a traditional seat position. Right arm segments are shown at intervals of 0.01 sec.
          Speed 4 km/h. Power 15 W. Stroke Frequency 53 cycles/min. Peak torque 17 Nm.

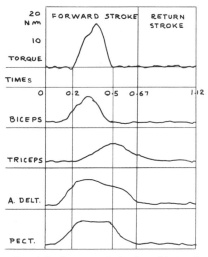

Fig. 3   Diagram showing muscle activity for one cycle for the subject and conditions shown in Figure 2.
          Values are estimated from EMG records of several runs.

Typically the upper arm is adducting as it is flexing at the shoulder. This is more apparent
with subjects having longer arms relative to trunk length.

FACTORS THAT AFFECT THE STROKE

The conditions that affect the stroke are speed, work load and seat position. At higher speeds, such as in racing, there is no time to grip the rim and torque is transmitted by contact usually by the thumb and fore-finger which become callused. In this situation, the torque curve is spike like with a very short duration and correspondingly high force. The hand motion begins well before rim contact and continues well beyond the end of torque input before beginning the recovery stroke.

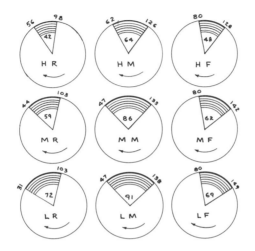

Fig. 4   Grasp and release points for 9 different seat positions: high, middle and low (H.M.L.) and rear, middle and forward (R.M.F.).

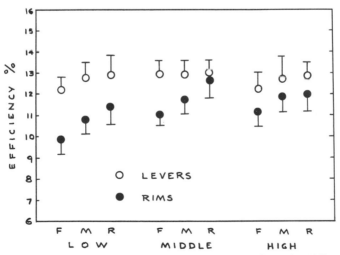

Fig.5   Mean and standard errors for efficiency for levers and rims at forward, middle and rear, and low, middle and high seat positions.

With higher work loads, as in ascending an incline, speed will be slower than normal and the torque input time may be increased while the recovery stroke may be quickened to minimise deceleration or prevent roll back. The stroke frequency is dependent in part upon the coasting characteristics of the wheelchair and terrain. On a smooth level surface with a low drag wheelchair, the torque input can be increased and the recovery time lengthened resulting in a slower frequency, say 30 cycles per minute, instead of the typical 60 cycles per minute. It is necessary only to maintain a work input equivalent to the work required. The work input per stroke is represented by the area under the torque curve. Seat position has a marked effect on the stroke pattern and to some extent on the efficiency of the propulsion. Figure 4 shows how the grasp and release position of the hand on the rim are affected by nine different positions of the seat with respect to the hub. Figure 5 shows the average results for 6 subjects when the seat position is changed forward and rearward and vertically. The traditional seat position, which is shown in Figure 2 corresponds to the middle forward position. The seat position affects also the torque curve, the motion of the limb segments and the muscle pattern.

EFFICIENCY

The propulsion efficiency is a good measure of how well a wheelchair suits an individual. As in other exercise routines, the efficiency is higher for higher work loads. It is readily determined on the dynamometer by comparing the power output with the metabolic energy rate measured by oxygen consumption in a steady state or aerobic condition using the equation:

$$1 \text{ litre } O_2/\text{min} = 349 \text{ W}$$

Since wheelchair propulsion on a smooth level surface with good tyres requires only about 10 watts at 3.6 km/h, efficiencies may be quite low, of the order of 3 to 5% for most subjects. However at higher work loads such as experienced on an upward incline the work load and efficiency can be greatly increased. The highest power output recorded at UVa was 1.56 watts per kilogram of body weight with an efficiency of 13.9%. It is interesting to compare this with Daedalus athletes (Nadel and Bussolari, 1988). who with foot pedals maintained an average output of 5.25 W/kg for four hours with efficiencies ranging from 18 to 34%. It is important to note that efficiency depends upon the mechanism and how well it is adapted to the individual. Using handrims, tests have shown that for moderate loads, gearing which reduces the input speed can increase efficiency. Although this may not be a practical solution, it does indicate that the drive mechanism is a factor worthy of consideration.

ALTERNATIVE PROPULSION SYSTEMS

Several studies have shown that alternative means such as cranks and levers result in much improved efficiency. Brubaker *et al.* (1984) have shown that single acting levers are not only more efficient than handrims, but are less dependent on seat position and provide a simpler means for gearing. This is shown in Figure 5 which compares levers and rims for various seat positions. It can be argued that levers provide a better means for utilising the arm and shoulder muscles, particularly for high loads where efficiencies as high as 20% have been recorded. It should be noted that less effort is required for gripping and for supporting the arm during the recovery stroke. This last factor propably plays a significant role using handrims with a low seat. Most applications using cranks or levers limit the control and manoeuverability of wheelchairs, although recent designs appear to have overcome this difficulty, allowing wheelies, curb hopping and other such manoeuvres, much the same as when using rims. The major application appears to be for those with impaired hand grip and arm strength. It will be interesting to see if such devices will replace the simplicity and low cost of handrims. Recent studies by Quinnan-Wilson *et al.* (1988) have shown that ski poles are considerably more efficient than either levers and rims. This raises some very interesting questions regarding the use of arm and shoulder muscles and presents new mechanical problems concerning the dynamic stability and braking of wheelchairs.

REFERENCES

Brubaker C E, McClay I S and McLaurin C A (1984). Effect of Seat Position on Wheelchair Propulsion Efficiency. *Proc. 2nd. int. Conf. Rehab. Engng.*, Ottawa, Canada, 12-14.
Coe, Paul L, Jr (1979). Aerodynamic Characteristics of Wheelchairs. *NASA Technical Memorandum 80191.*

Nadel, R and Bussolari, R (1988). The Daedalus Project Physiological Problems and Solutions. *Am. Sci.*, July-August, 350-351.
Quinnan-Wilson L, Hughes C J, Chen J J and Brubaker C E (1988). Efficiency Comparison of Wheelchair Propulsion Using Handrims, Lever Drives and Ski Poles. *Proc. ICAART 88*, Montreal, Canada, 300-301.

11

PRESSURE SORE PREVENTION FOR SPINAL INJURED

WHEELCHAIR USERS

J C Barbenel

INTRODUCTION

This 25th Anniversary publication is one of a regular series of books recording the proceedings of Strathclyde Bioengineering Seminars. That held in 1975 dealt with pressure sores, and the Seminar and the resulting publication (Kenedi *et al.*, 1976) were epoch making, initiating considerable research on the prevention and treatment of pressure sores, surveys of sore prevalence and, ultimately, the formation of the Society for Tissue Viability. Unfortunately there is little evidence that this research and increased interest in pressure sores has resulted in any major benefit to patients; the delivery of care and devices for the prevention of sores remains limited and largely unsatisfactory.

Spinal injury patients are at special risk of developing pressure sores and this problem has led to the development and organisation of systems for the provision of suitable wheelchair cushions to these patients. One such programme which was based in Phillipshill Hospital, Glasgow, will be described as an example of how to achieve the specific aim of providing patients at risk with suitable devices but also as a paradigm on how aids, devices and rehabilitation can be delivered to disabled people.

THE PROBLEM

Para-and tetrà-plegic spinal cord injured patients are at particularly high risk of developing pressure sores and skin care and sore prevention is an important part of their long term rehabilitation (Watson, 1983). The consequence of care failure and pressure sore development has been illustrated by Lawes (1984), who reported on the pressure sore history of 260 patients with spinal cord injury; he found that 30% of the patients had developed a sore in the spinal injury unit during the initial stages of treatment and that 25% of the paralysed patients had been readmitted with a sore over the five year period covered by the study. Additional data from the US Regional Spinal Cord Injury System for the period 1978-80 indicated that 40% of the 7000 spinal cord injured people in the study developed a pressure sore between their injury and initial rehabilitation discharge. After discharge about 30% of the rehabilitated patients per year developed a pressure sore (Young and Burns, 1981).

The development of a pressure sore will delay initial discharge of the patient after their injury and will lead to discharged people being readmitted to hospital. The cost of the additional patient load has been extensively discussed and is generally agreed to be considerable, both in financial and manpower terms. The development of a sore will also have consequential effects on loss of income but this is largely masked by the high unemployment rate of spinal cord injured people.

RATIONALE FOR PRESSURE SORE PREVENTION

Pressure sores are a complex form of injury which may have many possible initiating causes. The most common cause is, however, the prolonged application of loads to the skin, particularly forces acting normal to the skin surface, usually called "pressure".

Animal studies show that both the magnitude of the force and the time for which it acts are important and there is an inverse relationship between the force magnitude and time; damage will be produced by small forces acting for long times or for high forces acting for short periods. The general applicability of the animal studies has been confirmed by Reswick and Rogers (1976), who measured the pressures at the support surface interface of seated patients, some of whom developed pressure sores. The results (Figure 1) show, once again, the inverse pressure - time relationship. The absolute magnitude of the pressures in Figure 1 are open to some doubt and the authors suggested that they be treated as guide lines. The uncertain zone between acceptable and unacceptable tissue loading is of considerable importance and reflects the real difference between the response of different patients.

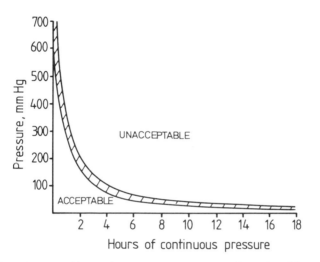

Fig. 1   Guidelines for acceptable interface pressures suggested by Reswick and Rogers (1976) for seated subjects; 100 mmHg = 13.3 kPa.

The established interrelationship between pressure and time required for the production of tissue damage provides a rational basis for sore prevention. The aim should be to reduce both the time for which the pressures act and the magnitude of the pressures.

For spinal injured patients who spend long periods in wheelchairs the duration of pressure action can be reduced by the patient making intermittent movements which relieve the pressure and allow the tissues to recover. For most patients the preferred method of pressure relief is partial raising from the chair by pressing down on the chair arms - a manoeuvre usually called a push-up. Patients who cannot push-up, particularly those who are quadriplegic, can produce intermittent pressure relief by changing posture, either by leaning forward or from side to side.

The training of patients to provide regular pressure relief is a key element in the process of long term rehabilitation of spinal injury patients but the preferred frequency and duration of pressure relief is largely a matter of personal opinion, ranging from 10 seconds relief every 10 minutes (Noble, 1981) to 1-2 minutes every hour (Watson, 1983). Measurements suggest, however, that relief movements are actually made at irregular intervals and are of variable duration (Fisher and Peterson, 1983; Barbenel *et al.*, 1984; Merbitz *et al.*, 1985).

The magnitude of the pressure is less amenable to simple modification. The pressures acting on the skin of a sitting patient are produced by the reaction forces generated by the proportion of the subject's weight supported by the seat. The total weight supported can be marginally altered by adjustment of the wheelchair, particularly of the foot rests. Since it is not possible to make a major reduction in the weight supported by the seat, the average interface pressure cannot be reduced to any great extent. Major alterations can only be achieved by redistributing the pressures to make it more uniform and to reduce the pressures over the sites at risk. These aims can be achieved by the provision of a suitable wheelchair cushion and the core of any pressure sore preventive programme for seated spinal cord injured patients must be the choice, provision and maintenance of a suitable wheelchair cushion.

PROVISION OF WHEELCHAIR CUSHIONS

There are many types of wheelchair cushions available but no single type is suitable for all patients. The primary objective in providing a cushion for spinal cord injured patients is to ensure that the pressures over sites at high risk of developing sores, particularly the ischial tuberosities, are as low as possible and preferably less than 40 mmHg (5.3 kPa). This must be achieved without overloading and endangering the tissues at the lower risk sites where the pressures are increased because of the redistribution. The methodology for achieving this objective is based on that described by Reswick and Rogers (1976) who used flexible foam cushions and a programme of provision which are best considered under two headings:

Equipment

For many patients it is possible to achieve a satisfactory pressure distribution on a simple rectangular block of high resilience, flexible polyurethane foam with a density of
30 - 40 kg/m^3. A thickness of 75 mm may be satisfactory for light patients but most will require a 100 mm thick block. Very heavy patients will need a 100 mm thick block of a denser foam.
Simple foam blocks are unsatisfactory for use with wheelchairs with sling seats which sag. The flexible blocks require a more rigid polythene foam base which is shaped to compensate for the sag. A cover will also be required. Biaxially extensible covers produced least modification to the load - deformation and pressure distributing properties of the foam (Ferguson-Pell *et al.*, 1981). Unfortunately no satisfactory waterproof two way stretch material is currently available and a one way stretch cover is used for incontinent patients who need the cover to be waterproof.
Contouring the surface of the foam block is a useful way of redistributing pressure from the ischial tuberosities to the lower risk regions of the femoral trochanters and thighs. The simplest contoured shape is produced by making a cut-out to remove a rectangular block of foam from the area beneath the tuberosities and coccyx; the polythene foam base should be similarly modified (Figure 2). The cut-out may either be left empty or filled with very low density foam.

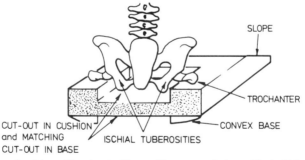

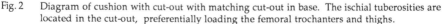

Fig. 2   Diagram of cushion with cut-out with matching cut-out in base. The ischial tuberosities are located in the cut-out, preferentially loading the femoral trochanters and thighs.

The dimensions of the cut-out may be most easily established by using a flexible bag containing polystyrene beads. The patient is asked to sit upon the bag and air is introduced to make the beads more fluid and to obtain an impression. The air is then evacuated to lock the beads together and make the impression less deformable. The bag retains an impression of the patient's sitting area with the locations of the tuberosities easily seen. The cut-out should be larger (by 1 cm) than indicated by the impression to allow for variability of patient posture.
A sloping front to the foam block will often allow easier transfer by the patient from the wheelchair.

Initially the bioengineer providing the cushion also manufactured it. Once the method and details of production had been established, the cushions were made by a commercial manufacture to a prescription, on a standardised prescription form.

Pressures generated at the cushion-skin interface may be measured by a variety of devices (Barbenel, 1983) but the pressure evaluator described by Mooney *et al.* (1971) has the advantages of simplicity and commercial availability (Talley Medical Limited, Borehamwood, Herts.). The device is available in a variety of sizes but a diameter of 100 mm appears to be most suitable and useful for routine clinical use.

Organisation

Before a cushion is provided it is necessary to ensure that the patient has a satisfactory wheelchair which is correctly adjusted. In addition the patient must be aware of the importance of regular pressure relief and self care of the skin.

The process of providing a cushion can be divided into several sequential steps.

Assessment - The patient should be generally assessed as to body build and weight. The posture should be evaluated to decide if asymmetry is likely to produce higher unilateral tissue loads, or symmetrically higher loading of the buttocks or thighs. The condition of the tissues in the sitting area should also be examined, particularly those overlying the bony prominences of the ischial tuberosities and femoral trochanters. The presence of atrophic tissue or scars from previous sores or surgery makes these sites particularly prone to breakdown and the control of the pressure in these areas is particularly critical.

Cushion Selection and Evaluation - The patient is provided with a polyurethane slab cushion, complete with base and cover. The wheelchair should be adjusted to compensate for the presence of the cushion and the posture of the patient confirmed as being satisfactory. The pressures beneath the ischial tuberosities and trochanters are then measured to check that the ischial pressures are less than 40 mmHg (5.3 kPa) and the trochanteric pressures no greater than 60 mmHg (8 kPa). If the pressures are satisfactory the trial cushion can be given to the patient or a copy prescribed.

If the pressures are not satisfactory a cushion with a cut-out is tried and the size and depth of the cut-out adjusted until the measured pressures are acceptable and once again the cushion either provided or prescribed.

If the cushions are prescribed it is essential that the cushions manufactured to the prescription are evaluated by interface pressure measurements. Flexible foams may show significant differences between batches of nominally the same material and mechanical properties may degenerate after prolonged storage. If the manufactured cushion does not provide satisfactory pressures another should be made, or a less dense foam used.

Although the pressures are judged to be satisfactory, there is considerable difference in response of the skin and superficial tissues of different patients. Whenever possible the patient should use the cushion, after which the skin of the loaded area is observed in order to confirm that there are no sites which show signs of trauma (e.g. prolonged hyperemic response). Alternatively temperature measurements or thermography may be used to evaluate the conditions of the tissue (Black and Reed, 1983). In the long term the patient should also monitor the condition of their skin themselves.

Aftercare - Flexible polyurethane foams have limited durability and after continual use tend to lose their initial resilience and stiffness, resulting in a degradation of performance with a danger of increased ischial pressures. As part of the wheelchair cushion service a follow up procedure was initiated, with patients returning six months after they had been provided with the cushion and three monthly thereafter. At the return visit the interface pressures were measured and a new cushion was provided if they had reached unsafe levels. Once again the pressures were measured on the newly provided cushion to confirm that it was satisfactory.

Records - A record keeping system was set up as an intergral part of the programme, facilitating regular patient recall. An important feature of the record system was that it contained data which made it possible to audit the outcome of the programme in order to establish its benefits and provide realistic cost estimates.

DISCUSSION AND CONCLUSIONS

The programme of wheelchair cushion fitting began as research, a phase which lasted for part of the first year and all the second year. Thereafter it was established as a service for spinal injury patients. During the fifth year the service was transferred to National Health Service funding and run by a senior nurse. The transition from research to service was not easy and it was found that research funds were more readily available than were service funds. This is unfortunate as successful research which benefits patients should be offered as a clinical service, particularly if it can be shown to be cost effective.

The records have allowed the outcome of the programme to be monitored and some results are presented in Table 1. The fall in the percentage of patients admitted each year with ischial sores is gratifying and the number of patients admitted actually fell despite there being an increase in the total patient load. Table 1 probably underestimates the benefits as some sacral sores are likely to be related to sitting. In addition average length of time in hospital for patients with ischial sores was 70 days for year 1 but fell to 7 days for year 5. The incidence of sores rose slightly on being transferred to NHS staffing. This might be due to the nurse requiring a training and familiarisation period but an element may be related to the complexity of the programme and the equipment required; perhaps more thought should have been given to simplification before transfer.

It was initially believed that the successful results were related to the use of a cushion with an ischial cut-out and only later was it realised that the structured system of providing the cushion had played a large part in the success of the programme.

Although this chapter refers only to wheelchair cushions, the ideas underlying the programme are applicable to the supply of many aids and appliances. The provision of cushions and other devices requires the accurate assessment of the patient, the prescription of an appropriate aid and the assessment of the suitability and usefulness of the aid once it has been provided. Follow up is extremely important as both patient and device may change. Indeed the provision of an aid is not the end of the process but is the first stage of a programme of continuing care.

Years after start of programme	Number of patients registered at Spinal Injury Unit	Percentage of patients admitted with an ischial presure sore	Number of cushions provided
1	188	6.9	-
2	220	6.4	-
3	240	2.5	46
4	287	2.4	75
5	316	3.8	90

Table 1 Fall in admissions of patients with ischial sores during first five years of wheelchair cushion programe. The percentage of patients admitted with ischial sores was significantly less during the years 3, 4 and 5 than in year 1 ($p<0.05$).

ACKNOWLEDGEMENTS

The author is delighted to acknowledge his collaboration with Dr I C Wilkie and Dr M W Ferguson-Pell, in the development of this wheelchair cushion programme, and with Mrs Morse and V Jamieson of the Pressure Sore Prevention Unit, Florence Street, Glasgow, after the Service had been taken over by the South Eastern District of the Greater Glasgow Health Board.
Financial support is acknowledged from the Scottish Home and Health Department and the Greater Glasgow Health Board.

REFERENCES

Barbenel J C (1983). Measurement of interface pressures, in J C Barbenel, C D Forbes and G D O Lowe (eds.), *Pressure Sores*, Macmillan Press, London, pp 67-78.

Barbenel J C, Davies R F and Ferguson-Pell M W (1984). Movements made by paraplegic patients in wheelchairs. *Care, Science and Practice*, **3**, 60-63.

Black R C and Reed L D (1983). The Use of Thermography in the prevention of pressure sores, in J C Barbenel, C D Forbes and G D O Lowe (eds.), *Pressure Sores*, Macmillan Press, London, pp 167-176.

Ferguson-Pell M W, Barbenel J C and Evans J H (1981). Biomechanical factors in the aetiology and prevention of pressure sores, in I Stokes (ed.), *Mechanical Factors and the Skeleton*, John Libbey and Co., London,
pp 205-214.

Fisher S V and Paterson P (1983). Long Term Pressure Recordings Under the Ischial Tuberosities of Tetraplegics. *Paraplegia*, **21**, 99-106.

Kenedi R M, Cowden J M and Scales J T (1976). *Bedsore Biomechanics*. Macmillan Press, London.

Lawes C J (1984). Pressure sore readmission for spinal injured people. *Care, Science and Practice*, **4**, 4-8.

Merbitz C T, King R B, Bleiberg and Grip J C (1985). Wheelchair push-ups. Measured pressure relief frequency. *Arch. Phys. Med. and Rehabil.*, **61**, 433-439.

Mooney V, Einbund M J, Rogers J E and Stauffer E S (1971). Comparison of pressure distribution qualities in seat cushions. *Bull. Prosthet. Res.*, **10-15**, Spring 1971, 129-143.

Noble P C (1981). *The Prevention of Pressure Sores in Persons with Spinal Cord Injuries*. World Rehabilitation Fund, New York.

Reswick J B and Rogers J E (1976). Experience at Rancho Los Amigos Hospital with devices and techniques to prevent pressure sores, in R M Kenedi, J M Cowden and J T Scales (eds.), *Bedsore Biomechanics*, Macmillan Press, London, pp 301-310.

Watson N (1983). Skin care and long-term rehabilitation, in J C Barbenel, C D Forbes and G D O Lowe (eds.), *Pressure Sores*, Macmillan Press, London, pp 95-102.

Young J S and Burns P E (1981). Pressure Sores and the spinal cord injured model systems. *S.C.I. Digest*, **3**, 18-25.

12

EXPERT SYSTEMS, A PROMISING TOOL FOR PATIENT

ASSESSMENT

M W Ferguson-Pell, M Cardi and D E Hurwitz

INTRODUCTION

Since World War II there have been remarkable accomplishments in the application of technology to assist people with disabilities. The startling growth in rehabilitation engineering is in part attributable to the capacity this discipline now has to directly alter the lives of people with disabilities. Many areas of biomedical engineering have accomplished enormous advances in diagnosis, treatment and replacement of malfunctioning body components. Yet one tends to associate such achievements with the efforts of teams of specialist engineers, systematically reducing problems to culminate in a solution that has implications for large numbers of people in medical need. At this point in its evolution rehabilitation engineering maintains a reputation for concentrating on the needs of individuals with problems that require highly custom solutions. Such solutions are normally developed by skilful and highly motivated engineers in association with a variety of clinical professionals who help to define specifications and evaluate outcomes. This situation in however changing rapidly, driven by heavy demand, limited availability of expertise and economic constraints.

The disabled person, or consumer of these services often has a different relationship to providers of his/her care than the acutely ill patient. The medical model normally associated with clinical decision-making tends to breakdown in the rehabilitation field. More often than not the consumer's experiences are as pertinent to defining an appropriate solution as the clinician's. The parameters being synthesised are invariably a mix of medical, social, economic and funtional issues with only a few being readily quantifiable. The problem solver therefore faces the challenge of mapping these factors for a given individual onto a set of available solutions, usually in the form of hardware.

Traditionally, solutions have been established through a process of trial, error, tenacity and native ingenuity. The principles employed in achieving an outcome are frequently intangible and have resulted in the field of rehabilitation engineering being academically undersubscribed. In the United States, at present there are only opportunities for students to gain a rehabilitation engineering parenthetic to a biomedical engineering degree. All too often such qualifications are associated with very limited or highly specialised clinical activities and do not prepare students for the real challenges of clinical rehabilitation engineering service delivery.

Fortunately, in association with the escalating demand for rehabilitation technology services two changes have taken place. Firstly, growing demand and increased consumer involvement have created a far more competitive marketplace than even only five years ago. For example the explosion of options and generally improved quality of wheelchairs in the last five years is the direct effect of consumer demand and participation coupled with sound Research and Development. Secondly as the commercial viability of rehabilitation technology products grows the need for engineers to provide custom systems built from scratch diminishes. Instead there is a need for clinically trained individuals such as physiotherapists, occupational therapists and speech and language pathologists to become competent in configuring and prescribing commercial systems. In the United States clinicians specialising in these activities are becoming known as "rehabilitation technologists". In broad economic terms this trend is encouraging, as it fosters the growth of a new industrial niche and helps to limit the reinventing of the wheel associated with programmes based on routine customisation.

A number of models for the engineer's role are emerging and more than one may be needed to meet specific needs. In the rehabilitation centre the rehabilitation engineer's role is likely to become that of a consultant to the primary clinical team and the associated rehabilitation technologists. Durable medical equipment dealers are finding an increasing need for rehabilitation engineers in providing product support to clinical specialists. In a number of cases, rehabilitation engineering practices have been established and may range from a centrally based office to a mobile van with materials and equipment for on the spot problem solving.

One of the major problems currently being faced in the delivery of rehabilitation technology services is making them systematic and teachable. When these problems are not addressed the programme becomes increasing dependent on the expertise of a small handful of experienced clinicians. Unless their expertise can be passed on, the programme is vulnerable to staff turnover, illness and retirement. It also becomes highly localised and although outstanding in its achievements may only be available to a small segment of the population.

Computer-based expert systems offer a very promising approach to reducing some of these problems. By definition they offer the opportunity to trap the expertise of one or more specialists into a computer program that then interprets decision-making steps as rules.

By carefully interacting with a non-expert the computer can help emulate the decision-making process of the specialist and thereby help to provide similar services. In this paper an example of such an expert system will be given and some of its attributes and drawbacks discussed.

THE PROBLEM

Many wheelchair users face the very serious problem of pressure sores. As described earlier in this section, they are debilitating, costly and demoralising. For many active wheelchair users they can impact on employability, family life and recreational activities. In many cases the solution is to provide a suitable wheelchair cushion. It is generally recognised that no one cushion type meets the needs of all and the clinician is therefore faced with the task of matching individual needs to a range of commercial or custom-fabricated options. Reswick and Rogers (1976) offered a pragmatic approach to this problem by establishing a pressure sore prevention clinic that was primarily directed to the needs of spinal cord injured patients. Tools were developed, including a pressure evaluator to provide a quantitative index of normal stress at the body/support interface. They even published pressure-time guidelines to help the clinician and patient match sitting duration and pressure relief activities to their risk score.

A number of similar pressure sore prevention clinics were started and most quoted anecdotal evidence to support the effectiveness of the clinics in reducing pressure sore incidence (Key and Manley, 1978; Ferguson-Pell *et al.*, 1981; Krouskop *et al.*, 1983; Garber, 1985). Despite the apparent success of this approach and numerous educational and scientific meetings to disseminate the concept, very few additional clinics were started. There is little doubt that clinicians and consumers alike most wanted "the ultimate cushion" and some excellent and successful products have resulted from this quest. For many spinal cord injured patients these products are effective. A combination of clinical judgement and measurement using an interface pressure evaluator have resulted in a lasting reduction of risk of breakdown for many patients. However, for a significant minority a more sophisticated approach is indicated and requires careful clinical evaluation, application of basic biomechanical principles and a process of iteration, testing different cushion designs until clinical goals have been met. For such a process to proceed efficiently it is important for the clinician to document each step carefully, ensure that the iterative process converges rapidly and that the solution is both clinically and cost effective. This process can then become dependent upon one or two individuals who may be unwilling to devote their professional efforts fully to wheelchair cushion prescription and pressure sore prevention.

A TECHNOLOGY BASED SOLUTION

Expert systems have long been considered a potentially useful concept for a wide range of industrial applications where complex, multifactorial decision-making has to be undertaken. Until recently personal computers have not been fast enough or have had insufficient resident memory to provide assistance with routine clinical decision-making. Furthermore the complexity of software development using symbol-manipulation languages such as LISP (Winston and Horn, 1981) combined with the need for specialised computer systems left the concept in the academic domain of artificial intelligence. In the last five years, not only have personal computer systems developed the capacity to undertake realistic problem solving using the expert systems approach, but authoring systems have been developed to allow non-programmers to develop applications software of their own.

During the transition of expert systems from an academic curiosity to "knowledge engineering" the authors developed a rudimentary expert system, written for the personal computer in BASIC. The program, CUSHFIT, is designed to assist the competent but non-specialist clinician through the process of prescribing a wheelchair cushion for spinal cord injured patients at risk of developing pressure sores.

The specifications for CUSHFIT included:
A robust and friendly user interface
Capacity to schedule patient's appointments and keep track of follow-up
Full documentation of observations and measurements made during evaluation
Provision of 'goal pressures' for each 'at risk' body site determined according to each individual's risk status
Graphics capability to provide rudimentary CAD for cushion customisation

Upon arrival in the clinic a brief clinical history is recorded by the clinician responding to prompts from the program. Many of the answers are not used by the program's 'inference engine' but certain information is important in assessing risk and eliminating unsuitable cushion options.

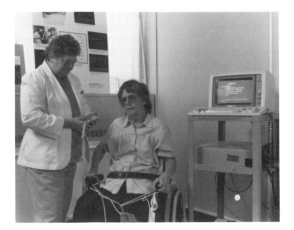

Fig. 1   CUSHFIT in operation during typical clinical evaluation.

Baseline information on how the patient interacts with two standard foam cushions is required by the program to determine the category of cushion most likely to succeed.

Central to the operation of the program is the need to identify a set of goals which can be used by the program to test potential solutions against. Clinical data, collected by the program, is used to establish goal interface pressures for the tissues overlying the ischial tuberosities, trochanters, sacrum and coccyx. The clinical factors are identified in table 1 and a typical set of goal pressures is given in table 2. Opportunity is provided for the client to comment on the performance of the cushion in terms of comfort, stability, heat dissipation, moisture build-up etc.

Parameter	Sample Response
Sitting time	6h/day
Pressure relief	infrequent
Skin condition	RIT = 3
	LIT = 1
	R. TROCH = 3
	L. TROCH = 1
	COCCYX = 1
	SACRUM = 1
Sensation	Absent
Prominence bony areas	Prominent

Table 1   Clinical factors Right (R) Left (L) Ischial tuberosity (IT) Trochanter (TROCH)

Acceptable goal pressures for: John Doe

RIT =	50 mmHg	(6.6 kPa)
LIT =	50 mmHg	(6.6 kPa)
R. TROCH	60 mmHg	(8.0 kPa)
L. TROCH	60 mmHg	(8.0 kPa)
COCCYX	20 mmHg	(2.7 kPa)
SACRUM	30 mmHg	(4.0 kPa)

OTHER

Table 2   Specimen acceptable goal pressures for 'John Doe'.

CUSHION	RIT	LIT	RTR	LTR	COCCYX	INSERT	EVAL
3 inch med density 3040	105	98	91	80	35	0	SB	t
4 inch med density 3040	75	70	68	60	30	0	SB	ok
2 in med T/2 in 3570	60	58	53	50	25	0	SI	ok
2330 ins/3040/3570 Cut-out dim: 6.5 x 6 in	48	50	62	60	20	0	SI	ok
PRESSURE GOALS	40	50	60	60	20		
DIFFERENCE FROM GOAL	8	0	2	0	0		

The recommended cushion is the 2330 ins/3040/3570
1=continue with computer   2=therapist selection   3=stop/select final cushion
Please make a choice?

Fig. 2   Chart which is updated for each cushion evaluated for a given client. Columns RIT... COCCYX indicate interface pressure measurements obtained. Blank column provides space for user-specified pressure data. INSERT column allows record of accessory insert to accommodate sling seat sag. EVAL indicates client feedback.

As the evaluation progresses, the clinician makes measurements of each cushion recommended by the program until the goal pressures are met, the clinician is satisfied with the interaction of the cushion with postural and functional needs, and the client is satisfied and comfortable. Figure 2 represents a status report on the computer monitor which allows the clinician to monitor the progress of the evaluation.

At all times the clinician is completely in charge of the decision-making process and can over-ride the recommendations made by the computer. Clearly the goal is to accomplish a satisfactory result in a minimum number of iterative steps. Our experience to date indicates an average of four cushions tested/evaluation which is less than the six we typically required when the clinic was operated by an experienced clinician without the assistance of CUSHFIT's structured approach.

The cushions accommodated by the program are classified into categories according to the types of materials used in their construction (figure 3). The way the program manages the selection of two of the cushion classification categories is of particular note.

```
              CUSHION SELECTION MENU
              _____


              1.   Foams and T-Foams

              2.   Cutouts

              3.   Gels

              4.   Special (ROHO, Jay)



       The recommended pathway is Foams and T-Foams

       Please select a cushion group using keys 1-4 ?
```

Fig. 3   Menu of cushion categories available.

Special Cushions

This category of cushions consists of specific commercially produced cushions designed for prevention of pressure sores. It includes products such as various models of the Roho and Jay cushions. The program as written will first test to see if conventional cushions employing foams or gels will meet the patient's needs. At the time of the program's development, funding for 'special cushions' was more scarce and required proof that conventional cushions were not satisfactory. More recently the superior durability of special cushions compared with foams, has been acknowledged and funding constraints have been relaxed.

Cut-Out Cushions

Foam cushions that have been modified by cutting away a rectangular section of foam beneath the ischial tuberosities are used in many clinics. A simple protocol is used to allow the clinician to input the position of the ischial tuberosities relative to the wheelchair. Other data from previous cushions evaluated by the program help to determine the design and materials recommended by the program for the cut-out cushion. The clinician is then able to use the drawing to communicate needs to a technician or for documentation purposes.

DISCUSSION

CUSHFIT provides a tool which assists with the systematic selection of cushions for wheelchair users at risk of developing pressure sores. The need for a tool of this type centres on the problem of expertise transfer that is crucial to a developing field in which procedures are evolving and the specialists who apply them are scarce. CUSHFIT also offers other benefits too. Accountability in medical care and decision-making places increasing demands on the clinician for documentation while demand for services continues to outstrip supply. Expert systems provide a mechanism for automatically documenting observations and decisions as the fitting process progresses. Data can be stored on disk for further analysis and/or printed out for inclusion in the medical record. The only time the CUSHFIT user has to take pen to paper is to sign his name authorising the prescription.

A particular asset of the system is the means to provide for the development of a large database for research purposes. One of the problems that has been encountered with pressure sore research is that the data collected at a single centre do not include enough cases to permit informative multifactorial analysis. Since expert systems collect data in a systematic and consistent fashion they offer the opportunity for multi-centre databases with large numbers of cases available for analysis.

CUSHFIT is a very simple and primitive example of how expert systems can be used to address important issues in patient assessment. It has been operating for approximately four years and has been well accepted by staff and patients alike.

ACKNOWLEDGEMENTS

The work reported in this chapter was funded in part by the Walter Scott and Lyons Foundations and the Paralysed Veterans of America SCRF award. Helen Hayes Hospital is operated by New York State Department of Health.

REFERENCES

Ferguson-Pell M W, Wilkie I C, Reswick J B and Barbenel J C (1981). Pressure sore prevention for the wheelchair-bound spinal cord injured patient. *Parapelgia*, 18, 42-51.
Garber S (1985). Wheelchair cushions for spinal cord injured individuals. *Am. J. Occup. Therp.*, 39, 722-725.
Key A G and Manley M T (1978). Pressure distribution on wheelchair cushions for paraplegics: its application and evaluation. *Paraplegia*, 16, 403-412.
Krouskop T A, Noble P C, Garber S L and Spencer W A (1983). The effectiveness of preventative management in reducing the occurrence of pressure sores. *J. Rehab. Res. Dev.*, 20, 74-83.
Reswick J B and Rogers J E (1976). Experience at Rancho Los Amigos Hospital, devices and techniques to prevent pressure sores, in R M Kenedi, J M Cowden and J T Scales (eds.), *Bedsore Biomechanics*, Macmillan, London, pp 301-310.
Winston P H and Horn B K P (1981). *LISP*, Addison-Wesley, Reading, MA, USA.

13

DELIVERY AND COSTING OF REHABILITATION

ENGINEERING

D N Condie

INTRODUCTION

Rehabilitation was defined by Mair (1972) as "the restoration of patients to their fullest physical and social capability". Rehabilitation Engineering (RE) might therefore be defined as "the application of engineering in combination with medicine and related sciences in the attainment of these objectives". it is therefore concerned with, although not exclusively, as will be described, the provision of equipment designed to compensate for or replace impaired natural functions.

THE DELIVERY SYSTEM

Assessment and Prescription

Any Service will only succeed if its customers know that it exists and are aware how to obtain access to it. It follows therefore that adequate publicity, a controversial word in medicine, and a clear-cut method of referral of patients are basic requirements. The purpose of the assessment procedure is to define the nature and consequence of the patient's impairments. Put in even more basic terms: What's wrong with the patient and as a result what can he or she not do?

The personnel involved in the assessment process should include a medical doctor, therapist(s) of an appropriate discipline, and a technical representative who for the purpose of this discussion will be referred to as the Rehabilitation Engineer. The role of the engineer during this stage is likely to depend on his particular scientific knowledge, eg in the subject of biomechanics, which will complement the skills of the medical and life scientists. This might include the responsibility for conducting specific investigative procedures as part of the assessment process such as gait analysis tests or tests of upper limb function. It must be stressed however that the roles of the team members should ideally overlap and all must be reasonably conversant with each other's speciality.

The prescription process will follow naturally a successful assessment; however, once again a logical step by step process is indicated, the first step being the definition of treatment goals or objectives. Obviously, these must take account of the patient's needs and aspirations, however they will be fundamentally dictated by the individual's physical and mental status. At all cost these must be realistic if the patient is not to suffer subsequent disappointment. This is an area where enthusiastic engineers must learn from the traditionally more cautious approach of the medical and therapy professions without necessarily "playing safe".

The second step is to consider the means of achieving the agreed objectives. The best answer is always not to burden the patient with cumbersome equipment if, for example, a simple surgical procedure can achieve the same end. All possible means of treatment must be considered - surgical, therapeutic, chemo-therapeutic and equipment both singly and in combination.

Assuming that it is agreed that the supply of equipment is indicated it is now possible to proceed to the selection of the specific item. Considerable progress has been made in recent years in the development of "simulators" which can be employed in the clinic to test the prescription and refine the characteristics of the prescribed device on a sort of trial and error basis while monitoring some vital function, eg contact pressure. At the simplest where stock items are involved the availability of demonstration equipment which can immediately be tried out albeit on a limited timescale will increase the accuracy and success rate of initial prescriptions. It would be impossible to overstress the importance of this the first element of the supply process since it is quite simply the keystone of any subsequent successful programme of treatment.

Production and/or Delivery

The organisation for the production and/or delivery of equipment will come under the control of the Rehabilitation Engineering staff. Options for the fulfilment of the prescription include the use or adaptation of stock items, in-house fabrication and/or subcontracted fabrication such as the system of orthotic contractors in the UK. There is a growing trend towards wider use of stock items which may have a limited degree of adjustability, eg modular devices. Whilst this approach is to be welcomed in general as cost effective, it must be stressed that optimal matching of the prescription must not be sacrificed simply to achieve savings.

The latter stages of the delivery of equipment will very frequently involve an element of training requiring the active involvement of therapists, other caring personnel such as nurses and, of course, the relatives of the patient. Unfortunately this requirement is often neglected or at best inadequate, sometimes due to inadequate liaison and on other occasions due to simple resource limitations. The best time to start making these arrangements is at the time of assessment and prescription.

Follow-up

Lastly but most certainly not least important is the follow-up element. All too often at least in the UK the delivery of a piece of equipment is seen as being the end of the process with disastrous consequences for the patient and the use of scarce resources. Three basic situations must be catered for in any follow-up system.

The initial review procedure is concerned with the verification of the accuracy of the prescription, the integrity and safety of the equipment and the adequacy of the training. No precise timing can be specified for this event which must be individually planned.

Subsequent reviews require to deal with two distinctly different requirements:

    technical considerations such as repairs and/or maintenance and
    clinical considerations such as the changing status or age and hence demands of the patients.

Concerning the former, scheduled "preventative" maintenance is certainly the ideal, however with "reliable" patients it may be accepted to use the "hot-line" approach for technical problems. Clinical reviews require to be individually scheduled and are time-consuming and expensive since they should involve the whole team.

These then are the three elements of the supply process; however, it must be apparent that for this process to function effectively a number of other structures must be created to complete the delivery system, some of which have been hinted at.

OTHER ELEMENTS OF THE DELIVERY SYSTEM

Patient Management

The efficient operation of a large department receiving several hundreds of referrals annually and with a standing population of some thousands, such as in Tayside, represents a complex management task.

The handling of the documentation and the resulting data necessary to programme these activities, even for a much smaller department, clearly will benefit from the use of a computerised management system.

Investigative Facilities

Mention has already been made of the use of gait analysis procedures and other forms of physical, physiological and functional testing as a complementary feature of assessment procedures. Clearly there are resource implications for the creation, staffing and operation of such facilities.

Information

The quality of the prescription emanating from the service will depend on the existence of effective intelligence regarding current products and techniques. This information must be available on a local basis, however it is becoming increasingly apparent that this requirement is likely to be most practically and economically achieved by "subscribing" to national and/or international databases.

Repair and Maintenance

Much of the repair and maintenance activities associated with RE equipment has traditionally been carried out by "in house" staff at a cost to new production levels. The introduction of increasingly complex equipment however is now creating a requirement for greater use of externally based maintenance services.

Evaluation and R & D

It appears obvious that it is essential for the Service organisation to have the opportunity to clinically evaluate new products and techniques in a scientific manner.

Equally it would appear desirable for the Service personnel to be intimately involved with local research and development (R & D) programmes. The principal implication of these requirements relates to the time spent by the service personnel when engaged on these activities which will diminish their routine service capacity.

Education and Training

Even assuming that the staff recruited to the organisation have received adequate basic training, it is certain that all categories of staff will require continuing education whether by formal courses or by informal visits and conference attendances.

COSTING OF SERVICES

Traditionally within the UK the cost of the provision of equipment has been seen as comprising the direct labour costs, consumable materials and components and standard overheads for administrative and building costs. Recognition of the demands of a "delivery system" such as the one previously briefly outlined requires a radical rethink of this policy. If one were to implement such a system it would be necessary to include in the cost of provision of an item of equipment the cost of all the related elements of the delivery system just outlined. The cost of each of these items will very clearly depend on the nature of the equipment being provided and the size of the service. An estimate based on the operational costs of the Tayside Orthotic Service, which is responsible for the supply of approximately 3,000 items/year, suggests that these additional costs could be as high as 80% of the direct supply costs.

Most overseas Rehabilitation Engineering practitioners will not be surprised at this conclusion since they have been accustomed to this type of costing for many years. In the UK, however, it is a new concept which NHS practitioners are only just coming to terms with but which has important implications for the future of RE services.

CONCLUSION

Research, development and manufacture of Rehabilitation Engineering technology are obvious avenues for the application of bioengineering expertise. It is submitted however that Bioengineers must also involve themselves in the systems for the delivery of technology, including the unpleasant subject of costing. Unless this role is accepted it is questionable whether these products will ever reach the patient, let alone achieve their full potential for benefiting society.

REFERENCES

Mair A (1972). Medical Rehabilitation. The Pattern for the Future. *Report of a sub committee of the Standing Medical Advisory Committee*, HMSO, London.

14

DELIVERY OF AIDS FOR THE DISABLED

P J Lowe

INTRODUCTION

A number of technical aids for the disabled can now be obtained either through statutory bodies or purchased privately from commercial organisations. However, there will always be those people whose problems necessitate an individual solution. The Northern Regional Medical Physics Department (based in Newcastle-upon-Tyne) has established the Regional Technical Aid Service to help these people. Over 900 referrals have been made to the Service. The majority of these have been resolved successfully.

STATUTORY PROVISION OF EQUIPMENT

Local Authority

Local authorities are required by the Chronically Sick and Disabled Persons Act, (1970) to provide assistance with home adaptations and any additional facilities designed to secure the person's greater safety, comfort or convenience. They also have the power to provide or help to obtain various environmental aids. In addition they are required to assist with radio, television, library or similar recreational facilities and specialised telephones. Various local authorities interpret the Act differently leading to considerable variation throughout the country in the types of aids provided.

Health Authority

In general, aids required in connection with medical and nursing care at home are supplied through the NHS. These include ripple beds, cushions, hoists and aids for incontinence.

Disablement Services Authority (DSA)

This is a temporary authority established to deliver wheelchairs and artificial limbs. In 1991, this aspect of technical aid provision will be undertaken by Health Authorities.

Manpower Services Commission

The Commission can provide, on free permanent loan, any aids necessary to enable disabled people to work. These include modifications to machines, special desks and seating, typewriters, telephone aids, reading and writing aids and Braille measuring devices.

Local Education Authority

Aids for use in connection with education may be supplied by these authorities.

PRIVATE PURCHASE

A number of devices are now available to help the disabled individual to improve his quality of life. Some of these are not considered essential by the statutory authorities. They can be purchased, often with the financial assistance of voluntary organisations, either by mail order or from the local chemist.

SOURCES OF INFORMATION

Disabled Living Centres

These centres have been established in various cities around the country together with mobile centres when that is the appropriate solution. They provide an area where various devices from a wide range of companies can be displayed and information on other aids not on display can be obtained. The devices can be used by visitors and trained staff are available to offer advice on the various items. Some centres, such as the Dene Centre in Newcastle, incorporate other services for the disabled under one roof. Typically these will include an information service (dealing with benefits etc.) and a continence advisory service.

Communication Aid Centres (CACs)

Six CACs were established by RADAR and the DHSS in 1983 to provide a resource of expertise on communication aids. They have a wide selection of devices on display and the staff are able to suggest suitable communication aids for a wide range of disabilities.

Catalogues (Equipment for the Disabled)

A number of books on devices are available and the disabled living centres will be able to help with suggestions but "Equipment for the Disabled" (Oxford Regional Health Authority) is a good basic reference.

Exhibitions

There are exhibitions every year covering new developments in aids and equipment for disabled people (NAIDEX). The major one is in London but others are organised on a regional basis.

THE REGIONAL TECHNICAL AID SERVICE

The Regional Medical Physics Department has considerable technical expertise that can be used to alleviate some of the more difficult problems faced by those people with disabilities. Prior to 1980, many of these problems had been solved on very much an *ad hoc* basis. Late in 1980, the Regional Technical Aid Service was established to make this provision a firm commitment of the Department's Bioengineering Section.

Any problem involving the provision of a technical aid for a disabled or elderly person can be referred to the Service provided that:

The aid or service requested is not available through normal commercial channels or other government departments eg DSA.

The referral has come from a health or social service professional.

Problems referred have involved communication devices, continence aids, wheelchair controls, bathing aids and aids to daily living. The referral may involve:

The development of a completely new device.

The modification of an existing device to suit an individual's particular needs.

The repair of a device when that repair cannot, for one reason or another, be completed by the original manufacturer or agent.

Referrals are made to the Service on a request form. The form asks for information on the client, the details of the specific request and the person making the referral. Most of this information is put onto a microcomputer database. This database was established in 1981 to hold information that is needed to answer specific questions related to the effective management of the Service. The major requirements of the database are:

To help in the running of fortnightly progress meetings. Each outstanding project is reviewed and priorities established taking account of urgency, current delay and resource demands.

To provide basic statistics on the Service. It is important to identify trends in the use of the Service whether they are area, referror or device based. Maintenance of basic statistics allows trends to be recognised and can lead to a plan of action to take account of that trend.

To find specific information.

A number of questions arise either within the Service or from outside colleagues. Finding the name of a particular client who was supplied with a particular device is typical of an internal enquiry. Outside enquiries may relate to problems encountered with particular brand name devices for the disabled.

Each project is handed over to a project leader. The project leader is responsible for managing the referral. This will normally involve organising visits, deciding on appropriate action and keeping the project active. All communications and visits are recorded on the reverse of the request form. This project diary provides a comprehensive record of activity that allows other staff to answer queries when the project leader is absent. All visits are recorded in a desk diary to make efficient use of travel. When the project is completed, the request form is filed under the client's name together with all relevant drawings and documentation. A letter is usually written to the referror explaining exactly what we have done and suggesting an appropriate form of follow-up. Sometimes the Service will follow-up the referral but there are occasions when it is more sensible for the referror to do this.

A significant proportion of referrals are for the provision of various types of switches usually for either wheelchair control or a communication aid. Consequently the Service now operates a weekly Switch Assessment Clinic.

In addition to its general role in the Northern Region, the Service provides scientific and technical support to the Newcastle Communication Aid Centre and to the local Paediatric Wheelchair Clinic. By linking with these and other services for the disabled, the Regional Medical Physics Department through its Regional Technical Aid Service is able to improve the overall provision of technical aids for disabled people.

FUTURE DEVELOPMENTS

Two recent reports, Review of Artificial Limb and Appliance Services (1986) and Royal College of Physicians (1986) are having a dramatic effect on the future provision of disability services. The major recommendation of both reports is the establishment of Regional Disability Units (RDU) in each Regional Health Authority (RHA). If established, these Units would assess and manage all types of complex and severe physical disability. Ideally, all facilities for the assessment of the severely disabled would be available under one roof. A small number of beds with full nursing support would also need to be provided. Employment rehabilitation centres might be nearby to assist with retraining and resettlement in employment. The Units would have an important role in teaching and research and, in this respect, a university link would be an advantage.

The RDU would co-ordinate the efforts of a wide range of different agencies (voluntary, Social Services and Health Service) to ensure an efficient and economical provision of disability services for severely physically disabled patients.

Until recently, the Artificial Limb and Appliance Centres (ALAC) had responsibility for the provision of artificial limbs, wheelchairs, surgical appliances, vehicles and artificial eyes. The McColl report (DHSS, 1986) has led to the restructuring of this service. A new health authority, the DSA has been created with geographical regions similar to those of the NHS. In 1991, each RHA will absorb this service and be directly responsible for it. The provision of wheelchairs will probably be devolved to District Health Authorities (DHA) except for severely disabled people requiring specialised controls or specialised seating. Other services currently provided by the DSA should be provided by the RDU.

The Clinical Director of a RDU would be a specialist in rehabilitation medicine. Clear guidelines on appropriate training have been laid down in the Royal College of Physicians handbook on training for consultants in rehabilitation medicine.

The specific functions of the RDU outlined by the RCP Report are:

The assessment of severely disabled patients, especially those with multiple problems.

Orthotics, prosthetics and difficult wheelchair problems. Appropriate workshops would be provided.

The Unit would contain a Disabled Living Centre, where a wide variety of equipment is available for inspection and trial.

A Regional Communication Aids Centre could be included.

The Unit might incorporate the management of certain specific clinical disorders, such as spinal injury.

In addition to these Regional Services, other disability services need to be provided. The RCP list of necessary generic services requiring a District policy is reproduced below. Generic services are those likely to be used by a variety of disabled people that are not obviously the responsibility of a particular medical speciality.

Disabled Living Centres.

Housing, housing modifications and re-housing.

The physically disabled school leaver.

Support services for younger severely disabled and handicapped people.

Driving for the disabled.

Sexual counselling.

Head injury services.

Visual impairment.

Hearing impairment.

Communication aids.

Wheelchairs.

Prosthetics and orthotics.

Urinary continence service.

Stoma care service.

Pressure sores.

Regional Health Authorities are currently reviewing their disability services bearing in mind the RCP Report and the integration of the DSA in 1991.

The Royal College of Physicians, following on from their report, are undertaking a study of "Health Services for People with Disabilities". One aim is to obtain base-line information about the provision of services, so that the review of the services in the early 1990's, called for in the RCP Report, can identify changes and progress since the publication of the report. The aim is a national survey of services for disabled people, aged 16 and over, in all Districts and Regions in England. The main items to be covered include: facilities, staff, services, management, liaison and collaboration, satisfaction, training and research.

The Institute of Physical Sciences in Medicine has published its own report (IPSM, 1988) which stresses the need for bioengineers and physical scientists in the provision of disability services and the steps required to ensure appropriate representation as these disability services develop.

The reports reviewed above have created an environment in which disability is at last receiving attention by health authorities. This means that each RHA and DHA has disability services high on its priority list, probably for the first time ever. They are assessing their current provision of these services and making their own recommendations for change in line with the proposals summarised above. It is important that Medical Physics Departments are involved in this process to ensure that the application of technology to solve the problems of disabled people is addressed in a professional manner.

REFERENCES

IPSM (1988). Physical Disability: the role of the Physical Scientist in the Health Service. *Clin. Phys. Physiol. Meas,* **9**, 81-84.

DHSS (1986). Review of Artificial Limb and Appliance Centre Services, I. McColl (ed.), HMSO, London.

Royal College of Physicians (1986). Physical Disability in 1986 and beyond. *J. Roy. Coll. Physicians Lond.,* **20**, 160-194.

15

THE EFFECTS OF DEFORMITY AND POSTURE ON BODY-

SEAT INTERFACE VARIABLES: A PROJECT SUMMARY

D A Hobson

INTRODUCTION

Over the last two decades considerable research emphasis has been placed on studying the effects of pressure on the buttock tissues of individuals with spinal cord injury. It is evident that the prolonged application of pressures above certain threshold levels will initiate a pathological process in the tissues that can lead to necrosis and ulceration (Kosiak, 1961; Exton-Smith, 1976; Reswick and Rogers, 1976). The exact mechanics of the process is not clearly understood. There is no evidence which suggests that a number of additional factors could be involved in the formation of pressure sores. For example, shear stresses and tissue distortion (Dinsdale, 1974; Chow and Odell, 1978; Bennett *et al.*, 1979) repeated loadings (Brand, 1976), impact stress (Patterson, 1984) temperature and humidity (Brattgard *et al.*, 1976), metabolic stress, nutritional status, age, body stature (Garber and Krouskop, 1982), and psychological factors (Richards, 1981) have all been implicated as possible contributing factors. In spite of this extensive research effort concise measures and guidelines have not resulted that can be used confidently for clinical decision-making. That is, there still seems to be unresolved questions related to the pressure sore problem.

This study investigated the possible contributions of pelvic and spinal deformity and body posture to the variables occurring at body-seat interface. Four variables were investigated involving two study groups; a normal control group (10) and a group of individuals with spinal cord injury (12). The variables are: spinal/pelvic alignment, pressures across the buttock support area, tangential shear at the support surface, and locations of centre of gravity. The last three variables were measured for both groups in nine standardised postures commonly assumed while sitting in a wheelchair. The spinal/pelvic alignment study involved a radiographic series taken in three of the nine standardised sitting postures.

The results indicate that pressure distribution and tangentially-induced shear forces are highly influenced by body posture. The results also indicate differences between the study groups in terms of pelvic alignment and displacement of the ischial tuberosities during changes of body posture. It is proposed that these findings have important implications relative to the design of future seating devices and in the clinical practice of pressure sore management.

OBJECTIVE

It was postulated that the abnormal sitting posture exhibited by individuals with spinal cord injuries can have a significant effect on the body-seat variables, which in turn can increase the probability of tissue breakdown and formation of pressure sores (Hobson, 1984). A second objective has been to investigate the changes in interface variables resulting from changes in sitting posture.

METHODS AND MATERIALS

The overall project involved four interrelated studies, all using the same standard sitting postures. The standard postures were obtained through use of a specially designed body position chair (BPC). The studies compared four variables between an able-bodied control group and a sample of subjects with varying levels of spinal cord injury. The study variables were:

> spinal/pelvic angles and displacements,
> interface pressure
> tangentially-induced shear forces, and
> the location of the centre of gravity.

Comparisons of pelvic alignment between subject populations were based on a radiographic series. The radiographs were taken in the BPC with the subjects positioned in three of the nine specified postures. The centre of gravity (CG) measurements were made using four load cells, which permitted computation of the CG of the body while seated in nine reproducible postures. Gross shear forces acting tangentially to the seat surface were measured using load cells mounted on the seat substructure of the BPC. Buttock/seat interface pressures were measured simultaneously using the Oxford Pressure Monitor. Analyses and comparison of results between the study samples have permitted conclusions to be drawn relevant to the stated objective.

Definitions of Standard Postures

An objective of the study was to detect differences in interface variables that may exist within and between the two study groups; and how these differences may change as a person moves from one body posture to another, i.e., from upright sitting to 30° of forward trunk flexion. Definitions of the nine standardised postures are as follows:

Position (P1M): Defined as the posture in the BPC with the pelvis placed as far posteriorly on the seat as possible, trunk and pelvis on the mid-line, and head in contact with the headrest. The seat surface is horizontal with the backrest reclined 10° to the vertical (100°). Arms are comfortably placed on the thighs and footrests are adjusted to take approximately 10% of the body weight.

Upright Trunk Bending - Left (P1L): Defined as the same posture in the BPC as P1M above, except that the trunk is flexed to the left until the left side contacts the elevated armrests (approximately 15° of lateral trunk flexion).

Upright Trunk Bending - Right (P1R); Defined as the same posture in the BPC as P1M, except the trunk is flexed to the right side.

Forward Trunk Flexion 30° (P2); Defined as the same posture in the BPC as P1M, except the head and trunk are flexed forward 30° along the anterio-posterior (A/P) mid-line

Forward Trunk Flexion 50° (P3): Defined as the same posture in the BPC as P1M, except the head and trunk are flexed forward 50° along the A/P mid-line.

Back Recline 110° (P4): Defined as the same posture in the BPC as P1M, except the backrest is reclined to the 110° position.

Back Recline 120° (P5): Defined as the same posture in the BPC as P1M, except the backrest is reclined to the 120° position.

Body Recline 10° (P6): Defined as the same posture in the BPC as P1M, except the whole seat assembly (seat and back) is tilted 10° in space.

Body Recline 20° (P7): Defined as the same posture in the BPC as P1M, except the whole seat assembly (seat and back) is tilted 20° in space.

Definitions and Radiographic Measurements

The angle and displacement measurements used to delineate spinal/pelvic alignment in two planes for both study groups are shown schematically in figure 1.

Definitions of Pressure Measurements and Analysis

Definitions of average, maximum and peak pressure gradients are given schematically in figure 2.

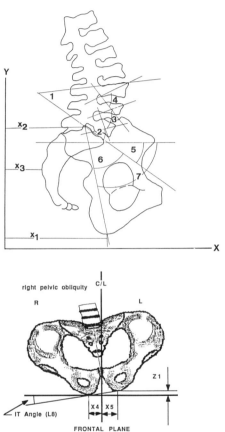

Fig. 1   Definitions of sagittal (upper) and frontal (lower) angles and distances used to define the orientation and displacement of the lower spine and pelvis.

RESULTS

Spinal/Pelvic Alignment

In general, a person with a spinal cord injury (SCI) will sit in their neutral posture with a posteriorly tilted pelvis. On average, the pelvis will be tilted 15° more than for able-bodied individuals (Pelvic angle (7) - figures 1 and 3).

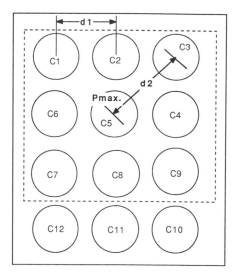

Fig. 2 Definition of pressure measurements.
Average pressure = Sum (C1.....C12)/12.
Maximum pressure = P_{max} (C1.....C12)
Peak Gradient = [P_{max} - P_{min} (C1.....C9)]/d1 or d2.

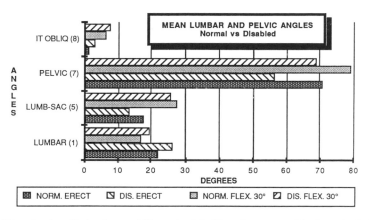

Fig. 3 Three lumbar (1), lumbosacral (5), and pelvic (7) angles measured in the sagittal plane; and IT obliquity (8) angle measured in the frontal plane. Values shown are for the upright posture (P1M) and 30° trunk flexion posture (P2), for both normal and disabled groups.

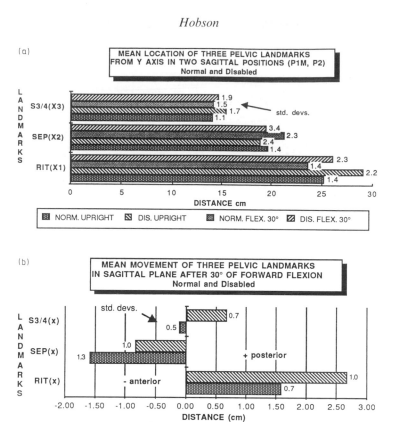

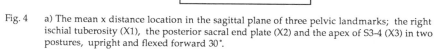

Fig. 4   a) The mean x distance location in the sagittal plane of three pelvic landmarks; the right
ischial tuberosity (X1), the posterior sacral end plate (X2) and the apex of S3-4 (X3) in two
postures, upright and flexed forward 30˚.
b) The mean movement in the sagittal plane of three reference pelvic landmarks (RIT, SEP,
S3/4) upon forward trunk flexion of 30˚ (P2) from the neutral posture P1M.

Forward trunk flexion to 30˚ causes forward rotation of the pelvis, with angular
motion also taking place mainly at the lumbro-sacral joint (Lumb-Sac (5) - figures 1 and 3).
The mean pelvic angles (L7) change from N = 71˚, D = 56˚ in the neutral posture to N = 79˚,
D = 68˚ in the flexed posture. Minimal motion takes place at the sacroiliac joint (6) - figure
1).

Sitting in a neutral posture with 100˚ backrest recline causes the lumbar spine, as
defined by spinal segments S1 - L3, to assume a lordotic angle of about 26˚ for people with
SCI and about 22˚ for the able-bodied. The difference of 4˚ between the groups was found not
to be significant (Lumbar (1) - figures 1 and 3).

The posterior pelvic tilt causes the ischial tuberosities of the person with a SCI to
be displaced anteriorly. On average, the ischial tuberosities are displaced 40 mm anterior
to those of individuals without a spinal injury (figure 4a, RIT (x_1)).

Forward flexion of the trunk to 30˚ from a neutral upright posture causes the ischial
tuberosities to move posteriorly. On average, the posterior displacement is 27 mm for
individuals with a SCI, and only 16 mm for individuals without a spinal injury, an
average difference of about 10 mm (figure 4b, RIT (x_1)). This small difference was not found
to be statistically significant.

Pressure Distribution

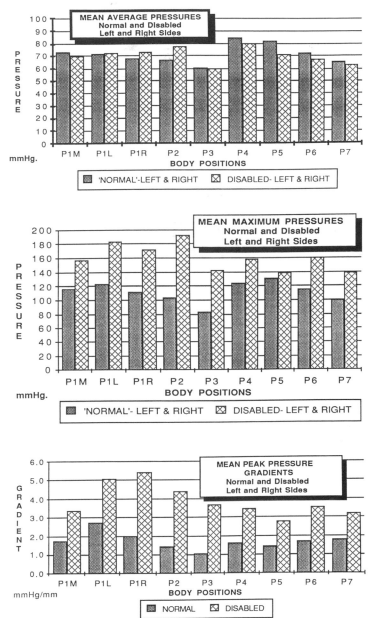

Fig. 5   Plot of pressure measurements for both study groups in the nine study postures; average pressures (top), maximum pressures (middle) and peak pressure gradients (lower).

Average pressure distribution between groups is not effected significantly by alterations of sitting posture. Forward flexion of the trunk causes the largest decrease within both groups from the average pressures in the neutral position. On average, a reduction of approximately 15% from the values in the neutral position occurs at 50° of trunk flexion (P3) (figure 5, upper).

Mean maximum pressures of individuals with a SCI are significantly higher than able-bodied individuals in all nine sitting postures studied. In the neutral posture, on average, the maximum pressure for a person with a SCI is 26% higher. The differences range from a low of 6% in the P5 position to a high of 46% in the P2 position (Figure 5, middle).

Postural changes can reduce the maximum pressures from those values occurring in the neutral P1M posture. These reductions are; lateral trunk bending 15° (32-38%), forward flexion to 50° (9%), backrest recline to 120° (12%), and full body tilt to 20° (11%) (figure 6, middle plot).

On average individuals with spinal cord injury have peak pressure gradients that are 1.5 to 2.5 times greater than non-injured people (figure 5, lower plot). Among the postures studied, maximum reduction of the peak gradients from those measured in the neutral posture occur after backrest recline to 120° (P5) (8%).

Tangential Shear Force

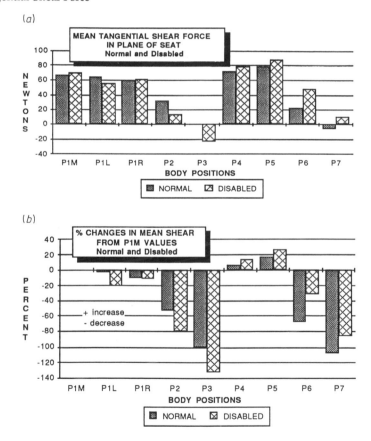

Fig. 6   a) Mean tangentially-induced shear forces as measured by the A/P load cells in the nine body postures for both normal and disabled subjects.
b) Percent changes in the mean values of the tangentially-induced shear (TIS) from values measured in the neutral position, P1M.

Tangential shear force (TIS) acts at the body-seat surface in all nine postures studied (figure 6a). With respect to the neutral posture, maximum reductions of TIS force occur upon trunk flexion of 50° (133% or a reversal of 33%) and upon full body tilt to 20° (85%; figure 6b). It is reasonable to assume that full body tilt beyond 20° will reduce the TIS to zero, and thereafter cause a reversal and an increase in TIS force in the opposite direction. The manner in which the TIS interacts with the normally induced shear stresses to produce potentially deleterious effects in the supporting tissues has not been clarified by this work.

Centre of Gravity

In general, the centre of gravity of individuals with a SCI is displaced posteriorly when compared to a normal sample. The difference in posterior displacement is dependent on the posture.

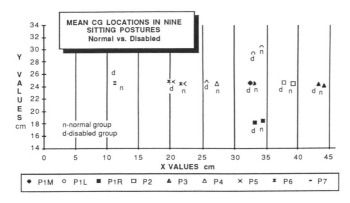

Fig. 7   Mean centre of gravity locations from references axes for disabled and normal groups in the nine sitting postures.

On average, the mean difference ranges from an insignificant amount in the P7 posture to a maximum of 18 mm in the P6 posture. Alteration of posture from the maximum forward flexed position of 50° (P3) to full body tilt of 20° (P7) causes the CG to move posteriorly a total distance of about 300 mm which is approximately the same for both groups (figure 7).

The relative location of the maximum pressure (Pmx) and the location of the CG line exhibit major differences as a result of postural changes. The exact relationship between these two variables, and how they may effect the physiological responses in the supportive tissues remains unknown.

IMPLICATIONS FOR CHANGES TO CLINICAL AND RESEARCH PRACTICES

Clinical Practice

Progressive management of the pressure sore problem is much broader than the selection and application of appropriate cushions and related mobility devices. It is the implementation of a philosophy of management that begins with the staff in the primary care facility, continues throughout the post-acute phase and eventually permeates the life of a person who has insensitive weight-bearing tissues. This management philosophy espouses the need for pressure relief regimens, personal hygiene, adequate nutrition, and individual responsibility and common sense in the pursuit of activities of daily living. Within this broader management approach the progressive rehabilitation team has a relatively small window of opportunity to assist in the selection of appropriate pressure relieving and other devices, and to impart practical concepts of self care. It is within this narrow window that the following suggestions are made for improvement to clinical practice.

Enhancement of Pressure Monitoring

If one accepts the research evidence that shear stress is a deleterious interface factor, second only to normal pressure, then clinical monitoring of pressure should be expanded to gain information on peak pressure gradients. Measurement tools exist commercially that readily allow monitoring of pressures at multiple sites across the buttock. Computation of pressure gradients is relatively simple. For example, knowledge of the location and magnitude of the peak gradients could assist in deciding between different cushions options, assuming all other factors being approximately equal.

It is evident that most active people in wheelchairs do not remain in one static posture. Results of this study indicate that changes in posture have a direct bearing on maximum pressure, pressure gradients, and pressure distribution across the supporting tissues. Evaluation practices should be adopted which can determine an estimate of the frequency of specific postures assumed by individuals. Pressure assessments should then be made in these postures. If possible, conditions of dynamic loading should also be determined during the evaluation process. It may not be possible to reduce the dynamic loads to safe thresholds, but it can have the benefit of making the person aware of daily situations that create peak loads on the tissues.

Management of Deformity

It is evident from this work and other studies that asymmetrical alignment of the pelvis, particularly in the frontal plane, can cause elevated interface pressure and indirectly shear stress. To date little emphasis has been given to preventing deformities of the spine and pelvis in the spinal cord-injured population. The conventional wheelchair seat and the interface materials used to distribute pressure have possibly done more to foster deformity than to prevent it. Assessment of cushion materials and wheelchair designs should include a critical analysis of their capabilities to prevent lateral and anterior drifting of the pelvis.

Seat backs should be analysed for height and possible dynamic features, which will allow a person to periodically extend their spine and realign their trunk and pelvis to postures that can periodically neutralise the deforming effects of gravity. The need for lateral trunk stability should be evaluated in order to provide the mid-line orientation necessary to minimise the onset of lateral spinal curvatures. Where deformities exist postures should be identified that will both reduce the interface stresses as well as counteract the deforming influences of gravity. Patients should be given the means and encouraged to assume these postures whenever it is practical to do so.

Tangentially-Induced Shear Force

As indicated throughout the study alterations of sitting posture can influence the nature of the stresses experienced by the supporting tissues. Presently designed wheelchairs provide trunk stability, especially for tetraplegic patients, by reclining the backrest. Most newer powered wheelchairs also have a reclining backrest as a powered option. Elaborate designs have been marketed that prevent relative movement between the person and the backrest when reclining takes place. Based on the results of this study, reclining the backrest creates the greatest increases in tangential shear forces at the seat surface. To the extent that pelvic sliding is resisted by the seat surface friction (TIS force), potentially damaging tangentially-induced shear stresses can be experienced by the supporting tissues. Also, backrest recline up to 120° has relatively little effect on reducing maximum pressure. The results suggest that a better approach to provide trunk and pelvic stability and reduce TIS force and maximum pressure is to tilt the whole body in space. On average, a tilt angle of 25-30° will reduce the TIS value to zero, while reducing maximum pressure values by 10-15% from those values experienced in a neutral upright posture (horizontal seat with 100° of backrest recline).

Centre of Gravity Information

Clinical tools for measuring and rapidly displaying the centre of gravity of the seated person in various postures are not readily available. It is not clear from this study how CG measurements can be cost-effectively used in a clinical environment. Until it is more clearly demonstrated how centre of gravity location and displacement interacts with the other interface variables, it is probably best that it remains primarily a research tool. One can speculate that one useful application may be the quantification of the stability of the wheelchair and occupant, when the occupant assumes various postures within the wheelchair. This information could be particularly useful when specialised seating inserts are placed in existing wheelchair frames. Monitoring of occupant CG may give clues regarding postural asymmetry and how this may change with time. This information could then be used to develop indications of increasing deformity or altered postures.

Research Practices

This study has suggested that several points regarding past research practices are worth noting. Significant resources have been expended on conducting pressure-related studies involving both animal and human subject. Various models and laboratory procedures have been developed, complete with a variety of methods for analysing and communicating the results. Unfortunately, the variability between the procedures and methods of data analysis and presentation is such that repetition of studies or comparisons of results are virtually impossible. This author makes a plea for standardisation of measurement models and disclosure practices, so that results can be more readily compared. It is only through cooperative efforts of this nature that consensus will be achieved and guidelines produced that are more meaningful in clinical environments.

Secondly, it was noted that in several studies normal subjects were used as readily available substitutes for individuals with spinal cord injuries. This study has demonstrated that rather marked differences exist between the two populations, particularly when the body-seat interface variables are being investigated. It is recommended that these types of substitutions be done with extreme caution since the results may not readily extrapolate across populations.

ACKNOWLEDGEMENTS

This work was sponsored by the Crippled Children Hospital Foundation. The author gratefully acknowledges the UTREP staff contributions by Stan Cronk, Glen Ellis and Beverly Wilson. Guidance provided by Professors Barbenel and Paul of the Bioengineering Unit has been invaluable. Assistance provided by Martin Ferguson-Pell related to the Centre of Gravity instrumentation is also acknowledged with appreciation.

REFERENCES

Bennett L, Kavner D, Lee B K and Trainor F A (1979). Shear vs. pressure as causative factors in skin blood flow occlusion. *Arch. Phys. Med. Rehabil.*, **60**, 309-314.
Brand P W (1976). Patient monitoring, in R M Kenedi, J M Gordon and J T Scales (eds.), *Bedsore Biomechanics*, Macmillan, London, pp 183-184.
Brattgard S O, Carlsoo S and Severinson K (1976). Temperature and humidity in the sitting area, in R M Kenedi, J M Cowden and J T Scales (eds.), *Bedsore Biomechanics*, Macmillan, London, pp 185-188.
Chow W W and Odell E I (1978). Deformations and stresses in soft body tissues of a sitting person. *J. Biomech. Engng.*, **100**, 79-87.
Dinsdale S M (1974). Decubitus ulcers: role of pressure and friction in causation. *Arch. Phys. Med. Rehabil.* **55**, 147-152.
Exton-Smith A N (1976). Prevention of pressure sores: monitoring mobility and assessment of clinical conditions, in R M Kenedi, J M Cowden and J T Scales (eds.), *Bedsore Biomechanics*, Macmillan, London, pp 133-159.
Garber S and Krouskop T (1982). Body build and its relationship to pressure distribution. *Arch. Phys. Med. Rehabil.*, **63**.

Hobson D A (1984). Seated Posture and its Implications on Pressure Sore Management. *Nat. Symp. Care, Treatment and Prevention of Decubitus Ulcers*, PVA Washington, D.C., 29-31.

Kosiak M (1961). Etiology of decubitus ulcers. *Arch. Phys. Med. Rehabil.*, **42**, 19-29.

Patterson R P (1984). Is Pressure the most Important Parameter?, *Nat. Symp. Care, Treatment and Prevention of Decubitus Ulcers*, PVA Washington, D.C., 69-71.

Reswick J B and Rogers J E (1976). Experience at Rancho Los Amigos Hospital with devices and techniques to prevent pressure sores, in R M Kenedi, J M Cowden and J T Scales (eds.), *Bedsore Biomechanics*, Macmillan, London, pp 301-310.

Richards J S (1981). Pressure ulcers in spinal cord injury: psychosocial correlates. *Spinal Cord Injury Dig.*, **3**, 11-18.

16

DISCUSSION: DELIVERY OF REHABILITATION

The session opened with *Paul* asking McLaurin where the load cell in his dynamometer was placed. *McLaurin* replied it was usually about the height of the axle of the wheelchair. The discussion then moved on to the measurement and definition of efficiency with *Malagodi* asking why the time during which the subject repositioned himself was included in the time base for calculating the efficiency. *McLaurin* explained that significant muscular effort could be required to reposition the arms, especially if the subject was seated low in the wheelchair. *Childress* asked why we did not see more lever operated chairs if they were more efficient and to what extent altering the subject by muscular exercise would be a better choice for improving efficiency than altering the wheelchair. *McLaurin* replied that unconventional means of propulsion such as levers were more expensive than rim propelled wheelchairs and, in particular the use of ski poles, was rather conspicuous. Both these effects mitigated against their more general use. He also suggested that part of the function of the rehabilitation centre was to develop the musculature of the paraplegic or tetraplegic patients.

Barbenel agreed with the suggestion by *Lazim* that, in seating, pressures should be preferentially exerted on skin which covers muscles.

Hobson then asked for further information on the expert system described by Ferguson-Pell. In particular he asked about the clinical acceptance of the "Cushfit" system and whether there had been any problems with therapists becoming familiar with computers and finally what experience had been gained outside the department in which Dr Ferguson-Pell worked. *Ferguson-Pell* replied that the acceptance of the system had been very good and that this might represent a problem with the clinicians using the system, who might watch the computer more than the patient, and a critiquing system might be preferable to an expert system. He also commented on the fact that personal liability as a program developer was a problem yet to be overcome. In view of the widespread interest which has been shown the goal of the developers of the program was essentially to produce a large intercentre data base which would allow further refinement of the system.

Fernie commented that the systems described for patient and device evaluation had utilised simple pressure monitoring techniques, although devices such as laser Doppler flow meters were available for assessing tissue condition. He asked the speakers to comment on the appropriateness and applicability of these devices and simple tests of tissue physiology. *Barbenel* agreed there had been considerable development in these devices which have been applied in the research context. This limited use was partly because of their complexity and partly because of their cost. The research findings using these devices had shown very wide inter-subject variability and had not led to simpler guide lines and safe standards of pressure for routine clinical use. *Ferguson-Pell* replied that he and his co-workers were investigating the possibility of systemic screening systems in which biochemical species in the blood could be used as an indicator of muscle breakdown and identify patients needing more detailed investigation. He also confirmed that there was a clinical need for simple devices to look at tissue condition in specific patients in a clinical environment.

Kenedi pointed out that what had been discussed was reactive and suggested, in the light of the future growth in the number of elderly people, preventive community health care would become of increasing importance; he asked the speakers if they saw this as a developing area and if so what sort of steps were being taken to make it affective. *Ferguson-Pell* replied that amongst his clinical colleagues there was a real wish to produce such preventive community care. Unfortunately one of the major problems they faced was the lack of professional manpower available to do this. It was necessary to determine how to attract people into the field of patient care and to build a stronger orientation towards community care than is apparent at present. *Barbenel* suggested that the best method of preventing pressure sores was not to allow the patients to become immobile and incontinent. He suggested the highest priority should be in keeping elderly patients active by teaching them suitable exercise regimes for example. There were models of how this might be done e.g. at the University of Dundee but a key problem was showing the benefits produced by the expenditure associated with such programmes.

M Clark asked how it would be possible to transfer the technology and ideas in preventive care to a wider range of patients. *Barbenel* suggested that the success of the work at Philipshill Hospital which he had described was in part due to the close involvement of clinical colleaques, both medical and nursing. He also suggested there was a need for further education particularly amongst nurses, to make them more receptive to the ideas of prevention and patient support which had been suggested. *Scales* pointed out one of the problems in Britain was a declining budget. He also asked whether rocking chairs could usefully be used by the elderly and suggested that wheelchairs could be redesigned to eliminate the sagging seat. *Barbenel* agreed on the need for redesigning wheelchairs, but expressed his ignorance about the use of rocking chairs although he believed that an alternative to the standard ward chair was required. *Kenedi* commented that Scales was quite correct in his comments about finance but part of the function of bioengineers was to provide better technology and extend the possibilities of diagnoses and treatment and open new areas of patient treatment. This was one of the reasons why there was an apparent funding crisis in the National Health Service. He suggested that investment in community health care would control the escalation of disease, age and other effects.

Kralj asked Hobson how he coped with artefacts when measuring intersegmental angles using X-rays. *Hobson* said that the body positioner chair had been developed in order to provide stable postures to minimise the artefacts.

McLaurin opened a wider discussion on the training and education of rehabilitation engineers by asking Condie what changes he envisaged in the next ten to twenty years because he believed that there would be a need for an increasing number of rehabilitation engineers and it was necessary to clarify what training they would require. *Condie* said that this was a very challenging question but he had been struck by the presentations which had been made in the session in which the speakers had described research within a wider context of service and delivery to the patients. He suggested that in the future he would like to see rehabilitation engineers becoming increasingly involved in the decision making processes and if necessary, becoming more political to ensure that this occurred. *Paul* commented that from the UK perspective there appeared to have been a boom in bioengineering in North America followed by disillusionment, and asked what lessons can be learnt from this in respect of training and implementation. *Hobson*, after clarifying the question, commented on an important meeting in relation to engineering education held in 1976 at which it had been predicted that within ten years there would be a need for 2000 rehabilitation engineers to meet the demand for rehabilitation services. He suggested that the crystal ball gazing had been fairly accurate and that ten years later state legislation has been enacted requiring that each state should have a technical service delivery system. This had led to considerable problems which had in fact been foreseen in the 1976 meeting although, at that time, no follow up decisions had been made. He further suggested that similar demands might arise in the United Kingdom and the problem should be addressed now. *Ferguson-Pell* commented that he believed rehabilitation engineers were engineers with postgraduate training who carried out research and development or evaluation but saw workers who described themselves as rehabilitation engineers performing what he would describe as technical functions e.g. building ramps. One aspect of this was the problem of appropriate education of rehabilitation technicians. *Kenedi* suggested what was really required was good engineers working in medicine because the problems in medicine are sophisticated and high level. He agreed with Ferguson-Pell but disagreed with Hobson and suggested that in the US some of the demand for rehabilitation engineers was politically motivated and represented a bandwagon effect. A major problem in the application of engineering in medicine was that society was now dynamic and rapidly changing and so were its demands. No rehabilitation engineer specially trained for that purpose would have the flexibility to fulfil these changing demands and one can guarantee that anyone trained in a specialised area such as rehabilitation would be out of date in ten years. *Simpson* commented on

Ferguson-Pell's definition of a bioengineer. He pointed out his own flexibility changing from bioengineer to orthopaedic bioengineer to clinical engineer and rehabilitation bioengineer. He supported Kenedi's suggestion that the important thing was the way you approach the job, the way you thought and the way you see medicine and the patient; these were the things which mattered. He suggested that a broad base and a broad understanding was required to successfully apply engineering in medicine.

Lowe said that in his opinion the starting place for clinical or rehabilitation engineering was with a good engineer. Sometimes it was desirable to have a non-bioengineer on the staff e.g. an electrical engineer because they could bring new viewpoints and new ideas which might well alter the direction of the work. Another important issue was that in the UK there was likely to be a regional disability service and not a regional rehabilitation service. Finally he stressed that when solving problems it was valuable to have the ideas of a group of people from different disciplines to more clearly define the problem and to solve it. *Condie* said he preferred not to think about rehabilitation engineers but rather to think about a rehabilitation engineering organisation and that this rehabilitation organisaton should have multidisciplinary staffing. He thought rehabilitation engineers should certainly be involved in research and development but should also be involved in the more basic service provision where they have a different role. *Fernie* warned of the danger of basing rehabilitation engineering on service delivery. It was important that rehabilitation engineers addressed real problems and obtained first hand knowledge but it bred the danger that these rehabilitation engineers could tend to create jobs for themselves, there being an obvious incentive to expand rehabilitation engineering departments and the role of rehabilitation engineers. Good rehabilitation engineering was expensive and should ideally aim to provide solutions which made the bioengineer unnecessary. Part of his function should be to develop systems that can be applied both competently and economically by others with the support of technicians. *Hobson* replied to Fernie's point. He believed that the strength of rehabilitation engineering lay in service delivery. It was in service delivery that rehabilitation engineering could show its worth and this was a better basis for growth than was government or state decisions to set up rehabilitation engineering services. *Murdoch* said he was convinced that anybody calling themselves a rehabilitation engineer should have a routine clinical responsibility. *Fernie* said he was not suggesting anything else but it did carry the dangers of excessive specialisation and inappropriate use of their skills, as outlined by Ferguson-Pell. *McLaurin* suggested that Condie had answered the question which he had asked on training - that the rehabilitation team and the engineer within that team was important. Part of the educational philosophy, in addition to developing a good engineer, should be to develop an attitude and a confidence to work as an active and productive member of the team.

Barbenel suggested that a key consideration in the growth of rehabilitation and bioengineering was to show that the service provided was of value. *Condie* agreed with this and stressed the difficulty of providing such estimates. Part of the problem was that some of the advantages of a rehabilitation service were hypothetical and nominal while the cost was real.

Childress asked about volunteer engineers within voluntary services for the patient and how these services are integrated with more formal services associated with hospitals. *Lowe* said that in the Northern Region there were such voluntary services but they tended to concentrate on leisure activities. Formal organisation could offer engineering expertise if that were lacking in certain cases.

17

LOADING OF ORTHOPAEDIC IMPLANTS

J B Morrison and P Procter

INTRODUCTION

The first detailed attempts to measure the dynamic loading of biomechanical structures can be attributed to Marey (1873) and Fischer (1898-1904). Marey correlated motion analysis with ground-foot forces and body accelerations, while Fischer developed a biomechanical model of human locomotion. It was not until the introduction of rapid response transducers by Cunningham and Brown (1952), that dynamic loading could be accurately measured. Their work contributed to two separate approaches: the prediction of internal loading from external forces using biomechanical models, and the direct measurement of prosthetic loading *in situ*.

DIRECT MEASUREMENTS OF IMPLANT LOADING

Hip Joint

Rydell (1965, 1966) attached strain gauges to the inner neck of two hip prostheses. The prostheses were calibrated prior to implanting, and analysis of dynamic joint loading was accomplished at six months post-operation by reconnecting to wires buried below the fascia. Maximum joint forces were 3.3 ± 0.3 body weight (bw) when walking and 4.3 ± 0.3 bw when running.

English and Kilvington (1979) and Davy *et al.*, (1988) also implanted instrumented hip prostheses (table 1). Each study comprised one patient walking at 0.7 and 0.5 m/s respectively. Maximum joint loadings were less than 3.0 bw in all activities with the exception of one legged stance, where a peak load of 3.6 bw was measured. Davy *et al.*, (1988) reported a maximum bending moment of 35 Nm at the base of the neck when walking, and a torsional moment of 22 Nm around the stem when ascending stairs.

An instrumented hip prosthesis was developed by Carlson (1971) which incorporated pressure sensors sealed within the endoprosthesis ball. Following *in vitro* studies by Rushfeldt *et al.*, (1981), a device of this type was implanted in one patient by Hodge *et al.*, (1986) in apposition to natural cartilage. Pressures measured by 10 sensors were found to be non-uniform, typically 0.5 to 7 MPa. A maximum pressure of 18 MPa was measured when rising from a seated posture.

Brown *et al.*, (1982) used telemetry to record the bending moments acting at the nail plate junction of implanted hip nails in three subjects. Maximum bending moments (in the plane of the implant) were 20 to 30 Nm when walking. The authors also noted high values of torque acting on the implant (2 to 12 Nm) indicating a substantial load component in the anterior-posterior direction.

AUTHOR	METHOD	HIP		KNEE		ANKLE	
		N	P	N	P	N	P
Rydell (1965)	D		3.3				
			1.8				
English and Kilvington (1979)	D		2.3				
Davy *et al.* (1988)	D		2.6				
Paul (1966)	B	5.0					
Morrison (1967)	B			3.0			
Poulson (1973)	B	5.7		3.2			
Harrington (1974)	B			3.5	2.4 (W)		
					2.6 (F)		
Paul and McGrouther (1975)	B		3.5				
Tooth (1976)	B	7.7	3.7	3.3			
Stauffer *et al.* (1977)	B					4.7	3.3
Hardt (1978)	B	5.7		2.7		3.5	
Crowninshield *et al.* (1978)	B	4.3					
Procter (1980)	B					3.9	
Brown *et al.* (1984)	B	4.8	4.4 (M)				
			3.2 (C)				
Rohrle *et al* (1984)	B	5.5		5.1		3.4	

B = Biomechanical Model: D = Direct Measure
N = Normal: P = Prosthesis

W = Waldius Hinge: F = Freeman-Swanson surface replacement
M = Muller: C = Charnley

Table 1 Maximum joint loading during level walking expressed as multiples of body weight.

BIOMECHANICAL ANALYSIS OF JOINT FORCES

Hip Joint: Normal

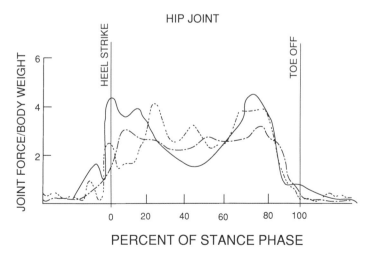

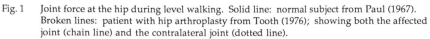

Fig. 1   Joint force at the hip during level walking. Solid line: normal subject from Paul (1967). Broken lines: patient with hip arthroplasty from Tooth (1976); showing both the affected joint (chain line) and the contralateral joint (dotted line).

In order to estimate joint loading from external measures, Paul (1965) developed a biomechanical model of the hip joint. Cine film and force-plate data provided input to the model. The indeterminate nature of muscle activity was resolved by "reduction" in which muscles having similar function were grouped and defined as a single activator. Results of nine subjects showed a characteristic curve of joint loading at the hip during walking (Paul, 1966, 1967). Peak forces occurred during the early and late stance phase (figure 1) and had mean values of 3.3 and 3.9 bw respectively.

This work was extended by Poulson (1973) who employed a two joint biomechanical model to investigate slow, normal and fast level walking, and using a ramp and stairs. The most stressful activities were fast walking (2.1 m/s) and using stairs, in which individual forces of 10 to 11 bw were calculated at the hip joint. The mean maximum forces in these activities are presented in table 2.

JOINT	HIP				KNEE			
AUTHOR	A	C	D	F	B	C	E	F
ACTIVITY								
Slow Walk		5.2		4.1		2.6		4.0
Normal Walk	3.3	5.7	4.3	5.5	3.4	3.2	3.1	5.1
Fast Walk		7.5		6.9		4.2		6.2
Up Ramp		5.9			4.0	3.5		
Down Ramp		5.2			4.0	4.6		
Up Stairs	3.4	7.2	7.4		4.3	4.5		
Down Stairs	2.8	7.1	3.8		3.8	5.9		
Running	4.3						10.1	
Rise from Chair			3.1					

A	Rydell (1966)	Implant: Direct Measure: One patient	
B	Morrison (1969)	Normal)	
C	Poulson (1973)	Normal)	
D	Crowninshield (1978)	Normal) Biomechanical Model	
E	Harrison & Nicol (1988)	Normal)	
F	Rohrle *et al.* *(1984)*	Normal)	

Table 2   Maximum joint loading at the hip and knee during different types of activity expressed as multiples of body weight.

A dynamic analysis of hip, knee and ankle joint forces during walking was reported by Hardte (1978) and Rohrle *et al.* (1984), who used optimisation techniques to determine muscle activity based on the criterion of minimum muscle force. Hardte expressed lack of confidence in the solution of muscle forces, which ignored physiological aspects of muscle activity. Crowninshield *et al.* (1978) imposed a maximum stress constraint which forced the optimisation procedure to distribute muscle forces more evenly among protagonists. These authors reported maximum hip joint forces of 3.5 to 5.5 bw during level walking (n = 5), compared with 7.4 bw ascending stairs and 3.1 bw when rising from a chair (n = 1).

Knee Joint: Normal

A biomechanical model of the knee was developed by Morrison (1967) which included muscle, ligament and joint contact forces. Analyses indicated three peaks of joint force during the stance phase of walking (figure 2). Morrison (1970) reported the centre of pressure at the joint surfaces to be located over the medial condyle during the stance phase, indicating that this structure carried the greater load. This was confirmed by Harrington (1974a) and Tooth (1976). Poulson (1973) reported joint force patterns similar to those of Morrison (1970). In a study of 22 subjects, Rohrle *et al.*, (1984) identified only two maxima during the stance phase of walking, and predicted greater joint forces than other investigators.

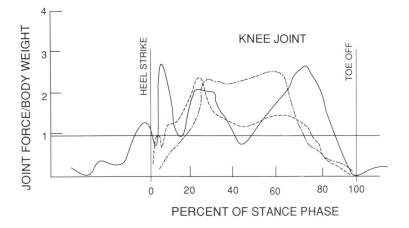

Fig. 2   Joint force at the knee during level walking. Solid line: normal subject from Morrison (1967).
Broken lines: patient with surface replacement prosthesis (chain line) and patient with
hinge prosthesis (dotted line) from Harrington (1974a).

Joint forces were also analysed at different walking speeds by Poulson (1973) and
Rohrle *et al.*, (1984), and during ascending and descending a ramp and stairs by Morrison
(1969) and Poulson (1973). These authors predicted higher joint forces during fast walking
and when using ramp or stairs. The mean maximum joint forces predicted by these authors
are shown in tables 1 and 2.

Ankle Joint: Normal

Procter (1980) modelled the ankle joint as three separate articular surfaces. During level
walking, maximum force occurred at the talocrural (Tc) joint during late stance phase, the
mean maximum of 7 subjects being 3.9 bw (Procter and Paul, 1982). The corresponding forces
at the talocalcaneonavicular joint were 2.4 and 2.8 bw at the anterior and posterior facets
respectively. Hardte (1978) and Rohrle *et al.* (1984) who modelled the ankle as a single
joint centre, reported joint loading similar to that of Procter and Paul at the Tc joint (table
1).

Hip Joint: Implant

Using the methods of Paul 1965 and Morrison 1967, several investigators have analysed
hip and knee forces following joint implant procedures (Harrington, 1974a, b; Paul and
McGrouther, 1975; Tooth, 1976 and Brown *et al.*, 1984).
Tooth (1976) analysed joint loading patterns of normals and patients with hip
arthroplasty. Maximum loading of hip implants was 3.7 bw compared with 5.4 bw acting at
the contralateral (normal) hip joint and 7.7 bw calculated in normal subjects. Patients were
elderly (76 years) and walked more slowly (0.9 m/s) than normals (1.5 m/s). Implant
loading was relatively constant during the stance phase, in contrast to the more dynamic
force-time characteristics seen in normals (figure 1). Similar joint force patterns were
reported by Paul and McGrouther (1975). Brown *et al.* (1984) analysed hip joint loading in
normals and patients having Charnley and CAD Muller hip joint replacements. Implant
loading displayed the same bimodal characteristics as normal joints, but maximum forces
were lower in the Charnley than in the Muller or normal joints (table 1).

Knee Joint: Implant

The joint loading characteristics of normal and prosthetic knee joints were assessed by Harrington (1974a). Maximum joint force was lower in subjects with knee implants (Waldius hinge and Freeman Swanson surface replacement) than in those with no disability (table 1). Joint loading patterns varied among patients having knee implants (figure 2), two of whom displayed reasonably normal results. Tibial shaft torques were less than 12 Nm, compared with up to 17 Nm reported in normals (Morrison, 1970). The centre of pressure of joint loading was located over the medial condyle during the stance phase in both normal and prosthetic joints.

Ankle Joint: Implant

Stauffer *et al.* (1977) analysed ankle joint loading during walking in normals and patients having total joint replacement. The joint was modelled in two dimensions as a simple hinge controlled by the anterior tibial and achilles tendons. Although forces were lower in the prosthetic joints than in the normal ankle (table 1) it was noted that the patients adopted a slower cadence.

APPLICATION OF BIOMECHANICAL DATA TO IMPLANT DESIGN

Joint Replacement

Both direct measurements and biomechanical analyses have played a role in the development of implant design. As each approach has distinct advantages and limitations, these techniques have been complementary in their contributions. Direct measures are restricted to a few studies, due to the difficulties of *in vivo* measurement and the range of expertise required. Patients have not always been fully active and the implanted transducers cannot be recalibrated at time of measurement. Measurements have concentrated on hip joint forces, and the range of activities and postoperative periods which have been investigated are limited.

Biomechanical analyses have been used to investigate a wider range of joints and activities in both normal and prosthetic joints, and thus provide more comprehensive design data. However, in the absence of any direct validation of the biomechanical models employed, these data remain predictions rather than true measures. The use of optimisation procedures to assign muscle forces provides a mathematical solution which ignores underlying physiological events. An alternative "physiological" solution has been published by Pierrynowski and Morrison (1985), but joint forces were not reported. Researches have generally been concerned with model development rather than implant design.

Early studies show joint forces predicted by biomechanical analysis (Paul, 1967) to be substantially greater than those measured directly from instrumented prostheses (Rydell, 1966). Subsequent works suggest that differences can be attributed to two factors: walking speed and joint replacement. There is a positive correlation between maximum load and walking speed (Poulson, 1973; Rohrle *et al.*, 1984) with implant patients walking slower (0.5 to 1.5 m/s) than normal subjects (1.0 to 2.0 m/s). Patients appear to favour the prosthetic joint even when free of pain, so that implant loading is lower than in the contralateral normal joint (Tooth, 1976). If these two factors are acknowledged, there is reasonable consistency among studies of normal and prosthetic joints.

While estimates of maximum joint loading have been helpful to implant development (Rybicki, 1982), proper design requires a four dimensional analysis of joint loading describing the variation of joint force vector with time. This information is provided for the hip by Paul (1966), Rydell (1966), Brown *et al.* (1984) and Davy *et al.* (1988), although major differences exist between joint force vectors reported by these authors. Similar data have been provided at the knee by Morrison (1970) and Harrington (1974a) and at the ankle by Procter (1980). Such data has yet to be fully incorporated in the design of implants (Clarke *et al.* 1980). Brown *et al.* (1984) have analysed the interactions between implant geometry, dynamic loading and stress distribution. In their analysis

Brown *et al.* (1984) confirm the importance of axial torque as a contributing factor to fatigue fracture and stem loosening.

Trauma Implants

In the treatment of trauma, an implant may begin by transmitting all the load from one bone segment to another, as in a comminuted fracture of the femur treated by an intermedullary locking nail. In hip fractures a substantial proportion of the load may be borne by the bony structures from the outset, and as healing proceeds progressively less load transfers through the implant. Hence the biomechanical loading of such structures is subject to a greater degree of variance and uncertainty than total joint replacement.

The proliferation of designs for trauma implants (Tronzo, 1974; Browner and Cole, 1987) far outstrips any supportive biomechanical analysis of loading characteristics. Development has been based mainly on static analysis of material properties and clinical experience. This experience has shown that it is necessary to design for three dimensional dynamic loading and that axial torsion in particular is significant. Figure 3 shows a device used for treating hip fractures which features distal screws that control axial torsion as well as telescoping of the fracture.

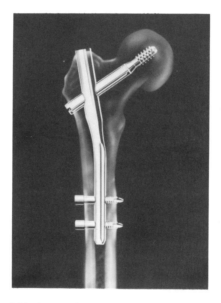

Fig. 3   Gamma locking nail hip fracture device.

Bone and implant stress distributions have been analysed using finite element methods (Levine and Stoneking, 1980; Rohlmann *et al.* 1980; Bucholz *et al.* 1987), but accuracy is limited by lack of knowledge of tendon and ligament forces and failure to consider the full range of dynamic loading. Failure of intramedullary implants have been reported by Weinstein *et al.* (1981) and Bucholz *et al.* (1987). Schneider *et al.* (1988) describe the development of an instrumented interlocking nail, and point out that no *in vivo* information on the initial loading or the history of unloading during healing exists for this type of implant.

The advent of bioresorbable materials which can be used in treating fractures poses a further challenge (Katz, 1986; Vainionpaa, 1987). The problems associated with these materials are that they may lack initial strength and that their properties can decay too rapidly as the fracture heals. While some devices carry only one third of the load borne by the composite of bone and implant (Brown *et al.* , 1982), this is untrue of complex unstable fractures. In order to determine the loading patterns of fracture implants, biomechanical models must be extended to investigate the dynamic loading of bone structures other than joint surfaces, and a wider spectrum of instrumented implants must be developed.

REFERENCES

Brown R H, Burstein A H and Frankel V H (1982). Telemetering in vivo loads from nail plate implants. *J. Biomechanics*, **15**, 815-823.

Brown T R M, Nicol A C and Paul J P (1984). Comparison of loads transmitted by Charnley and CAD Muller total hip arthroplasties. *Engineering and clinical aspects of endoprosthetic fixation*, Proc. Inst. Mech. Engng., C210/84, 63-68.

Browner B D and Cole J D (1985). Current status of locked intramedullary nailing: a review. *J. Orthop. Trauma*, **1**, 183.

Bucholz R W, Ross S E and Lawrence K L (1987). Fatigue fracture of the interlocking nail in the treatment of fractures of the distal part of the femoral shaft. *J. Bone Jt. Surg.*, **69A**, 1391-1399.

Carlson C E (1971). A proposed method for measuring pressure of the human hip joints. *Expl. Mech.*, **11**, 499-506.

Clarke I C, Gruen T A W, Tarr R R and Sarmiento A (1980). Finite element analysis of total hips versus clinical reality in B R Simon (ed.), *Conf. on finite element analysis in biomechanics*, Tucson, Arizona, 487-510.

Crowninshield R D, Johnsson R C, Andrews J G and Brand R A (1978). A biomechanical investigation of the human hip. *J. Biomechanics*, **11**, 75-85.

Cunningham D M and Brown G W (1952). Two devices for measuring the forces acting on the human body during walking. *Proc. Soc. expl. Stress Anal.*, **IX** (2), 75-90.

Davy D T, Kotzar G M, Brown R H, Heiple Jr. K G, Goldberg V M, Heiple K G, Berilla J and Burstein A H (1988). Telemetric force measurements across the hip after total arthroplasty. *J. Bone Jt. Surg.*, **70 A** (1), 45-50.

English T A and Kilvington M (1979). In vivo records of hip loads using a femoral implant with telemetric output. *J. Biomed. Engng.*, **1**, 111-115.

Fischer O (1898-1904). Der Gang des Menschen. *Abh d Koenigl Saechs Gesellsch' d. Wissensch Math. Phys. d.*

Part 1 (1898) (with Braune)	**21**, 151
Part 2(1899)	**25**, 1
Part 3 (1900)	**26**, 85
Part 4 (1901)	**26**, 469
Part 5 (1903)	**28**, 319
Part 6 (1904)	**28**, 531

Hardte D E (1978). Determining muscle forces in the leg during normal human walking - an application and evaluation of optimization methods. *Trans. Am. Soc, Mech. Engng. J. Biomech. Engng.*, **100**, 72-78.

Harrington I J (1974a). Knee joint force in normal and pathological gait. M.Sc. Thesis, University of Strathclyde, Glasgow.

Harrington I J (1974b). The effect of congenital and pathological conditions on the load action transmitted at the knee joint. *Total knee replacement*, Proc. Inst Mech. Engng., C204/74, 1-7.

Harrison D W and Nicol A C (1988). Knee joint loads during running and turning activities, in *Biomechanics in Sport*, Mechanical Engineering Publications Ltd, London, pp 13-18.

Hodge W A, Fijan R S, Carlson K L, Burgess R G, Harris W H and Mann R W (1986). Contact pressures in the human hip joint measured in vivo. *Proc. natn. Acad. Sci. USA*, **83**, 2879-2883.

Katz J L (1986). Bioengineering and implant surgery: some new directions, in R Kassowsky and N Kossovsky (eds.), *Materials Sciences and Implant Orthopaedic Surgery*, Martinus Nijhoff Publ. Dordrecht, Netherlands, pp 397-406.

Levine D L and Stoneking J E (1980). A three dimension finite element based, parametric study of an orthopaedic bone plate, in B R Simon (ed.), *Conf. on finite element analysis in biomechanics*, Tucson, Arizona, 713-728.

Marey E J (1873). De la locomotion terrestre chez les bipedes et les quadrupedes. *J de L'Anat et de la Phys.*

Morrison J B (1967). The forces transmitted by the human knee joint during activity. Ph.D. Thesis, University of Strathclyde, Glasgow.

Morrison J B (1969). Function of the knee joint in various activities. *J. Biomed. Engng.*, **4** (12), 573-580.

Morrison J B (1970). The mechanics of the knee joint in relation to normal walking. *J. Biomechanics*, **3**, 51-61.

Paul J P (1965). Bioengineering studies of the forces transmitted by joints. II. in R M Kenedi (ed.), *Biomechanics and Related Bioenginering Topics*, Pergamon, Edinburgh, pp 369-380.

Paul J P (1966). The biomechanics of the hip-joint and its clinical relevance. *Proc. Roy. Soc. Med.*, **59** (10), 943-948.

Paul J P (1967). Forces transmitted by joints in the human body. *Proc. Inst. Mech. Engng.*, **181** (3J), 8-15.

Paul J P and McGroutheer D A (1975). Forces transmitted at the hip and knee joint of normal and disabled persons during a range of activities. *Acta. Orthop. Belg.*, **41** (Suppl 1), 78-88.

Pierrynowski R M and Morrison J B (1985). A physiological model for the evaluation of muscular forces in human locomotion: theoretical aspects. *Math. Biosci.*, **75**, 69-101.

Poulson J (1973). Biomechanics of the leg. Ph.D. Thesis, University of Strathclyde, Glasgow.

Procter P (1980). Ankle joint biomechanics. Ph.D. Thesis, University of Strathclyde, Glasgow.

Procter P and Paul J P (1982). Ankle joint biomechanics. *J. Biomechanics*, **15** (9), 627-634.

Rohlmann A, Bergmann B and Kolbel R (1980). The relevance of stress computation in the femur with and without endoprosthesis, in B R Simon (ed.), *Conf. on finite element analysis in biomechanics*, Tucson, Arizona, 549-566.

Rohrle H, Scholten R, Sigolotto C, Sollbach W and Kellner W (1984). Joint forces in the human pelvis-leg skeleton during walking. *J. Biomechanics*, **17** (6), 409-424.

Rushfeldt P D, Mann R W and Harris W H (1981). Improved techniques for measuring in vitro the geometry and pressure distribution in the human acetabulum-II. Instrumented endoprosthesis measurement of articular surface pressure distribution. *J. Biomechanics*, **14** (5), 315-323.

Rybicki E F (1982). The role of finite element models in orthopaedics, in R H Gallagher, B R Simon, P C Johnson and J F Gross (eds.), *Finite Elements in Biomechanics*, John Wiley and Sons. Ltd, New York, pp 181-194.

Rydell N (1965). Forces in the hip joint II, Intravital Studies in R M Kenedi (ed.), *Biomechanics and Related Bioengineering Topics*, Pergamon, Edinburgh, pp 351-357.

Rydell N (1966). Forces acting on the femoral head-prosthesis. *Acta Orthop. Scand.*, Suppl 88.

Schneider E, Genge M, Wyder D, Mathys R and Perren S M (1988). Telemetrized interlocking nail for in vivo load determination in the human femur. *Proc. Eur. Telemetry Conf.*, 272-296.

Stauffer R N, Chao E Y and Brewster R C (1977). Force and motion analysis of the normal diseased and prosthetic ankle joint. *Clin. Orthop. Related Res.*, **127**, 189-196.

Tooth R (1976). The biomechanics of arthrodesis and arthroplasty in the human leg. Ph.D. Thesis, University of Strathclyde, Glasgow.

Tronzo R G (1974). Hip nails for all occasions. *Sym. Fractures of the Hip Part I, Orthop. Clin. N. Am.*, **5** (3), 479-491.

Vainionpaa S (1987). Biodegradation and fixation properties of biodegradable implants in bone tissue - an experimental and clinical study, Multiprint, Helsinki.

Weinstein A M, Clemon A J T, Starkebaum W, Milicic M, Klawitter J J and Skinner H B (1981). Retrieval and analysis of intramedullary rods. *J. Bone Jt. Surg.*, **63 A**, 1443-1448.

18

INTERFACE FAILURE DYNAMICS: OSTEOCLASTIC AND

MACROPHAGIC BONE LOSS

I C Clarke and P Campbell

INTRODUCTION

The optimal interface for fixation is generally considered to be direct contact between bone and implant (Charnley, 1979; Albrektsson et al., 1981; Linder et al., 1983; Zweymuller, 1988). However, in most instances the surgeon has had to accept an interpositional fibrous tissue layer between implant and bone, regardless of fixation methods. We can therefore consider implant failure as loss of fixation, which was the end result of progressive loss of bone.

This chapter will discuss this phenomenon of bone loss in relation to three destructive events: wear debris, micromotion and "stress shielding". We shall explore common pathways to implant loosening, whether the initial clinical result was achieved by cement, press-fit or porous-ingrowth fixation. Much of the bone-loss mechanisms will be deduced from a review of the cemented implant literature, which in turn will be used as a counterpoint for the discussions on press-fit smooth and press-fit porous implant designs. We shall also emphasise the time scale leading to clinical failure, i.e. short-term (0-3 years), intermediate (3.1- years) and long-term (10-20 years).

Wear Debris, Macrophages and Bone Loss

Bone loss can be mediated by either:

osteoclasts, multinucleated cells with resorb bone in both normal and pathological situations and

macrophages, phagocytic cells which, when activated by physical or chemical messengers, appear to be capable of direct bone resorption, as well as being able to stimulate osteoclasts to resorb bone. The three stimuli known to activate the macrophage are foreign particulate matter, bacteria and cell death. Since the wear particles from implants cannot be digested by macrophages, the debris will be released when the cells die and repeated phagocytosis, cell death and progressive tissue necrosis continues.

The presence of particulate debris in membranes around cemented implants has been well described (Mirra et al., 1976, 1982; Revell et al.,. 1978; Willert et al., 1981; Bullough et al., 1988), and includes cement, polyethylene and metal. Particulate size is important to the activation of macrophages since only particles which are small enough to be phagocytosed will stimulate a histiocytic response (table 1). There is experimental evidence of equally adverse histiocytic tissue responses to fine particulates of PMMA (Goodman et al., 1988; Horowitz et al, 1988), polyethylene and metals (Rae, 1986; Howie and Vernon-Roberts, 1988;). However, in the clinical literature, there is no consensus as to which material is responsible for the most severe histiocytic response. For example, Mirra et al., (1976) and Revell et al., (1978) reported that polyethylene resulted in a more marked histiocytic response, compared to metal or cement, while Pazzaglia et al., (1988a) claimed that metal particulates provoked a macrophage response more readily than either acrylic or polyethylene. It may be that in each of the clinical failure types, there was a marked propensity for an excess of one or more materials which led the authors to conclude that one material was more reactive than another. In contrast, there seems to be some agreement that acrylic debris does not evoke the same marked response as polyethylene or metals (Mirra *et al.*, 1976; Pazzaglia *et al.*, 1988a; Malcolm, personal communication, 1988). However, it may be that the volume and particle sizes of the debris have an overwhelming effect on the body's defense mechanisms, much more so than the material itself.

DEBRIS	SHAPE	PARTICULATE SIZE	CELL TYPE	CELL SIZE
Metal	Particles[a]	$< 0.1 \ \mu m$	Macrophages	NS
Metal	Sphere/rod[b]	$1-4 \ \mu m$	Macrophage	$20-40 \mu m$
Metal	Aggregate[b]	Moderate-massive	Giant cells	$100-400 \mu m$
UHMWPE	Fibril[b]	$< 10 \ \mu m$	Macrophage	$20-40 \mu m$
UHMWPE	Fibril[b]	$10-40 \ \mu m$	Giant cell; E/C, M/P	$100-400 \mu m$
UHMWPE	Fibre[b]	$> 40 \ \mu m$	E/C, G/C	
UHMWPE	Flake[b]	$> 500 \ \mu m$	In scar tissue	

Note: Details taken from Pazzaglia *et al.* (1988a)[a] and Mirra *et al.* (1976)[b].
    E/C = Extracelullar, G/C = Giant Cell, M/P = Macrophage, NS = Not Stated

Table 1 Summary of materials, particulate sizes and cell types.

Implant retrieval studies have shown wide distribution of particulates throughout the joint tissues, which suggests that a transport mechanism must be at work. Bocco and Charnley (1977) initially speculated on a pumping mechanism capable of moving macrophages and/or particulate debris to even distant interfaces of the joint. Polyethylene from the hip socket has been identified in pathological fractures of the distal femur (Pazzaglia and Byers, 1984) and inside the femoral head of double cup arthroplasties (Campbell, unpublished). The exact transport mechanism is unclear, but this means that either macrophages and/or wear debris produced at one interface can induce an osteolytic response and component loosening in other areas, even remote from the initial loosening site.

Loosening of the Femoral Stem

The femoral stem exhibits a greater propensity for pain and shows higher (1.4 - 50 times) failure rates than the acetabular side (Gudmundsson *et al.*, 1985). The spectrum of bone loss spans from the initial interfacial radiolucent zone, (Charnley *et al.*, 1968; Gruen *et al.*, 1979), to progressive radiolucencies, cyst formations (Blacker and Charnley, 1978; Johnson and Crowninshield, 1983; Huddleston, 1988; Rimnac *et al.*, 1988) and finally to massive "granulomatous pseudotumours" (Griffiths *et al.*, 1987). A clinical and radiographic review of collared and collar-less stem types of cemented THRs illustrated that osteolytic cysts were the highest predictors of failure, i.e., "Overall, 80% of those demonstrating osteolytic cyst formation had come to revision" (Bannister, 1988).

Proximal bone loss is generally identified as loss of the medial cortex, or a general cancellization of the proximal femur (figure 1). Bocco and Charnley (1977) reported that "a universal finding was a change in the texture of the bone of the calcar, resulting in the cortical layer of the medial border of the calcar tending to assume a cancellous structure" (figure 1). Sarmiento *et al.*, (1988) reported survivorship statistics of calcar resorption: this was evident in 15% of cases at six years, 47% at ten years, and by 15 years had risen to 83%. Griffiths *et al.*, (1987) noted that the most common cystic changes were either adjacent to the femoral calcar or the central medial aspect of the stem.

Carlsson *et al.*, (1983) observed localised endosteal resorption, scalloping and cysts in 47% of revised stems, 76% of which were progressive at 1 - 13 years follow up (average 5 years). Virtually 80% of distal femoral cysts occurred where the prosthesis tip penetrated through the cement sheath and juxtaposed the endosteal wall. One possible explanation is that the cyst was a result of the metal debris which was produced due to bone/stem (type

III) micromotion. An alternative explanation, and one which we prefer, was provided by Willert (personal communication) i.e., such defects in the cement sheath create cement particles by cement/stem micromotion (type I) which then elicit a histiocytic response.

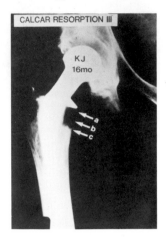

Fig. 1   Radiograph of cemented total hip showing disappearance of the medial cortical wall due to a histiocytic response to wear debris.

Blacker and Charnley (1978) further commented that "the process of resorption was sometimes preceded by cystic areas". Several investigators have suggested that such changes result from cement breakdown (Harris *et al.*, 1976; Willert *et al.*, 1981), excessive local stresses (Carlsson *et al.*, 1983), micromotion (Harris *et al.*, 1976; Scott *et al.*, 1985; Eftekhar and Nercessian, 1988). However it is particularly interesting that histological studies have correlated proximal bone loss with higher patient activity and evidence of acetabular cup wear (Johnson and Crowninshield, 1983; Hierton *et al.*, 1983a, b; Schmalzried and Finerman, 1988).

A very compelling explanation is that increased patient activity levels produced more wear and exacerbated the histiocytic osteolysis of the proximal femur !

Johanson *et al.*, (1986) described 30 mm of calcar resorption with pain onset at ten years and histology confirmed the presence of polyethylene wear and histiocytes in the resorbed area. Similarly, Pazzaglia *et al.*, (1988b) noted in four cases that, over a three to nine year period, there was progressive resorption of the medial cortex associated with either a grey or a white caseous material containing stainless steel of CoCr particles.

Interestingly, Bocco *et al.*, (1977) speculated that a good proximal bone-cement junction would help minimise the entry of polyethylene particles into the interface. In the clinical reviews of the collarless Exeter stem by Fowler *et al.*, (1988) and Bannister (1988), the collarless prosthesis had the least cystic problems and the lowest loosening rate. It is believed that this is due to the self-wedging effect of the collarless stem inside its cement sheath i.e., as the cement creeps (Ebramzadeh *et al.*, 1985), the stem keeps wedging it against the femur's endosteal surface, thereby "sealing off" the proximal canal against the ingress of wear debris. Carlsson *et al.*, (1983) measured this increase in stem-cement gap radiographically as 0.6 mm/year for the first three years, decreasing to 0.24 mm/year from 3-5 years. This gap has been observed at implant retrieval as a 50-100 μm thick membrane between stem and cement (Fornasier and Cameron, 1976; Pazzaglia *et al.*, 1988b).

Acetabular Component Loosening

Loosening of cemented acetabular components has long been associated with membrane formation, bone loss and wear debris (Bocco *et al.*, 1977; Blacker and Charnley, 1978; Eftekhar and Nercessian, 1988). McCoy *et al.*, (1988) noted that 85% of the Charnley cups exhibited wear and proximal femoral bone loss after fifteen years. Annual wear rates in males averaged 0.3 mm per year whereas females averaged 0.1 mm. Rimnac *et al.*, (1988) and Silverstein (1986) described progressive cystic changes over a three to nine year period which eventually resulted in bone destruction and revision. Wrobleski *et al.*, (1988) noted that as cup wear increased beyond 4 mm, radiographic evidence of cup migration increased from 60 to 100% of such cases. This data coupled with the evidence from the femoral side, would appear ample evidence that the intermediate to long term loosening of total joints is to a major extent controlled by wear mediated osteolytic processes.

Non-Cemented Implants

Bone ingrowth, stress shielding and the problem of proximal bone loss

There has been a marked shift away from the use of bone cement and there are now many designs of porous coated implants for bone ingrowth fixation. Pilliar (1980) theorised that "the use of a porous-coated implant into which bone would grow should provide a similar method of transfer of stress into this region of bone, thereby stimulating healthy bone retention". It was hoped that with suitable surface characteristics, initial implant to bone contact and adequate bone stock, that bone ingrowth would be both achieved and sustained for long term fixation.

A number of animal studies have detailed the bone reaction around porous coated stems (table 2). The common finding was distal fixation by bone ingrowth but major loss of the more proximal bone by three to six months, with cases in fully coated stems where the implants beaded surfaces appeared through the resorbed femoral cortices. In extreme examples the femoral cortices had been entirely resorbed. However, the press-fit model of the same design demonstrated little or no bone changes (Turner *et al.*, 1985; Sumner and Turner, 1988;).

AUTHOR	POROUS IMPLANT TYPE	RESULTS
Turner '86	THR Stem	Proximal to distal cortex lost
Bobyn '87	THR Stem	Antero-medial cortex lost
Chao '83	THR Stem	Anterior cortex lost
Chen '82	THR Stem	Decreased femur width, increased cortical porosity
Miller '76	Segmental replacement	Anterior femoral cortex lost

Table 2 Summary of canine models describing massive proximal bone loss around porous coated femoral implants.

It is therefore interesting in view of the animal models that clinical results have also demonstrated major bone loss. Engh and Bobyn (1988) demonstrated short-term "subtle changes but indicative of substantial loss of bone", i.e., one to two years. This bone loss became more extensive using canal-filling prostheses, extensive porous coating and in the presence of bone ingrowth. In fact these authors have stated that if there is no proximal bone loss, there cannot be bone ingrowth! The converse of this has been proven true in the clinical series of Cameron (1986) and Brown and Ring, (1985) - those implants which obtained only fibrous tissue ingrowth exhibited little or no bone loss.

Reviewing intermediate (6 years) to long-term clinical series (10 years) of porous coated stems, it is a curious historical fact that five of the pioneering devices have been more or less abandoned due to problems associated with massive osteolysis, pain, stem fracture and extreme difficulties of revision (Judet, 1978; Brown and Ring, 1985; Cameron, 1986; McClelland, 1987; Letournel, 1988; Lord, 1987 - personal communication).

Porous implants which failed at ten years exhibited "granulatomous thinning of the cortices" and debris laden black tissue at the resorbed interfaces (Brown and Ring, 1985; Letournel, 1988). This would appear somewhat akin to the massive osteolysis labelled as "cement disease" around failed long-term cemented implants (Jones and Hungerford, 1987). However, in the intermediate to long-term series of porous-coated AML prostheses (which utilise a smaller stem and bead size than currently used; Engh and Bobyn, 1988), it is believed that the bone changes at 1-2 years will be non-progressive and the ten year results look promising so far.

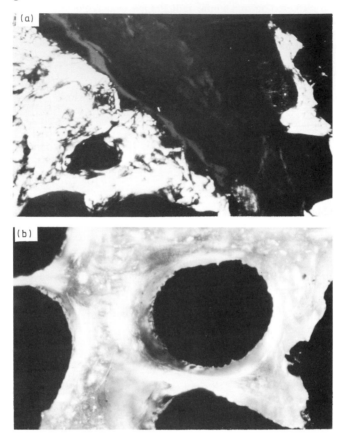

Fig. 2   a) Polarised light micrograph of a ground selection of ingrown but failed porous surface replacement showing an invading histiocytic membrane containing metal and polyethylene debris.
b) Higher power showing polyethylene and metal debris at the interface, i.e., well within the femoral head.

In terms of problems arising from serious contamination of the implant-bone interfaces, some concern must be expressed regarding the potential for porous coatings to fret, fatigue, fragment and cause wear problems. Buchert *et al.*, (1986) had two short term (2-3 months) revisions due to detached CoCr metal beads. Cheng and Gross (1988) documented loose CoCr beads on postoperative X-rays in five PCA knees, by 13 months the number had increased to 23. Callaghan *et al.*, (1988) also reported loose beads in PCA components, 36% of cups and 55% of stems. With 41% of patients complaining of some pain at two years, the authors concluded that "the pain in the thigh, the progressive radiographic changes and the loosening of beads are of concern". This concern was echoed by Cameron (1986) and Ryd *et al.*, (1985). Another consequence of non-cemented stems is wear between the endosteal wall and the prosthesis itself. As Collier *et al.*, (1988) noted, "a significant proportion of all retrieved femoral hip prostheses demonstrated burnishing of the smooth portion of the stem".

Analysis of retrieved porous coated components suggest that bone ingrowth is not being widely achieved. Cook *et al.*, (1988) described no bone ingrowth in one third of 90 retrieved THR implants, about 2% ingrowth in one third and the remainder had at most 10% ingrowth. Both Cook *et al.*, (1988) and Collier *et al.*, (1988) found bone ingrowth in patients with pain, indicating that bone ingrowth alone is not sufficient to guarantee pain relief. In fact, "prostheses retrieved for pain showed a somewhat higher incidence of bone ingrowth than those retrieved for all other reasons" (Collier *et al.*, 1988).

In terms of the wear-particle induced osteolysis, retrieval analysis of porous coated femoral and acetabular surface replacements from the UCLA series has shown that excellent short-term bone ingrowth can be achieved, but that failure can occur due to wear induced histiocytic response. Out of 12 short to intermediate-term failures to date (24 to 55 months post-operatively), four fractured through femoral neck cysts which comprised histiocytic membranes loaded with metal and polyethylene debris (figure 2). The joint tissues were grossly black and the titanium alloy bearing surfaces were burnished revealing about three square centimetres of wear scar. For metal particles believed to be about 0.1 μm in size and assuming an average wear scar depth of only 1 μm, this would be equivalent to about 3×10^6 metal particles. It is to be expected that such release of metal particulate would have an aggravated effect on the polyethylene wear rates as well. This points out the seriousness of the wear issue when dealing with implant systems in younger patients with higher activity levels.

Non-Cemented Implants:

Press-fit and osseointegration concept

As Jones and Hungerford (1987) stated, "failure of the press-fit prostheses used prior to the introduction of bone cement fixation was virtually never associated with the massive osteolysis that is currently being seen with the cemented implants". Ring (1978) described the results of 1797 press-fit CoCr straight stems as 95% pain-free with annual revision rates of only 0.5% and little or no bone loss. He emphasised that his revision results were more successful than typical cemented revisions.

A histological review (Kozinn *et al.*, 1986) of retrieved press-fit Moore type implants (up to 23 years follow-up) demonstrated that, although there were fibrous membranes and micromotion was probably present, these stems remained "functionally stable" with little or no bone loss.

The recent concept of osseointegration "direct - on the microscopic level - contact between living bone and implant". (Albrektsson *et al.*, 1981) has been demonstrated with titanium alloy (Ti-6-4) implants. They illustrated osseointegration of 34 Ti-6-4 bone screws removed from various skeletal sides. Linder *et al.*, (1988) reported that the bone appeared to be held to the titanium oxide film by a 20-50 nm thick film of ground-substance "glue", and that osseointegration was well established by five months and was "attained regularly, even in porotic bone and in patients with rheumatoid arthritis".

Lintner *et al.*, (1986) and Zweymuller *et al.*, (1988) reported osseointegrated titanium femoral stems at three weeks to 73 months follow-up, noting broad "osteoid seams" at 3-4 months, and osseointegration at 5-10 months. They reported that "from the first month after surgery, the presence of newly-formed osseous tissue growing on the titanium surface could be observed" and that "this incorporation can be observed over a period of 2-3 years".

Thus there would appear to be short to long-term evidence that non-cemented "press-fit" devices can function with reasonable clinical success without the risk of coated surfaces contaminating the fragile bone-implant interfaces.

DISCUSSION AND CONCLUSIONS

It would appear that failure of a cemented hip in the short-term (1-3 years) has to be due to a combination of poor design, poor materials choices, sub-optimal surgical technique, or excessive patient activity. True interface loosening of the "well-cemented" implant appears as an intermediate to long-term phenomenon, sometimes taking anywhere from six to 18 years to progress to failure. In our opinion, the most powerful mechanism inducing bone loss is a wear-mediated osteolytic process which normally takes ten years to become noticeable but can be triggered much earlier by higher patient activity-levels or poor materials combinations.

In our wear/loosening model (figure 3), it seems likely that creep of the medial cement sheath may result in the femoral stem flexing more into varus and creating three wear effects:

micromotion and wear between the metal stem and the cement (type I),

appearance of the typical superio-lateral radiolucent zone between stem and cement (Carlsson *et al.*, 1983), and

a fibrous membrane inside the cement sheath (Fornasier and Cameron, 1976).

The resulting cement debris gets pumped around the joint (figure 3), thereby exacerbating the polyethylene wear (Pazzaglia *et al.*, 1988a; Willert, 1988 -personal communication). The resulting polyethylene and cement particles then activate the macrophages and get pumped back down the femoral canal to cause more destruction and loosening at the cement-bone interface. With continuing patient activity and longer in-service requirements, this process can proceed from radiographic signs of lucent zones to cystic changes and to "granulatomous pseudotumours" (Griffiths *et al.*, 1987) over the space of one to ten years or more. Thus the phenomenon of "cement disease" may well be a triple play, i.e., wear between cement:stem (type I), ball:cup (type II) and cement:bone (type III, figure 3).

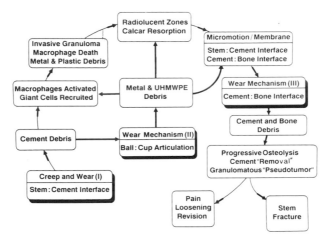

Fig. 3   Triad of wear mechanisms between cement:stem, ball:cup and cement:bone interfaces.

Wear of the polyethylene cup and associated metal ball wear can clearly cause considerable magnitude of bone loss without the need for an initial "prosthesis-loosening" theory (Scott *et al.*, 1985). For example, Jasty *et al.*, (1986) reported on localised bone resorption adjacent to rigidly anchored femoral stems. Nor do we believe it is necessary to hypothesise that the cement must first crack before the cement particles can be released (Stauffer, 1982; Schmalzried and Finerman, 1988). The evidence supports cement creep with subsequent micromotion at the cement:stem interface.

It is difficult to assess the cement particle size and shape distribution because:

the cement dissolves out of the processed tissue section so the smaller sized particles may go unnoticed, and

the cement-stem and cement-bone interfacial wear mechanisms are beyond our quantitative description.

However, the danger of the polyethylene particulates can be illustrated with respect to the study by Mirra *et al.*, (1976) who stated that the majority of the polyethylene wear debris was less than 40 μm in size (Figure 4). This would encompass types I to IV with the particle size classification that we use (table 3). Also from Mirra's study, we assumed a typical length:width:thickness ratio for the polyethylene particle of 10:2:1 to simplify calculations. Using the 0.1 and 0.3 mm/year wear rates identified by McCoy *et al.*, (1988) for Charnley prostheses, it is possible to predict the average number of particles released from the acetabular cup each year of use as a function of the assumed size and shape of the particles.

This analysis showed that polyethylene particles 40 μ to 10 μ size would be released in amounts up to 4×10^9 particles per year with some small variation due to ball size (table 3). However, the biggest effect results from the decrease in particle size. Going down to the 5 μ polyethylene size indicated by Mirra's study, the volume of particulate would rise to 15×10^9 and 46×10^9 per year for females and males (figure 4), respectively (with the limiting assumption that they all have the same size and shape). For particles of the order 1 μ or less, the numbers reach into the trillions of particles per year which would be more typical of metals perhaps than polyethylene (table 1). With numbers like these, one wonders how surgeons can continue to claim that wear is not a problem!

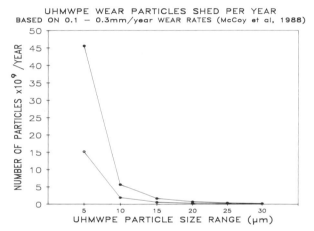

Fig. 4   Correlation between polyethylene particle size and quantity of particles released per year in normal hip function with 0.1 - 0.3 mm of wear per year, based on 5 μ - 40 μ sized particles which have been reported in macrophages .

DEBRIS TYPE	SIZE RANGE
I	< 0.1-1.0μm
II	1-10μm
III	10-20μm
IV	20-40μm
V	40-100μm
VI	100-200μm
VII	200-400μm
VIII	> 400μm

NO. POLYETHYLENE PARTICLES SHED EACH YEAR

BALL DIA	22mm	28mm	32mm	
FEMALES	38M-1.9B	62M-3.1B	80M-4B	
MALES	114M-5.7B	185M-9.3B	240M-12B	

$M = 10^6$ Particles, $B = 10^9$ Particles

Table 3: Range of particulate sizes encountered in retrieval studies and estimates of the amount of debris sged on an annual basis.

It is rather puzzling that many studies indict the loosening of the polyethylene cup as the first major indication of things going wrong, yet the femoral stems are revised in higher numbers. One explanation would be that both the stem and cup interfaces are affected by wear-mediated osteolysis, but by the nature of the geometry and loading conditions, radiolucent zones become more apparent in the acetabulum without necessarily resulting in clinical symptoms. In contrast, the pumping mechanism drives the particulate laden macrophages equally into the femoral interface where bone lysis and implant micromotion more readily transmit symptoms of pain.

Porous-coated stems and cups are very much the centre of attention today. The main thrust of the bone ingrowth studies was to "provide a similar method of transfer of stress into this region of bone, thereby stimulating healthy bone retention" (Pilliar, 1980). However, as the animal models, clinical results and retrieval studies show, there is actually bone loss from the beaded areas, frequently referred to as the "stress-shielding" phenomenon to describe extensive areas of proximal bone loss (Engh and Bobyn, 1988). As Galante (1988) pointed out for his short-term studies (12-30 months), the "bone remodelling processes that we have observed are more pronounced at this earlier stage than those that have been reported with cemented femoral components". This bone loss may be either osteoclastic remodelling due to "stress-shielding" or a histiocytic response to metallic and/or polyethylene debris, or both. Unfortunately, there is a lack of good histological data on the process of stress shielding, but Turner *et al.*, (1985) stated that osteoclasts were noted in the resorbing area.

It has been suggested that stress shielding will be non-progressive (Engh and Bobyn, 1988). However, this ignores any long-term osteolytic mechanisms of bone destruction brought about by the release of wear particles. In addition, the inclusion of younger active patients may hasten the onset of this destructive process in the intermediate term.

It is a sobering thought that the long-term porous stem clinical studies of Drs Lord, Judet, Letournel and Ring were terminated predominantly due to granulomatous reactions, bone destruction and pain. We may have to coin the term "bead disease"for future comparisons to the "cement disease" term referenced by Jones and Hungerford (1987).

With the increasing complexity of coatings on implants, the standard metallic femoral balls may be compromised by the release of particulates promoting an aggressive three-body abrasive wear. As the titanium surface replacement failures demonstrated, achieving bone ingrowth fixation is one thing, sustaining it in the face of an aggressive histiocytic response is quite another. The use of ceramic balls (table 4) may be mandated not only to eliminate the metallic particulate and corrosion products but also to reduce the polyethylene wear rates (Zweymuller *et al.*, 1988).

It seems clear that the great value of a cemented stem is a very consistent, pain-free result for short and intermediate term results. Press-fit and porous-ingrowth stems eliminate the threat of "cement disease" but may not have as good pain scores as the cemented procedure. The advantage of the press-fit stem is that it minimises the contamination at the implant-bone interface, i.e., no cement layer or porous coatings. However, it has yet to be confirmed if event the "osseointegrated" press-fit interface can maintain fixation and bone stock in the long term (Ring, 1978; Kozinn *et al.*, 1986; Zweymuller *et al.*, 1988). The next question will be for the young patient who gets to the long-term follow-up period and is still reasonably active; which of the three fixation systems, cemented, press-fit or porous ingrowth, will offer the best resistance to the granulomatous invasion, best maintenance of bone and best prognosis for revision? There seems little doubt that the press-fit design used with wear resistant ceramic balls (Semlitsch *et al.*, 1977) will provide a significant clinical benefit to the patient.

	CONTROL	UHMWPE WEAR ON CERAMIC (28-32 mm)
Paschen '86	Muller	2-3 times reduction
Okumura *et al.* '88	Charnley	Almost half
Dorlot *et al.* '88	Muller	50% reduction
Oonishi *et al.* '88	T-28	2.5-3 times reduction
Kuroki *et al.* '88	Muller	Least UHMWPE debris
Griss, 1984	Muller	3 times reduction
Sieber & Weber, 1983	Weber	4 times reduction

Table 4 Summary of clinical and radiographic studies of wear advantages of ceramic total hip joints.

REFERENCES

Albrektsson T, Branemark P I, Hansson H A and Lindstrom J (1981). Osseointegrated titanium implants - requirements for ensuring a long-lasting direct bone-to-implant anchorage in man. *Acta Orthop. Scand*, **52**, 155-70.

Bannister G (1988). Mechanical failure in the femoral component in total hip replacement. *Orthop. Clin. N.A*, **19** (3), 567-573.

Blacker G J and Charnley J (1978). Changes in the upper femur after low-friction arthroplasty. *Clin. Orthop.*, **137**, 15-23.

Bocco F, Langan P and Charnley J (1977). Changes in the calcar femoris in relation to cement technology in total hip replacement. *Clin. Orthop.*, **128**, 287-295.

Brown I W and Ring P A (1985). Osteolytic changes in the upper femoral shaft following porous-coated hip replacement. *J. Bone. Jt. Surg.*, **67 B**, 218-221.

Buchert P K, Vaughn B K, Mallory T H, Engh C A and Bobyn J D (1986). Excessive metal release due to loosening and fretting of sintered particles on porous-coated hip prostheses. *J. Bone Jt. Surg.*, **68 A**, 606-609.

Bullough P G, DiCarlo E F, Hansraj K K and Neves M C (1988). Pathologic studies of total joint replacement. *Orthop. Clin. N.A.*, **19** (3), 611-625.

Callaghan J J, Dysart S H and Savory C G (1988). The uncemented porous-coated anatomic total hip prosthesis. *J. Bone Jt Surg.*, **70 A** (3), 337-346.

Cameron H U (1986). Six year results with a microporous-coated metal hip prosthesis. *Clin. Orthop.*, **208**, 81-83.

Carlsson A, Gentz C F and Linder L (1983). Localized bone resorption in the femur in mechanical failure of cemented total hip arthroplasties. *Acta Orthop. Scand.*, **54**, 396.

Charnley J (1979). Low friction arthroplasty of the hip: theory and practice. Springer-Verlag, New York.

Charnley J, Fallacci F M and Hammond B T (1968). The long-term reaction of bone to self-curing acrylic cement. *J. Bone Jt. Surg.*, **50 B**, 822-829.

Cheng C L and Gross A E (1988). Loosening of the porous coating in total knee replacement. *J. Bone Jt. Surg.*, **70 B** (3), 377-381.

Collier J P, Mayor M B, Chae J C, Surprenant V A, Surprenant H P and Dauphinais L A (1988). Macroscopic and microscopic evidence of prosthetic fixation with porous-coated materials. *Clin. Orthop.*, **235**, 173-180.

Cook S D, Thomas K A and Haddad Jr R J (1988). Histologic analysis of retrieved human porous-coated total joint components. *Clin. Orthop.*, **234**, 90-101.

Dorlot J M, Christel P and Meunier A (1988). Alumina hip prostheses: Long term behaviour. *Bioceramics 88*, 32.

Ebramzadeh E, Mina-Araghi M, Clarke I C and Ashford R (1985). Loosening of well-cemented total hip femoral prosthesis due to creep of the cement, in A C Fraker and C D Griffin (eds.), *2nd Sym. Corrosion and Degradation of Implant Materials, ASTM STP 859*, 373-397.

Eftekhar N S and Nercessian O (1988). Incidence and mechanism of failure of cemented acetabular component in total hip arthroplasty. *Orthop. Clin N.A.* **19** (3), 557-566.

Engh C A and Bobyn J D (1988). The influence of stem size and extent of porous coating on femoral bone resorption after primary cementless hip arthroplasty. *Clin. Orthop.*, **231**, 7-28.

Fornasier V L and Cameron H V (1976). The femoral stem/cement interface in total hip replacement. *Clin. Orthop.*, **116**, 249-252.

Fowler J L, Gie G A, Lee A J C and Ling R S M (1988). Experience with the Exeter total hip replacement since 1970. *Orthop. Clin. N.A.*, **19** (3), 477-489.

Galante J O (1988). Clinical results with the HGP cementless total hip prosthesis, in R Fitgerald Jr (ed.), *Non-Cemented Total Hip Arthroplasty*, Raven Press Ltd, New York, pp 427-431.

Goodman S B, Fornasier V L and Kei J (1988). The effects of bulk versus particulate polymethylmethacrylate on bone. *Clin. Orthop.*, **232**, 255-262.

Griffiths H J, Burke J and Bonfiglio T A (1987). Granulomatous pseudotumors in total joint replacement. *Skeletal Radiol.*, **16**, 146-152.

Griss P (1984). Four to eight year postoperative results of the partially uncemented Lindenhof-type ceramic hip endoprosthesis, in E Morscher (ed.) *The Cementless Fixation of Hip Endoprostheses*, Springer-Verlag, New York, pp 220-224.

Gruen T A, McNeice G M and Amstutz H C (1979). Modes of Failure of Cemented Stem-type Femoral Component A Radiographic Analysis of Loosening. *Clin. Orthop.*, **141**, 17-27.

Gudmundsson G H, Hedeboe J and Kjaer J (1985). Mechanical loosening after hip replacement. *Acta Orthop. Scand.*, **56**, 314-317.

Harris W H, Schiller A L, Schaller J M, Freiberg R A and Scott R (1976). Entensive localised bone resorption in the femur following total hip replacement. *J. Bone Jt. Surg.*, **58 A**, 612-617.

Hierton C, Blomgren G and Lindgren U (1983). Factors leading to rearthroplasty in a material with radiographically loose total hip prostheses. *Acta Orthop. Scand.*, **54**, 562-565.

Hierton D, Blomgren G and Lindgren U (1983). Factors associated with calcar resorption in cemented total hip prostheses. *Acta Orthop. Scand.*, **54**, 584-588.

Horowitz S M, Frondoza C G and Lennox D W (1988). Effects of polymethylmethacrylate exposure upon macrophages. *J. Orthop. Res*, **6**, 827-832.

Howie D W and Vernon-Roberts B (1988). The synovial response to intraarticular cobalt-chrome wear particles. *Clin. Orthop.*, **232**, 244-254.

Huddleston H D (1988). Femoral lysis after cemented hip arthroplasty. *J. Arthropl.*, **3** (3).

Jasty M J, Floyd W E, Schiller A L, Goldring S R and Harris W H (1986). Localised osteolysis in stable, non-septic total hip replacement. *J. Bone Jt. Surg.*, **68 A** (6), 912-919.

Johanson N A, Callaghan J J, Salvati E A and Merkow R L (1986). 14 year follow-up study of a patient with massive calcar resorption. A case report. *Clin. Orthop.*, **213**, 189-196.

Johnson R C and Crowninshield R D (1983). Roentgenologic results of total hip arthroplasty. A ten-year follow-up study. *Clin. Orthop.*, **181**, 92-98.

Jones L C and Hungerford D S (1987). Cement disease. *Clin. Orthop.*, **225**, 191-206.

Judet R, Siguier M, Brumpt B and Judet T (1978). Prothese totale de hanche in poro-metal sans ciment. *Rev. Dhir. Orthop. et Reparatrice de L'appareil moteur*, **64** (2), 14-21.

Kozinn S C, Johanson N and Bullough P (1986). The biological interface between bone and non-cemented femoral endoprostheses. *ORS, 32nd Ann, Meet., New Orleans, Feb. 17-20*, p 350.

Kuroki Y (1988). A histological evaluation in revision ceramic polyethylene hip prosthesis. *Bioceramics 88*, p 35.

Letournel E (1988). Failures of biologically fixed devices: causes and treatment in R Fitgerald Jr (ed.), *Non-Cemented Total Hip Arthroplasty*, Raven Press Ltd, New York, pp 318-350.

Linder L, Albrektsson T, Branemark P I, Hansson H A, Ivarsson B, Jonsson U and Lundstrom I (1983). Electron microscopic analysis of the bone - titanium interface. *Acta Orthop. Scand.*, **54**, 45-52.

Linder L, Carlsson A, Marsal L, Bjursten L M and Branemark P I (1988). Clinical aspects of osseointegration in joint replacement. A histological study of titanium implants. *J. Bone Jt. Surg.*, **70 B** (4), 550-555.

Lintner F, Zweymuller K and Brand G (1986). Tissue reactions of titanium endoprostheses. Autopsy studies in four cases. *J. Arthroplasty*, **1** (3), 183-195.

McClelland S, James R L, Simmons N S and Shelton M L (1987). A well-fixed but clinically unacceptable porous-ingrowth hip prosthesis. *Surg. Rounds Orthop.*, Jul, 23-37.

McCoy T H, Salvati E A, Ranawat C S and Wilson Jr PD (1988). A fifteen-year follow-up study of one hundred Charnley low-friction arthroplasties. *Orthop. Clin. N.A.*, **19** (3), 467-476.

Mirra J M, Amstutz H C, Matos M and Gold R (1976). The pathology of the joint tissues and its clinical relevance in prosthesis failure. *Clin. Orthop.*, **117**, 221-240.

Mirra J M, Marder R A and Amstutz H C (1982). The pathology of failed total joint arthroplasty. *Clin. Orthop.* **170**, 175-183.

Okumura H, Kumar P, Yamamuro T, Ueo T, Nakamura T and Oka M (1988). Socket wear in total hip prosthesis with alumina ceramic head. *Bioceramics 88*, 33.

Oonishi H, Igaki H and Takayama Y (1988). Comparisons of wear of UHMW polyethylene sliding against metal and alumina in total hip prostheses - wear test and clinical results. *World Biomat. Cong., 3rd, Kyoto, Apr 21-25*, 337.

Paschen U (1986). In vivo report: Comparison of friction characteristics between Al_2O_3 ceramic and polyethylene in hip prostheses. Technical paper Sulzer, publisher.

Pazzaglia U and Byers P D (1984). Fractured femoral shaft through an osteolytic lesion resulting from the reaction to a prosthesis. A case report. *J. Bone Jt. Surg.*, 66 B (3), 337-339.

Pazzaglia U E, Ceciliani L, Wilkinson M J and Cell'Orbo C (1988a). Involvement of metal particles in loosening of metal-plastic total hip prostheses. *Arch. Orthop. Traum. Surg.*, 104, 164-174.

Pazzaglia U E, Ghisellini F, Barbieri D and Ceciliani L (1988b). Failure of the stem in total hip replacement. A study of aetiology and mechanism of failure in 13 cases. *Arch. Orthop.Traum. Surg.*, 107, 195-202.

Pilliar R M (1980). Bony ingrowth into porous metal-coated implants. *Orthop. Rev.*, 9 (5), 85-91.

Rae T (1986). The biological response to titanium and titanium-aluminium-vanadium alloy particles. II. Long-term animal studies. *Biomaterials*, 7 (1), 37-40.

Revell P A, Weightman B, Freeman M A R and Roberts B V (1978). The production and biology of polyethylene wear debris. *Arch. Orthop. Traum. Surg.*, 91, 167-181.

Rimnac C M, Wilson Jr P D, Fuchs M D and Wright T M (1988). Acetabular cup wear in total hip arthroplasty. *Orthop. Clin N.A.*, 19 (3), 631.

Ring P A (1978). Five to fourteen year interim results of uncemented total hip arthroplasty. *Clin. Orthop.*, 137, 87-95.

Ryd L, Lindstrand A, Rosenquist R and Selvik G (1985). Tibial component fixation in knee arthroplasty. *Clin. Orthop.*, 213, 141-149.

Sarmiento A, Natarajan V, Gruen T A and McMahon M (1988). Radiographic performance of two different total hip cemented arthroplasties. A survivorship analysis. *Orthop. Clin. N.A.*, 19 (3), 505-515.

Schmalzried T P and Finerman G A M (1988). Osteolysis in aseptic failure, in R Fitgerald Jr (ed.), *Non-Cemented Total Hip Arthroplasty*, Raven Press Ltd, New York, pp 303-318.

Scott Jr W W, Riley Jr L H and Dorfman H D (1985). Focal lytic lesions associated with femoral stem loosening in total hip prosthesis. *AJR*, 144, 977-982.

Semlitsch M, Lehman M, Weber H, Doerre E and Willert H (1977). New prospects for a prolonged functional life-span and artificial hip joints by using the material combination of polyethylene/aluminium oxide ceramic/metal. *J. Biomed. Mat. Res.*, 11, 537-552.

Sieber H P and Weber B G (1983). HDPE ceramic total hip replacement: the St. Gallen "in vivo" and "in vitro" experience in P Vincenzini (ed.), *Ceramics in Surgery*, Elsevier Scient. Pub. Co., Amsterdam, pp 287-294.

Silverstein G D (1986). Letter re "Focal lytic lesion associated with acetabular loosening in total hip prosthesis". *AJR*, 146 (2), 423-424.

Stauffer R N (1982). Ten-year Follow-up Study of Total Hip Replacement. *J. Bone Jt. Surg.*, 64A, 983-990.

Sumner D R and Turner T M (1988). The effects of femoral component design features on femoral remodelling, in R Fitgerald (ed.), *Non-Cemented Total Hip Arthroplasty*, Raven Press Ltd, New York, pp 143-157.

Turner T M, Urban R M, Rivero D P, Sumner D R and Galante J O (1985). A comparative study of porous coatings in a weight bearing total hip arthroplasty model. *ORS, 31st Ann. Meet., Las Vegas, Jan 21-24*, p 169.

Willert H G, Buchhorn G and Bucchorn U (1981). Tissue response to wear debris in artificial joints, in A Weinstein, D Gibbons, S Brown and W Ruff (eds.), *Implant Retrieval: Material and Biological Analysis*, US Dept. Commerce, Washington.

Wroblewski B M (1988). Wear and loosening of the socket in the Charnley low-friction arthroplasty. *Orthop. Clin. N.A.*, 19 (3), 627-630.

Zweymuller K A, Lintner F K and Semlitsch M F (1988). Biologic fixation of a press-fit titanium hip joint endoprosthesis. *Clin. Orthop.*, 235, 195-206.

19

TOTAL HIP PROSTHESIS DESIGN

A CASE HISTORY

P Lawes, R S M Ling and A J C Lee

In 1938 WIles implanted six total hip prostheses in London. This is the earliest publication of hip joint arthroplasty in which both of the femoral and acetabular bearing surfaces were replaced. The state of the art improved in 1951 when McKee took the Thompson concept of an intramedullary stem and incorporated it in his own total hip implant. The major leap forward came in the early 1960's when John Charnley (1961) introduced the use of acrylic bone cement, polyethylene acetabular cups, small bearing surface diameters and a new operative approach. Over the last 25 years the Charnley concept has evolved and there are many competing designs but the large majority of hip arthroplasties performed today continue to use acrylic bone cement for implant fixation, polyethylene for the acetabular bearing surface and small bearing surface diameters of between 22 and 32 mm. This paper analyses qualitatively four features of the Charnley concept and how they have evolved over the last 25 years. The four features are firstly, stem section and strength, secondly range of motion between the two prosthetic components, thirdly, the bearing surface diameters and fourthly the seating, fixation and location of the femoral component.

As the design of total hips evolved through the 40's, 50's and 60's there was no established database on loadings at the hip joint. It was therefore understandable that the intramedullary stems of Thompson (1959), McKee (1966) and Charnley (1961) were designed with very little consideration being given to fatigue strength. Fatigue failures of hip joint intramedullary stems began to show in considerable quantities in the mid 1970's, typically after the device had been implanted for seven or eight years. The cause of these failures was a combination of the use of acrylic cement, the misuse of acrylic cement, the much larger number of implants being performed, the chosen cross-sections of intramedullary stems, the stress raisers on the surface of the stems and the metallurgical condition of the materials employed which controlled their yield and fatigue strengths. As a response to this much publicised clinical problem, designs evolved improving the use of cement, increasing stem section geometries, removing stress raisers, reducing the bending moments applied to the stems, improving metallurgical strength and introducing fatigue testing both for new and the then-current designs. Of these changes the improvement in cement handling, the strengthening of the materials and the elimination of stress raisers can now be seen as design improvements. However, the increase in stem section geometries necessarily reduced the cement mantle section and it has been reported (Learmonth et al., 1988) that whilst the change from the Charnley Mark I to the Charnley Mark II stem did reduce fatigue failures of stems the loosening rate through cement mantle fracture increased. Furthermore the move to valgus necks to reduce the bending moments resulted in problems of dislocation and nerve impingement.

The design of the Exeter hip incorporated a stem of constant taper whereby the section reduced to a 4 mm diameter distal tip. This design was generated around 1969 well before the publicity of stem fatigue failures. This is fortunate as the design would certainly have been unacceptable to a large portion of the orthopaedics industry some ten years later. However, 18 years of clinical results now show that an alternative solution to the fracture problem is to increase the cement mantle section, reduce the distal stem section and raise the percentage of load being transferred proximally between the stem and the femoral shaft.

Fatigue testing protocols developed through the late 70's all aimed to simulate poor cement distribution, distal stem fixation and high bending moments in the middle third of the stem. This was a logical way to progress as it was noted that a test with good proximal cement support was unlikely to fail even the weakest of stem designs at that time.

Fatigue testing is now described in an International Standard. However, design evolution with some trial-and-error development has reduced the incidence of stem fatigue failures to a level where it is no longer considered to be a clinical design issue. The range of motion of total hip prostheses was also not an issue even as recently as ten years ago, but as attention has turned from the femoral component to the acetabular component explanations are being sought for the loosening and migration of acetabular cups particularly in the ten-year-plus subjects.

Wroblewski has postulated that impingement of the neck on the rim of the cup is a major cause of cup loosening. The snap-fit acetabular designs of the 70's are now less popular. The Charnley design has recently reduced the diameter of the neck of the femoral component in order to increase the range of motion. Other designs, including the Exeter, have increased the range of motion through choosing a larger head diameter than the 22 mm Charnley dimension, also reducing the neck cross-section but, in addition, trimming back the rim of the acetabular cup. The concerns about reducing the coverage of the femoral head by the acetabular cup are associated with the risk of dislocation. As there is no established database on the magnitude and orientation of hip joint forces immediately post-operatively (when dislocation has most commonly occurred clinically) these new design modifications are based on clinical experience. It remains to be seen whether the reduction in the incidence of impingement between the neck and cup does lead to a reduced risk of cup loosening and migration.

Low friction torque resulting from the adoption of a stainless steel versus polyethylene bearing combination and a 22 mm diameter was one of the corner stones of the Charnley concept. Some considerable scientific work was reported to demonstrate the importance of this design in relation to wear rates and loosening torques. In spite of the convincing nature of the technical support, visually the 22 mm head appeared to be very small and 32 mm became an equally popular design diameter for no great technical reason.

The passage of time has shown that whilst these two dimensions each have their own advantages and disadvantages there is no overwhelming clinical reason for choosing one in preference to the other. The clinical merits of low friction arthroplasty have justified the choice of polyethylene as a bearing material but not the head diameter of 22 mm.

For the future there is likely to be reduced usage of 22 mm headed hips and for new designs to be introduced with larger head diameters. it is anticipated that this will happen in order to increase the range of joint motion, both when the components are brand new and also as penetration of the head into the polyethylene cup occurs.

It is also anticipated that the use of 32 mm head diameters will decline and designs will be introduced with smaller head dimensions. Following the recent popularity of metal-backed cups with replacement polyethylene bearing liners, the choice of 32 mm femoral heads in patients whose natural acetabular diameters are less than 44 mm can leave the polyethylene liner uncomfortably thin in concentric cup designs and worse in eccentric designs. Over time a standardised dimension of around 26 or 28 mm might be predicted to become common-place.

The debate over the use of calcar collars has been continuing for the past fifteen years or more and has remained a highly controversial issue for longer than probably any other feature of total hip design. There are no signs of diminishing interest at present.

The femoral component designs of Thompson (1959), Austin-Moore (1959) and McKee (1966) through the 1950's established the need for calcar collars throughout the industry. These designs were all used without cement and therefore some simple mechanical device was necessary to locate the femoral component at the top of the femur. When bone cement was introduced McKee made no significant changes to his own femoral component, but added the use of cement to the existing design in order to stabilise the stem within the canal. Charnley with his brand new design of the early 60's, kept a small collar at the top of the stem which bears primarily on the cement mantle but can, depending on cement thickness, bear directly on the cut medial surface of the bone. The rationale for such a collar was presumably to take most of the vertical load on the hip along the medial trabecular column as is evident in the natural hip joint. But the precedents for collars had long been set and there was probably some subjective comfort derived from the existence of a mechanical stop which would prevent the prosthesis from subsiding along the femoral canal. The Exeter Group, in 1969, took a very different approach in producing their design with no step, shoulder or collar in the design at all (Lee and Ling, 1981). All loads were to be transferred into the femoral shaft via a tapering stem. Furthermore, the stem was polished so that all load transfer was effected by the taper and nothing by interlocking at the metal cement interface. Later, design revisions included a change to a matt finish surface for manufacturing and cosmetic reasons.

Eighteen years of clinical experience have now shown the collarless feature to be at least as effective as the collared designs in the short-term, but superior in the long-term in as much as the absence of a collar permits the stem to engage in the cement mantle initially and then re-engage as movement of the cement (either through creeping or cracking) occurs. Furthermore, the polished stem has been re-introduced in order to minimise all resistance to re-engagement of the femoral stem in the cement mantle.

The tapering stem was originally designed into the Exeter Hip primarily as a point of mechanical principle. Load could be transferred via a collar or via a tapering stem but both were not necessary. The collar was redundant rather than harmful. However after eighteen years of clinical experience it is now evident that the avoidance of the collar is far more significant than was ever expected and the mechanical principles which come into play were not those on which the original design was based. The consequences of a truly collarless stem, that is which has no shoulders, collars, protrusions, holes or any surface roughness that give direct purchase with the cement mantle are as follows:

There is no impediment to engagement between the tapering spigot and the cement mantle immediately post-operatively. Where there is any mechanical stop or resistance to taper engagement (for instance a collar at the top of the stem) as the cement shrinks on curing there can be a gap created at the stem/cement interface thereby causing a loose stem from the outset.

In the longer term where cement cracks or creeps as a response to bone remodelling or degeneration, the tapering stem can re-engage within the cement mantle typically causing 1-3 mm of leg shortening, but without disturbing the cement bone interface which, if it occurs, can lead to progressive loosening at the cement/bone interface. Re-engagement at the stem/cement interface causes a reestablishment of equilibrium rather than progressive deterioration.

Reengagement of the stem in the cement mantle causes the stem to move slightly into valgus. With collared designs loosening causes the stems to tip into varus increasing bending moments and again leading to progressive deterioration.

The uninterrupted tapering stem with its flexible distal section and more rigid proximal section causes greater proximal loading of the bone and guarantees proximal support of the femoral implant, thereby reducing the likelihood of distal stem support, cantilever loading and stem fatigue fracture.

The proximal loading of the Exeter stem is not physiological. There is substantial radiographic evidence to show the remodelling of the medial trabecular column to adapt itself to the new mechanism of load transfer.

No femoral component which depends on an intramedullary stem for its fixation, stabilisation and load transfer can claim to induce physiological load transfer patterns in the femur.

In conclusion the initial intramedullary stem designs were not produced with very great technical input. The technical rationale behind the re-design of stems was in part positive, but in part misleading. Range of motion between the prosthetic components was not identified as an important design feature in the early designs and whilst the importance has recently been recognised, the biomechanical data base does not enable one to predict the likelihood of joint dislocation. The technical rationale behind the choice of head diameter is now seen from clinical experience to have been overplayed when 22 mm heads were introduced.

Finally, the unique design of the Exeter femoral stem has shown that a different approach to loading of the cement and the bone, whilst not physiological has given extremely good results in long term clinical usage.

Whilst there is certainly a place for the Engineer in the evolution of total joint arthroplasty, the last twenty five years of experience certainly show that the proof of new designs and claims of improved performance depend on long term clinical follow-up and not laboratory simulation or theoretical modelling. Greater care is needed in the choice of experimental and theoretical models, and in researching carefully the boundary conditions for such models. Without this, there is the risk that the application of science to the evolution of implant design will, from time to time, discard sound and innovative ideas on the one hand and justify the unjustifiable on the other. If implant design is not to be dominated by clinically subjective reasoning then the quality of the application of science must be improved by the creation and utilisation of a much superior biomechanical data base.

REFERENCES

Charnley J (1961). Arthroplasty of the hip - a new operation, Lancet, Jan-June, 1129.

Learmonth I D, Dall D M and Miles A W (1988). A Comparison of First and Second Generation Charnley Femoral Stems in Total Hip Replacement. *55th Ann. Meeting Am. Ass. Orthop. Surg.*, Atlantia Georgia.

Lee A J C and Ling R (1981). Total Hip Prosthesis: Mechanics of Evolution of Optimal Design, in D N Ghista (ed.) *Biomechanics of Medical Devices*, Marcel Dekker, New York, pp 325-369.

McKee G K and Watson-Farrar (1966). Replacement of the arthritic hip by the McKee-Farrar prosthesis. *J. Bone Jt. Surg.*, 48 B, 245-259.

Moore A T (1959). The Moore self locking vitallium prosthesis in fresh femoral neck fractures. Instructional Course Lectures, *Am. Acad. Orthop. Surg.*, C V Mosby Co., St Louis, 16, 309.

Thompson F R (1959). Prosthesis indications in fresh fractures and basic considerations affecting choice of prosthesis. Instructional Course Lectures *Am. Acad. Orthop. Surg.*, C V Mosby Co., St Louis, 16, 299.

Wiles P (1957). The surgery of the osteoarthritic hip. *Br. J. Surg.*, 45, 488.

20

MECHANICAL ANALYSIS OF STEM LENGTH IN HIP

PROSTHESES

P M Calderale and C Bignardi

INTRODUCTION

The load-bearing joints of the human body (e.g. the hip, knee and ankle) are heavily loaded. The aim of replacing these joints with prostheses is to restore function and eliminate pain. Research aims at developing a long lasting prosthesis capable of adapting rapidly to the bone interface, thus reducing post-operative pain, and which permits mechanically correct revision and easy removal. As the shape of the bone elements is optimised for the function which they must perform, and the bone remodels itself in relation to the stresses to which it is subjected, inserting a prosthesis induces a remodelling in the bone structure. For the hip joint, the most frequent cause of failure is the aseptic loosening which can occur after a brief period for the femoral stem and which generally takes after a longer period for the acetabular cup.

Aseptic loosening is caused by biological and mechanical reasons. As regards the biological causes, that which is probably most important in the long term is the reaction to particles of debris caused by wear between the head of the femoral component and the acetabular cup: this reaction involves a progressive destruction at the interface bone-prosthesis by granulation tissue.

As regards mechanical causes, the problem lies in the conditions at the bone/prosthetic stem or cement/bone interface and, more generally, in the structure of the mechanical system constituted by the prosthesis and bone, with or without cement. The stem must transmit relatively high loads (axial, bending and torsional) to the bone, but the exact mechanism whereby these loads are transmitted is not fully understood.

The structural problems involving the femoral component derive from the fact that the bone/prosthesis system couples elements having different stiffnesses, and from discontinuity in the stiffness at the tip of the stem. It would seem clear that this connection is loaded in fatigue and that aseptic loosening has fatigue failure characteristics; in fact, loosening is progressive and the time involved shortens as stresses increase. Consequently, the stress pattern is closely linked to prosthesis life. We also believe that high stresses, and stress peaks in particular, are the principal cause of pain, which the majority of patients experience, especially during the first post-operative year. These stresses, which are higher and in any case differ with respect to those of the original bone, bring about bone remodelling.

PROBLEMS

The high percentage of failures in hip arthroplasty and the great variety in hip prosthesis designs on the market show that there is a need to clarify the mechanical nature of the problem; in particular, the redistribution of stresses caused by inserting a prosthetic stem in the bone structure must be evaluated biomechanically.

There are two approaches to designing stem shape: the first involves optimising stem length and the pattern of its cross-sectional moments of area, while the second involves optimising only the length.

Mainly for non-cemented prostheses, the shape and dimensions of the cross-section are affected by the shape and dimensions of the bone. We follow current trends in believing that it is necessary to ensure good mating between stem and bone in order to achieve tridimensional stability and a sufficiently large contact surface.

In view of the structural problems associated with surgical damage to bone (the likelihood that the bone structure will resist and adapt to the new stress distribution increases with the amount of bone which is preserved), the many surgical implications of stem length (possibility of insertion) and the fact that our study, for the moment, focuses on non-cemented prostheses whose cross dimensions are determined by femur geometry, we have conducted a static parametric analysis in order to investigate the influence of stem length on stress distribution in the femur. This analysis is the continuation of previous preliminary studies (Calderale and Garro, 1985; Calderale and Scelfo, 1987a, b).

METHODS

Various experimental and theoretical methods have been used in orthopaedic biomechanics.

Experimental analyses may be conducted *in vivo* or *in vitro*. *In vivo* analyses use instrumented prostheses, while one of the first approaches to experimental analyses was that of using both transmission and reflection photoelasticity.

Many of the analysis techniques considered are not appropriate in the context of this study (Calderale *et al.*, 1980a, b).

Experimental methods are of little help to us given that the performance of prostheses of different length on the same patient cannot be assessed *in vivo*, while the results may be of little significance due to *in vitro* difficulties of various kinds (e.g. the fact that prostheses of widely varying length cannot be implanted with the same bone resection, the lack of connection between prosthesis and bone, damage to the cancellous bone during the successive insertion and removal of instrumented prostheses, etc).

Analytical theoretical studies may be classified by the type of load considered. These may include bending alone, the coupling of axial and bending loads (which are the subject of most two-dimensional analyses), axial load alone or more complete analyses including torsional loads. These studies are based on closed form equations which are written for geometrically simplified models, i.e., which schematise the stem-bone system with constant parameters beams (Calderale *et al.*, 1980, 1979a, b, c; Huiskes, 1980).

Analytical theoretical analyses have now disappeared from the scene even though it was once believed that they would lead to results which could be used for design with a minimum of effort and expense. In reality, however, it was soon realised that the approximations which these methods involved were not valid because they did not allow for the physical complexity of the problem.

A better approximation may be achieved by using numerical procedures such as the finite element method. This is practically the only option still open. We believe this method to be the most suitable for implant design coupled with clinical verification.

Even here, the analyses conducted may be more or less approximate. Analyses may be two-dimensional or three-dimensional.

The distribution of bone characteristics may be assumed to be homogeneous, or its real variations within the bone element considered may be allowed for.

Connection between prosthesis and bone may be considered to be rigid, or the particular nature of the bone-stem interface may be better simulated, for example, by defining different laws of relative movement for the bone-stem interface nodes or by using gap elements to which stiffness and friction characteristics may be assigned.

The biomechanics of the stem-bone connection is not yet precisely understood. We can hypothesise an initial bone-to-prosthesis anchorage stage (during which bone-stem relative micromotions may take place) followed by subsequent spontaneous structural optimisation stages wherein stresses cause adequate bone remodelling if strength limits are not exceeded.

Using complex elements in the model and assigning characteristics which are considered to be likely on the basis of intuition alone without experimental backup means introducing causes of error which cannot be quantified but which are in any case considerable. At the moment, we believe it to be better to use a simple model, especially for parametric analyses whose scope is to understand the type of phenomenon.

Many authors have utilised two and three-dimensional FE models for parametric analyses and for comparing different types of hip implants (Andriacchi *et al.*, 1976; McNeice *et al.*, 1976; Yettram *et al.*, 1980; Crowninshield *et al.*, 1980; Huiskes, 1980; Tarr *et al.*, 1982; Calderale and Garro, 1985; Calderale and Scelfo, 1987; Rohlmann *et al.*, 1987). Among others Calderale *et al.* have studied the effect of stem length in uncemented hip prostheses on bone-stem stresses, particularly in the areas of the interface which were found to be more highly stressed. Rohlmann *et al.* have evaluated the influence of the essential design and material properties in cemented hip prostheses on the stresses in prosthesis, cement and surrounding bone.

In this work starting from former experiences some details of the mesh were modified; we have rigidly bonded the stem to the surrounding bone - separating nodes when bone ultimate stress is approached with margins which take into account the accuracy obtainable through calculations of this type - and, in view of the great variability in the bone's mechanical properties, we have made a simple distinction between cortical and cancellous bone.

Furthermore, we have considered the bone as an isotropic material with linear behaviour. Our analysis of the results has shown the stresses to be sufficiently far from the average values for failure.

The utilised model represents a left-hand human femur with a prosthesis implant. It is three-dimensional and consists of 847 nodes, 532 eight-node elements and 517 plate elements.

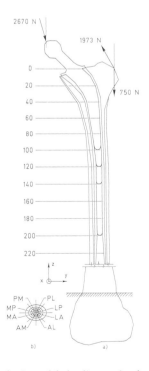

Fig. 1   a) Diagram of femur-prosthesis model: loading mode, four stem lengths, reference system.
b) Diagram of typical section of model. Elements are identified by their anatomical position:
LA (lateral-anterior), LP (lateral-posterior), AL (anterior-lateral), AM (anterior-medial),
MA (medial-anterior), MP (medial-posterior), PM (posterior-medial), PL (posterior-lateral).

Mechanical properties assigned to the elements are as follows: E = 210000 MPa for the prosthesis, E = 15500 MPa for the cortical bone and E = 500 MPa for the cancellous bone.

We used a Vax 8800 computer with the Marc code (version K3) for the calculation procedure and the Mentat code (version 5.3) for pre and post processing operations.

Four stem lengths (100 mm, 120 mm, 140 mm and 200 mm) were selected covering the wide range available on the market. The prosthesis-bone system was loaded and constrained as shown in figure 1a (Paul, 1967, 1971: Rohlmann *et al.*, 1983).

The prosthesis head is subjected to the resultant joint force, while the great trochanter is subjected to the muscle force of the abductors and the force that simulates the tension banding effect of the iliotibial tract of the fascia lata on the femoral shaft. This loading mode simulates the stance phase of normal walking.

We did not feel that it was appropriate to use more complex loading modes for this parametric analysis (Calderale and Scelfo, 1987a, b).

RESULTS AND DISCUSSION

This study analyses the stresses occurring at the bone-prosthesis interface for the different stem lengths under the hypothesized constraint and load conditions. In the model, the interface consists of a cylindrical layer of elements coupled to the prosthesis elements; in transverse section, the prosthesis is surrounded by a ring of eight elements identified with the following letters: LP (lateral-posterior), LA (lateral-anterior), AL (anterior-lateral), AM (anterior-medial), MA (medial-anterior), MP (medial-posterior), PM (posterior-medial) and PL (posterior-lateral), PM (posterior-medial) and PL (posterior-lateral) (figure 1 b).

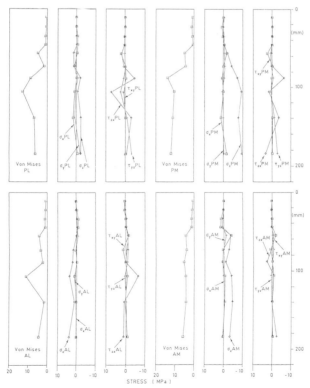

Fig. 2   Stem length 100 mm: σ_x, σ_y, σ_z, τ_{xy}, τ_{yz}, τ_{zx} and von Mises stresses for the elements LA, LP, MA, MP are plotted as a function of z coordinates.

For each stem length, we plotted curves for σ_x, σ_y, σ_z, τ_{xy}, τ_{yz}, τ_{zx} and of von Mises stress for each element (LP, LA,...) along the stem, extending the analysis to some elements located beyond the stem tip (figure 2 and figure 3).

The stress values in the various cases are not high, while stresses were found not to vary significantly with changes in stem length. For this reason, the results for just one length are shown in full.

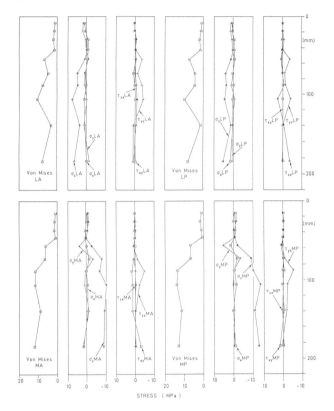

Fig. 3   Stem length 100 mm: σ_x, σ_y, σ_z, τ_{xy}, τ_{yz}, τ_{zx}, and von Mises stresses for the elements PL, PM, AL AM are plotted as a function of z coordinates.

In all four cases examined, the maximum stress was found at the tip of the stem. Consequently, to illustrate the influence of length on maximum stress we have plotted the values of the six stresses and the von Mises stress corresponding to the stem tip for the four stem lengths for each element (LP, LA, Al,...) (figure 4 and figure 5). This peak is also justified even if we consider only the sharp stiffness discontinuity which takes place.

It should be noted that von Mises stress patterns are chiefly determined by the pattern of σ_z stresses. Shear stresses would not appear to be significant; however, it should be noted that the stress values obtained depend on fineness of mesh. Consequently, a more accurate evaluation of the peaks would involve a local increase in mesh fineness. With the mesh used, however, we believe that we have identified the basic lines of the phenomenon sufficiently for clinical application.

The diagrams of figure 4 and figure 5 show that stress at the stem tip does not vary significantly with stem length. This stress is higher in the 200 mm prosthesis given that in this case the stem tip involves an area of the femur in which the torque due to the hypothesised loads is greater.

That stresses remain uniform for bone coupled to prostheses of different length may be justified by the fact that the bone, which constitutes a kind of variable stiffness elastic foundation on which the stem rests, causes a redistribution of stresses.

This leads us to conclude that, contrary to past belief, the use of a long stem provides no advantages. The prosthesis-bone coupling, in fact, was erroneously schematised as a lever supported in two points, leading to the conclusion that long stems were more suitable for containing stresses near the stem tip.

This evaluation is further borne out by the fact that, as length of stem increases, it may be considered as fixed within the cortical bone, given that its distal portion is interfaced in a narrow zone of the medullary channel and in an area in which the bone has

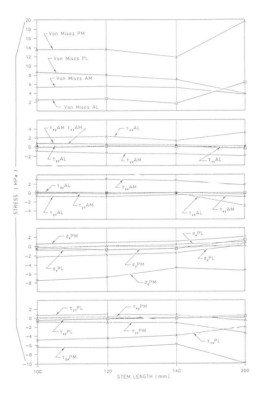

Fig.4   Values of σ_x, σ_y, σ_z, τ_{xy}, τ_{yz}, τ_{zx} and von Mises stresses for the elements MA, MP, LA, LP at tip of stem as a function of stem length.

cortical characteristics. This situation is particularly dangerous as it interferes with the physiological distribution of the loads from the femoral head, which would thus be transferred directly into the diaphysis, unloading the metaphyseal zone, (possibly leading to bone resorption) and creating high fatigue stresses due to prosthesis tilting within its seat during gait. This cyclic stress fatigues the bone-prosthesis connection; where it is sufficiently high, it cannot be balanced by bone reconstruction (bone healing).

Even if the situations involved are rather different from the structural point of view, our previous and current results are not in contrast with those reached by Rohlmann *et al.* (1987).

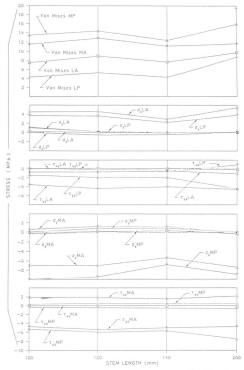

Fig. 5   Values of σ_x, σ_y, σ_z, τ_{xy}, τ_{yz}, τ_{zx} and von Mises stresses for the elements PL, PM, AL, AM at tip of stem as a function of stem length.

CLINICAL CONSEQUENCES

The possibility, or rather the advisability, of using a short stem makes it possible to adopt a high resection, including one immediately under the head of the femur. This makes it possible to preserve the original strength of the proximal femur's shell structure and to rest the prosthesis collar on an area which consists essentially of cortical bone.

In the case of revision, a high resection also makes it possible to perform a subsequent conventional resection at the base of the femoral neck in order to insert a new prosthesis implant. Carrying out a new resection also facilitates stem removal, as it uncovers part of its surface.

CONCLUSION

A mechanically correct implant design must allow for the biological component of the structure. Basically this is a question of avoiding excessive deviations in the stress from the physiological flow lines, or in other words of reducing bone remodelling to the minimum. Likewise, it is necessary to avoid subjecting the most delicate area of the prosthesis system, i.e., the bone-stem interface, to dangerous stresses. Biomechanical stress analysis can use finite element mathematical models. State of the art bone-prosthesis system modelling methods permit us to produce models which are suitable for average evaluations from the structure point of view, which are of considerable utility in identifying the orders of magnitude of the stresses and strain involved, and which allow us to better understand phenomena of which we formerly had only an intuitive grasp.

Parametric analyses may be performed adopting isotropic and linear bone characteristics. For a more detailed understanding, further developments in modelling must come closer to the real bone structure, and must consider its local characteristics in order to permit a more precise assessment of the bone-prosthesis interface conditions.

In the specific case which we have addressed using this method, we believe it possible to conclude that the use of a short stem in a hip arthroplasty is advantageous for biomechanical and surgical reasons.

These concepts have been considered in designing hip arthroprosthesis (Pipino and Calderale, 1987).

REFERENCES

Andriacchi T P, Galante J O, Belytschko T B and Hampton S (1976). A stress analysis of the femoral stem in total hip prostheses. *J. Bone Jt. Surg.*, **58 A**, 616-624.

Calderale P M, Gola M M and Gugliotta A (1979). New theoretical and experimental developments in the mechanical design of implant stems. *Pauwels Symposium*, Freie Univ. Berlin, 199-208.

Calderale P M, Gola M M and Gugliotta A (1979). Theoretical and experimental analysis of stem-femur coupling: new improved procedures. *II Eur. Soc. Biomech. Meet.*, Strasbourg, 1-14.

Calderale P M, Gola M M and Gugliotta A (1979). Mechanical stress distribution in stem implants. *Biomechanics VII*, 3A, Warsawa, 373-379.

Calderale P M, Gola M M and Gugliotta A (1980a). Systematic approach to the experimental and theoretical study of hip prosthesis coupling. I World Biomat. Congr., Baden near Vienna; *Advanced Biomaterials*, 3, J. Wiley & Sons, Chap. 17, pp 113-119.

Calderale P M, Gola M M and Gugliotta A (1980b). Experimental results for stem-femur couplings loaded in bending. *AL II-LEA Simp. Nat. Tensometrie*, Cluj-Napoca, Romania, 3-11.

Calderale P M and Garro A (1985). Theoretical analysis of the coupling between hip prosthesis stem and femur. *Proc. int. Conf. Experimental Mechanics*, Beijing, China, 1091-1096.

Calderale P M and Scelfo G (1987a). Mathematical models of muscolo-skeletal systems. *Engng. Med.*, **16** (3), 131-146.

Calderale P M and Scelfo G (1987b). A mathematical model of the locomotor apparatus. *Engng. Med.*, **16** (3), 147-161.

Calderale P M (1987). Biomechanical design of load bearing arthroprostheses. *Olgonopolska Konferencja Biomechaniki*, Gdansk, Poland, 22-24.

Crowninshield R D, Brand R A, Johnson R C and Milroy J C (1980). An analysis of femoral component stem design in total hip arthroplasty. *J. Bone Jt. Surg.*, **62 A**, 68-78.

Huiskes R (1980). Some fundamental aspects of human joint replacement. *Acta Orthop. Scand.*, 185 Suppl. 109-200.

McNeice G M, Eng P and Amstutz H C (1976). Finite element studies in hip reconstruction, in P V Komi (ed.), *Biomechanics V-A*, University Park Press, Baltimore, 394-405.

Paul J P (1967). Forces transmitted by joints in the human body. *Proc. Inst. Mech. Engrs.*, 181, Part 3J, 8-15.

Paul J P (1971). Load actions on the human femur in walking and some resultant stresses. *Expl. Mech.*, **11**, 121.

Pipino F and Calderale P M (1987). Biodynamic total hip prosthesis. *Ital. J. Orthop. Traum.*, **XIII** (3), 289-297.

Rohlmann A, Mossner U, Bergmann G and Kölbel R (1983). Finite element analysis and experimental investigation in a femur with hip endoprosthesis. *J. Biomechanics.*, **16**, 727-742.

Rohlmann A, Mossner U, Bergmann G, Hees G and Kölbel R (1987). Effects of stem design and material properties on stresses in hip endoprostheses. *J. Biomed. Engng.*, **9**, 77-83.

Tarr R, Clarke I C, Gruen T A and Sarmiento A (1982). Predictions of cement-bone failure criteria: three dimensional finite element models versus clinical reality of total hip replacement. Finite element in biomechanics (R H Gallagher *et al.*, eds), John Wiley, New York, pp 345-359.

Yettram A L and Wright K W J (1980). Dependence of stem stress in total hip replacement on prosthesis and cement stresses. *J. Biomed. Engng.*, 2, 54-59.

21

THE EFFECT OF YOUNG'S MODULUS ON IMPLANT/BONE

INTEGRATION

G L Maistrelli, A G Binnington, V L Fornasier, I J Harrington, K McKenzie and V Sessa

INTRODUCTION

It is now a well known fact that the presence of a stiff prosthesis in the proximal femur will alter the normal physiological state of the surrounding bone. According to Wolf's law this will lead to remodelling of the cortex and trabeculae. Several *in vitro* studies using strain gauge analysis have confirmed that following implantation of a stem, the strain levels in the cortical bone are shifted distally bypassing the proximal femur (Oh and Harris 1978; McBeath *et al.*, 1979, Lewis *et al.*, 1984). It can be expected that this so called "stress protection" will eventually lead to bone resorption. Other *in vitro* studies have also suggested that a more physiological state in the cortical bone can be restored by reduction in the stiffness (Torr *et al.*, 1979; Lewis *et al.*, 1981 and 1984; Manley *et al.*, 1982 a, b and 1983).

However, despite the *in vitro* evidence to date, no comparative morphometric animal studies have been done to support the theoretical advantages of a low modulus or isoelastic implant. The purpose of our discussion was to assess and quantify the biological effect of the modulus of elasticity on bone remodelling in implanted canine femurs.

MATERIALS AND METHODS

Six right, cementless total hip arthroplasties were performed in six adult male, mongrel dogs of mass greater than 25 kg. All procedures were performed under general anaesthetic using standard sterile technique and I.V. antibiotic coverage.

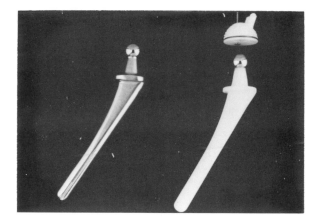

Fig. 1   Left stainless steel stem, right polyacetate stem with polyethylene cup.

The anterolateral approach was used in all cases. Half the animals had a stainless steel press fit femoral stem inserted while the other half had a reinforced polyacetate stem. Both implants were collared and of indentical size and geometry. Polyacetate is a high strength polymer with a modulus of elasticity of approximately 5 000 MPa. Stainless steel has a modulus of 200 000 MPa with bone ranging between 1 and 15 000.

Postoperatively the dogs were allowed unrestricted activity and were exercised daily. Radiographs were taken postoperatively at 3 and 6 months. The animals were sacrificed at six months. Force plate analysis of the gait was carried out prior to sacrifice of all the dogs. The pelvis and entire operated right and control left femurs were removed and dissected from the soft tissues. Contact X-rays of the specimens were then taken. The specimens were fixed and dehydrated in graduated alcohol acetone solutions and embedded in Spurr's low viscosity medium. Both the test and control femurs were then cut at 4 levels along the entire length of the prosthesis in one mm cross sections, using the Isomet low speed saw. Contact X-rays of the slices were taken and the slices were then sonicated, mounted and ground to 50-100 microns. The slices were then strained with toluidine blue for microscopic examination and quantitative measurements were performed using the Video Plan digitiser. The numerical data gathered were stored in the computer (MacIntosh Apple 2) and later analysed for statistical significance using the Student's t test.

For the analysis of our quantitative data, all the cross sections at each level were divided into 4 quadrants, anterior, posterior, medial and lateral. Representative areas from each quadrant in the test and the control femurs were analysed. For the evaluation of the bone density we calculated the percentage of the area of compact bone which was occupied by pores or resorptive cavities. Each implanted femur was compared with the contralateral femoral control. Along with the measurements of the percent porosity in compact bone, we also calculated the mean cortical thickness versus the control and the mean fibrous membrane thickness.

RESULTS

Force Plate Analysis

Force plate analysis revealed that the dogs were adequately weight bearing on both the right operated limb and the left control limb, keeping in mind that dogs will normally bear 40% of their weight on their hind legs. In comparing the two study groups there was a slight difference in favour of the lower modulus isoelastic group but the differences were not statistically significant (p = 0.01).

Radiographic Analysis

We also reviewed the post mortem contact X-rays to look for changes in the calcar and tip regions of our specimens. Calcar changes were classified according to three parameters: no change in the calcar, rounding off of the calcar and finally resorption of the calcar. Both groups showed changes in the calcar region but resorption was present only in the metal group. We also looked at the endosteal new bone formation around the implant tip. All implants exhibited new bone formation around the tip but only in the metal group were extensive changes evident.

Morphometric Analysis

We looked at the cross sections again to evaluate the difference in bone density. In the contact fine detail X-rays of the sections, we detected a qualitative increase in bone porosity in the metal implanted femurs compared to the polyacetate group. Morphometric analysis of the bone density revealed that the mean percent porosity in the low modulus group was 1.3% while the loss in bone density in the metal group reached a mean 7.8%. The difference was statistically significant. (p<0.01).

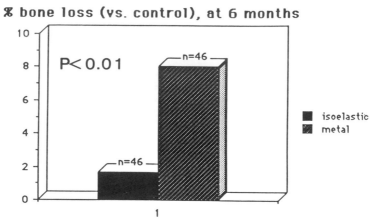

Table 1   % bone loss (vs. control) at 6 months.

CONCLUSIONS

We therefore conclude:

All prosthetic femurs in this study showed increased cortical bone porosity versus controls.

However, there was significantly less bone porosity in the low modulus group versus the high modulus group. (p<0.01).

There were no significant differences in mean cortical thickness and mean fibrous membrane thickness. (p = 0.01).

We realise that the follow-up is still short and in time further changes in remodelling may occur. Longer follow-up studies of specimens are currently underway.

This paper has been presented to the Canadian Orthopaedic Research Society Annual Meeting in Ottawa, June 5, 1988. It was awarded the founders Medal Award given by the Research Society for the most outstanding research project.

This paper was also presented to the University of Toronto Research Day held in November of 1987 and awarded the Robin Sullivan Award for the best research paper.

REFERENCES

Lewis J L, Askew M, Wixson R, Kramer G M and Tarr R R (1984). The influence of prosthetic stem stiffness and of a calcar collar on stresses in the proximal end of the femur with a cemented femoral component. *J. Bone . Jt. Surg.*, **66A**, 280-286.

Lewis J K, Kramer G M, Wixson R L et al., (1981). Calcar loading by titanium hip stems. *Trans. Orthop. Res. Soc.*, **6**, 75.

Manley M T, Gurtowski J, Stern L et al., (1983). A biomechanical study of the proximal femur using fullfield holographic interferometry. *Trans. Orthop. Res. Soc.*, **8**, 99.

Manley M T, Pachtman N, Stern L (1982). Strain levels in the proximal femur as a function of projected medial stem area. *Trans. Orthop. Res. Soc.*, **7**, 250.

Manley M T, Stern L, Gurtowski J (1982). Performance characteristics of total hip femoral components as a function of prosthesis modulus. *Bull. Hosp. Jt. Dis.*, **43**, 130-137.

McBeath A A, Schopler S A and Seireg A A (1979). Circumferential and longitudinal strain in the proximal femur as determined by prosthesis type and position. *Trans. Orthop. Res. Soc.*, **4**, 36.

Oh I, Harris W H (1978). Proximal strain distribution in the loaded femur. *J. Bone. Jt. Surg.*, **60A**, 75-85.

Torr R R, Lewis JL, Jaycox D et al., (1979). Effect of materials, stem geometry, and collar-calcar contact on stress distribution in the proximal femur with total hip. *Trans. Orthop. Res. Soc.*, **4**, 34.

22

THE BALGRIST SELF-LOCKING ARTIFICIAL HIP SOCKET

FOR IMPLANTATION WITHOUT BONE CEMENT

A Schreiber, H A C Jacob and B Hilfiker

INTRODUCTION

Soon after the advent of the Austin Moore femoral head prosthesis in the fifties, it became obvious that there was urgent need of a corresponding matching component for the acetabulum, since so-called acetabular migration was becoming a frequent complication. Various means of fixation, such as spikes and screws were attempted, but with discouraging results, and it was not until Charnley introduced acrylic bone cement in 1961 that a satisfactory method of fixing an acetabular socket to the pelvic bone was found.

Acrylic cement as a medium for attaching prostheses to bone was not universally accepted and some, like Ring (1968), are known to have always frowned upon it. Mittelmeier (1976), developed an absolutely new concept for attaching the artificial socket to the pelvic bone without the use of bone cement by using a tapered screw thread on the outside of the socket itself. The material he used was ceramic which, from our point of view, because of its high stiffness and poor wear properties under unfavourable loading conditions (Schreiber et al, 1976) did not appeal to us. However, as Endler introduced a similar socket design, but using high density polyethylene instead (Endler et al., 1983), we began to use this type of socket in 1980. It soon became evident, however, that cutting a tapered screw thread in the acetabulum, especially when the bone had become sclerotic, was difficult and the danger of stripping threads in the process was always imminent. Also, the risk of the ventral part of the acetabulum breaking out with the possibility of injuring the large blood vessels or the femoral nerve always existed.

In order to overcome the disadvantages of the Endler polyethylene screw-type socket, we developed our own self-locking socket for cementless implantation in our biomechanical research laboratory in 1982.

DESCRIPTION OF THE BALGRIST ARTIFICIAL HIP SOCKET

Basically, we have chosen a design for which there is no need to cut a thread in the pelvic bone, still without compromising on firm fixation. The socket consists of two parts, an outer split ring and a tapered insert. Initially both components were made of ultra-high molecular weight polyethylene. At present, the outer split ring is made of titanium while the tapered insert remains polyethylene (figure 1 a, b). We have chosen a cone instead of a hemisphere for the outer contour since we firmly believe that the surrounding bone finds better purchase on the former shape. We still rely on the cortical bone that is retained towards the periphery (Jacob et al., 1976) but also call upon the exposed cancellous material with its fast remodelling characteristics to anchor the prosthesis shortly after implantation. The split ring has the shape of a truncated cone and its outer surface has circumferential fins and longitudinal grooves. The split allows the ring to expand and increase its circumferences. Driving in the insert causes expansion of the split ring, and the circumferential fins and longitudinal grooves are pressed into the surface of the prepared recess in the acetabulum. This leads to a primary fixation of the prosthesis, the grooves enabling the bone tissue to ingrow, resulting in secondary anchoring while still permitting

131

further settling as well. After the outer ring has been inserted into the acetabulum, the tapered insert is driven in.

Fig. 1   a) The Balgrist hip socket with outer split ring made of titanium.
           b) The compartments separated.

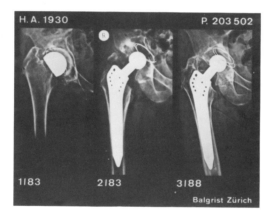

Fig. 2   A five year follow-up after replacement of a loosened ICLH prosthesis showing the contrast wires that mark the positions of the polyethylene outer ring and the insert. Notice the settling of the insert that usually occurs within six months postoperatively.

It is advisable not to drive the insert fully home but to have it stand proud of the outer ring by about 3 mm. This permits the outer ring to re-tighten its seating in the surrounding bone during the critical phase of bone remodelling that follows soon after implantation (figure 2). This self-locking possibility is a unique characteristic of this socket design. On reaching its final position, the tapered insert comes up against a stop, engaging at the same time with spikes that ensure absolute rotational immobility of the polyethylene insert within the titanium split ring. Any ball head with a diameter of 32 mm is suitable for articulation with the so-called BALGRIST socket. The currently used socket types have diameters of 50, 56, 60 and 64 mm. It must be expressly mentioned that this self adjusting socket is particularly applicable for dysplastic acetabuli since it holds bone grafts in place, with the required amount of precompression, so that the use of screws, plates, etc. is no longer necessitated.

However, sometimes additional support is required and we therefore have devised a special revision socket for such cases (figure 3).

Fig. 3   The special revision Balgrist socket with additional peripheral support.

CLINICAL RESULTS

We began to use the Balgrist hip socket from March 1982 onwards. Up to now, we have implanted over 389 such sockets. 283 of them were of polyethylene only, while a further 62 had the outer surface of the split ring coated with a layer of pure titanium. The remaining 44 had the split ring made entirely of titanium while the tapered insert continued to be of polyethylene.

The size of socket used most is the 56 mm one, in 58% of the cases, being followed by the 50 mm socket that has been employed in 26%, and the 60 mm one was used least, only in 16% of the implantations.

Of the 389 sockets that have been implanted since March 1982, 18 (4.6%) have been replaced.

In 6 cases (1.5%) technical errors were clearly responsible for inadequate primary stability. 12 sockets were revised because of loosening in which 4 were due to infection (1%) and 8 were aseptic (2.1%). On three occasions the socket was revised because of a loosened femoral prosthetic component and suspected low-grade infection, in spite of firm anchorage of the socket within the bone.

CONCLUSIONS

The special features of the BALGRIST artificial hip socket are:

The simple and safe procedure required to prepare the acetabulum.

The unique possibility for the socket to maintain pre-stress within the surrounding bone during the critical period of bone-remodelling, thereby ensuring absolute immobility at the bone-prosthesis interface.

Since no screw threads are required and by virtue of its self-adjusting and self-locking nature, the socket is particularly applicable for dysplastic acetabuli since it is capable of holding bone grafts in place without the use of screws, plates, etc. For the same reasons it might be mentioned that it does not call for the high technical skill otherwise generally mandatory for non-cemented implants.

The early results we have just reported with 2.1% aseptic loosenings, are indeed most encouraging. Whether those outer split rings made of polyethylene, as initially used, are doomed to suffer the same fate as other non-cemented polyethylene sockets that were, however, known to have moved within the bone, thus producing polyethylene debris, is a question that can only be decisively answered in about the next two years. Up to now, six and a half years after implantation, they are performing very well. In the meantime, since November 1987, we have gone over to the exclusive use of titanium split rings and now expect even better performance from these.

REFERENCES

Endler M, Endler F and Plank H (1983). Experimental and early clinical experience with an uncemented UHMW polyethylene acetabular prosthesis, in E Morscher (ed.), *The cementless fixation of hip prostheses*, Springer-Verlag, pp 191-199.

Jacob H A C, Huggler A H, Dietschi C and Schreiber A (1976). Mechanical function of subchondral bone as experimentally determined on the acetabulum of the human pelvis. *J. Biomechanics*, **9**, 625-627.

Mittelmeier H (1976). Anchoring hip endoprosthesis without bone cement, in M Schaldach and D Hohmann (eds.), *Engineering in Medicine, Advances in artificial hip and knee joint technology*, Springer-Verlag, pp 386-402.

Ring P A (1968). Complete replacement arthroplasty of the hip by the Ring prosthesis. *J. Bone Jt. Surg.*, **50 B**, 720-731.

Schreiber A, Huggler A H, Dietschi C and Jacob H A C (1976). Complications after joint replacement - long-term follow-up, clinical findings, and biomechanical research, in M Schaldach and D Hohmann (eds.), *Engineering in Medicine, Advances in artificial hip and knee joint technology*, Springer-Verlag, pp 187-202.

23

ANTERIOR KNEE PAIN - A BIOMECHANICAL

ASSESSMENT

D A MacDonald and I G Kelly

INTRODUCTION

Anterior knee pain is common, but its aetiology is uncertain, and its clinical and radiological assessment is unreliable (Fairbank *et al.*, 1984; Dowd and Bentley, 1986).

Abnormal patello-femoral loading has been implicated as the cause of anterior knee pain (Wiberg, 1941; Insall *et al.*, 1976; Hungerford and Barry, 1979), but none has quantified this load.

Patello-femoral loading has been investigated in normal subjects, in cadavers and in mathematical models, establishing normal parameters, but none of these models has been applied to patients with anterior knee pain.

We have developed a technique for calculating patello-femoral contact force (P.F.C.F.) at varying angles of knee flexion which can be used in the clinical situation. The P.F.C.F. has been correlated with the arthroscopic findings of the patello-femoral joint in an attempt to obtain a quantifiable assessment of patients with anterior knee pain.

BIOMECHANICAL MODEL

A free body diagram of the patello-femoral joint is shown in figure 1.

The P.F.C.F. can be calculated by simple vector analysis of the force and line of action of the quadriceps tendon (Q_T) and patellar tendon (P_T).

Patellar tendon force (P_T) can be calculated by taking moments about the femoro-tibial contact point.

$$P_T d = R + WC \cos \theta$$

$$P_T = \frac{R + WC \cos \theta}{d} \qquad (1)$$

The extensor moment (R) is measured at varying angles of knee flexing (θ) using a force transducer based on electrical resistance strain gauges incorporated in a specially designed chair (figure 2).

The subject sits in the chair, with the shank strapped to the test arm. The test arm is adjusted to position the axis of rotation level with the apparent centre of knee joint rotation. The test arm is fixed at five angles of knee flexion (30°, 45°, 60°, 75° and 90°) and at each angle the subject performs three maximum extensor efforts. The mean extensor moment at each angle is taken.

Prior to each test standard verbal instructions are given and the subject has two practice extensions. At the conclusion of each test a sustained extension is performed at 60° to obtain a fatigue curve. An assessment of sincerity of effort is made from the repeatability and the shape of the fatigue curve.

The shank weight (W) and the centre of mass (C) are calculated from the body weight and shank length utilising the New York University factors (Drillis and Contini,

1966). The amount of knee flexion (θ) is taken as the position of the test arm. This assumption holds true for small angles of flexion as long as the limb is correctly positioned.

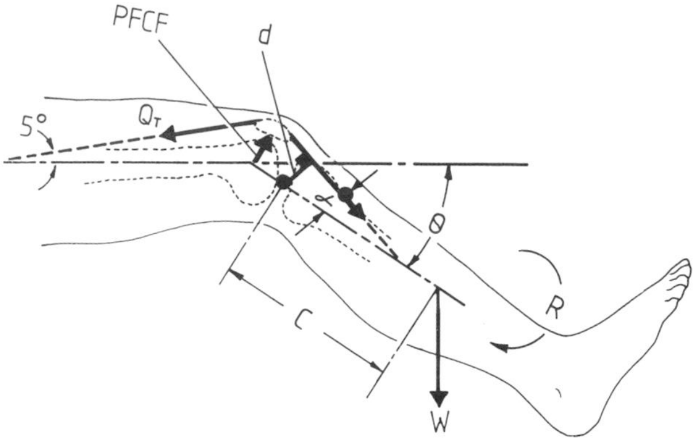

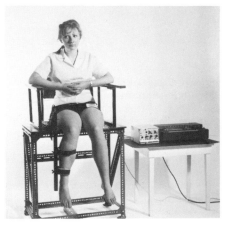

Fig. 1   Free body diagram of the patello-femoral joint.
          R = Extensor Moment
          W = Shank Weight
          Q$_T$ = Quadriceps Tendon Force
          P$_T$ = Patellar Tendon Force
          PFCF = Patello-Femoral Contact Force
          C = Distance from Centre of Rotation to the Centre of Mass of the Shank
          d = Patello-Femoral Moment Arm
          θ = Angle of Knee Flexion
          α = Angle of Patellar Tendon to Long Axis of Tibia

        The patellar tendon moment arm (d) has been calculated by previous workers for varying angles of knee flexion in normal subjects (Morrison, 1967; Ellis *et al.*, 1980; Nisell, 1985). They have shown that the moment arm remains relatively constant for the range of knee flexion studied (30°-90°). In this study the patellar tendon moment arm was measured from standard lateral radiographs taken at 45° of knee flexion using a specially designed frame to ensure constant position and a magnification factor of 1.11.

Fig. 2   The Test Chair

Previous workers have assumed that the patellar tendon force equals the quadriceps tendon force. Ellis *et al.* (1980) and Nisell (1985) have shown that this is not true. At increasing angles of knee flexion the patellar tendon force decreases relative to the quadriceps force (Q_T). From their work, the ratio $k = P_T/Q_T$ can be estimated for varying angles of knee flexion (table 1).

Knee Flexion Angle (deg)	30	45	60	75	90
k	1	0.90	0.80	0.75	0.65
α (deg)	15	10	5	0	-5

Table 1 Variation of k and α with knee flexion angle where $k = P_T/Q_T$ and α = angle of patellar tendon to long axis of tibia.

The angle of Q_T to the femoral axis is assumed constant at $5°$, throughout the range of knee flexion tested. This assumes tendofemoral contact does not occur until greater than $90°$ (Perry, *et al.*, 1975). The angle of action of P_T relative to the long axis of the tibia has been previously calculated (Morrison 1967; Missell, 1985) (Table 1).

Thus, knowing Q_T, P_T and the respective lines of action, the patello-femoral contact force can be calculated from Pythagorus' theorum.

$$(PFCF)^2 = (\text{Horizontal Vector})^2 + (\text{Vertical Vector})^2$$
$$\text{Horizontal Vector} = Q_T \cos 5 - P_T \cos(\alpha + \theta) \qquad (2)$$
$$\text{Vertical Vector} = Q_T \sin 5 + P_T \sin(\alpha + \theta) \qquad (3)$$

By substituting P_T/k for Q_T and P_T from (1) into (2) and (3) and rationalising then

$$PFCF = [R/d + (WC/d)\cos\theta]\left[1 + \frac{1}{k^2} - \frac{2}{k}\cos(\alpha + \theta + 5°)\right]^{\frac{1}{2}}$$

Despite the simplifying assumptions, this model provides a method of calculating patello femoral forces that is applicable to the clinical environment. When applied to the data of Huberti and Hayes (1984), obtained by using pressure sensitive film in cadavers comparable values are found.

SUBJECTS AND PATIENTS

18 volunteers were studied to establish a normal range (10 males, mean age 27 years, range 20-36, 8 females mean age 29 years, range 19-34). One volunteer was tested on seven consecutive days to establish reliability.

40 knees in 37 patients were tested prior to arthroscopy for a variety of knee pathologies (22 males, mean age 33, range 17-55, 15 females, mean age 23 years, range 16-29). 29 knees had symptoms and signs of anterior knee pain as defined by Ficat *et al.* (1979). At arthroscopy the patellar cartilage was inspected and probed for softness. The cartilage was described as either normal or showing evidence of chondromalacia patellae. Statistical analysis is by students unpaired 't' test.

Results

Two females failed the sincerity test and are excluded.

Patients with anterior knee pain and chondromalacia patellae have lower patello-femoral contact forces than those with normal articular cartilage. This difference is statistically significant at all angles of knee flexion except $30°$ in the female patient (figure 3), but only significant at the $45°$ and $75°$ test angles in the male patients (figure 4). In patients with chondromalacia the reduction in contact forces compared to the normal group is significant at all angles of knee flexion, except $30°$ in male patients ($p < 0.05$).

These reductions showed no correlation with clinically observed quadriceps wasting, and the test is not merely a non-specific index of knee pathology since patients with other knee pathologies had normal contact forces.

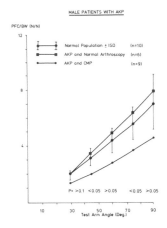

Fig. 3   Patello-femoral contact forces (PFCP) expressed as a multiple of body weight (BW) in female patients with anterior knee pain (AKP) and normal arthroscopy and with AKP associated with chondromalacia patellae (CMP) compared to the normal group.

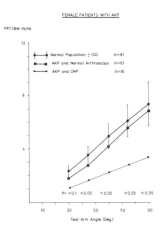

Fig. 4   Patello-femoral contact forces (PFCF) expressed as a multiple of body weight (BW) in male patients with anterior knee pain (AKP) and normal arthroscopy and with AKP associated with chondromalacia patellae (CMP) compared to the normal group.

CONCLUSION

These results show a previously unreported lower patello-femoral contact force in those patients with chondromalacia patellae. Patients with normal articular cartilage have normal or in some cases high contact forces.

This technique gives a quantitative assessment of anterior knee pain, previously lacking in clinical practice.

We are now able to provide an objective assessment of the results of the many surgical procedures designed to affect patello-femoral loading and thus improve the management of patients with the symptoms of anterior knee pain.

ACKNOWLEDGEMENTS

This work was funded by the Medical Research Fund of the University of Glasgow.

REFERENCES

Drillis R and Contini R (1966). Body Segment Parameters in *Technical Report 116 03* New York University School of Engineering and Science.

Dowd G S E and Bentley G (1986). Radiographic Assessment in Patellar Instability and Chondromalacia Patellae. *J. Bone Jt. Surg.*, **68 B**, 297-300.

Ellis M I, Seedhom B B, Wright V and Dowson D (1980). An Evaluation of the Ratio Between the Tensions along the Quadriceps Tendon and the Patellar Tendon. *Engng. Med.*, **4**, 189-194.

Fairbank J C T, Pynsent P B, Poortuliet J A V and Philips H (1984). Mechanical Factors in the Incidence of Knee Pain in Adolescents and Young Adults. *J. Bone Jt. Surg.*, **66 B**, 685-693.

Ficat R P, Phillippe J and Hungerford D S (1979). Chondromalacia Patellae - A System of Classification. *Clin. Orthot.*, **144**, 55-62.

Huberti H H and Hayes W C (1984). Patellofemoral Contact Pressures. The Influence of Q Angle and Tendofemoral Contact. *J. Bone. Jt. Surg.*, **66 A**, 715-724.

Hungerford D S and Barry M (1979). Biomechanics of the Patellofemoral Joint. *Clin. Orthot*, **144**, 9-15.

Insall J, Falvo K A and Wise D W (1976). Chondromalacia Malacia. *J. Bone Jt. Surg.*, **58 A**, 1-8.

Morrison J B (1967). The Forces Transmitted by the Human Knee Joint During Activity. *Ph.D Thesis*, University of Strathclyde, Glasgow.

Nisel R (1985). The Biomechanics of the Knee. *Acta. Orthop. Scand.*, Suppl 216.

Perry J, Antonelli D and Ford W (1975). Analysis of Knee Joint Forces During Flexed Knee Stance. *J. Bone Jt. Surg.*, **75 A**, 961-967.

Wiberg G (1941). Roentgenographic and Anatomic Studies on the Femero Patellar Joint with Special Reference to Chondromalacia Patellae. *Acta. Orthop. Scand.*, **12**, 319.

24

ADVANCES IN TV-BASED MOTION ANALYSIS SYSTEMS

E H Furnée

SUMMARY

Progress in measurement instrumentation is not synonymous to Progress in Bioengineering, but it helps in some places. This chapter recalls some early history in tv-to-computer motion analysis instrumentation, and points to several beneficial instances of cooperation. The present features of our PRIMAS precision movement analysis system are highlighted. Details follow of noise, precision and resolution performance. Using a figure of merit Q for spatio-temporal resolving power (or relative noise reduction), some current opto-electronic systems are compared, with no uncertain conclusion. A recent application is given by the 2-D estimation of the polar locus of rotation and translation in planar mandibular motion.

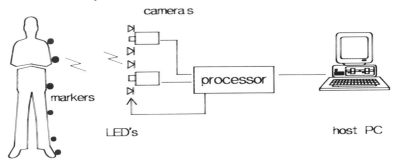

TV-based motion analysis system

Fig. 1   TV-based motion analysis system.

HIGHLIGHTS OF EARLY DEVELOPMENT

1967   Delft University: TV --> IBM video-digital coordinate-converter for multiple moving markers at 50 tv-fields/s (Furnée, 1967)

      1970's   Strathclyde University: TV <--> DEC system
               (Paul *et al.*, 1974)
            Dundee Limb Fitting Centre (Jarrett)

                    1980's VICON TV <--> DEC motion analysis system

1974   Stroboscopic illumination of reflective markers and feed-in to digital tape unit, or to HP minicomputer (Furnée *et al.*, 1974)

1984   Real-time estimation of marker centroids from complete contours:

    data reduction (feed-in to HP desktop with GP-IB)
    enhanced resolution
    output of marker size and shape, additional to centroid coordinates (Furnée,
    1984)

PRESENT FEATURES OF PRIMAS SYSTEM

1986   Dedicated, electronically-shuttered, CCD matrix cameras at 100 Hz

    enhanced contrast of retro-reflective markers (daylight operation)
    high geometric linearity and stability
    high speed, 100 field/s
    high resolution, 1:9 000
    low measurement noise and high precision, typical 1:10 000
    by   . tight coupling, at all clock levels, of camera and system
           . marker centroid processing, in real time
    data reduction, by real-time centroid processing, which allows:
    output interfacing to modest IBM-AT (compatible) host PC's
    multi-camera 3-D system
    (Furnée, 1986)

1988   strip-scan option: enhanced speed of 200 half-height field/s
    High Technology Holland measurement-quality matrix cameras
    ultra low noise and high precision, 1:18 000 hor. and 1:14 000 vert.
    (Furnée, 1988 a,b)

PRIMAS PRECISION AND RESOLUTION TESTS

The precision or repeatability of measurement devices is commonly expressed as one standard deviation of measurement noise, divided by the measurement range.
    We tested a matrix distribution of 20 stationary markers of various sizes, and a diagonal of 30 small markers. We took some 250 consecutive samples and calculated standard deviation from the mean coordinate for each marker. For each marker size, the results were averaged over the 20 or 30 markers, respectively, and these are shown in table 1. Field height is approximately 280 tv-lines, and sizes of the circular markers are expressed in tv-lines. The measured marker centroid coordinates, and their standard deviations, are expressed in units, chosen such that the measurement range (horizontal and vertical) is about 28 000. Coordinate output is in 15 bit binary code. Averaged over all marker sizes standard deviation is 1.5 hor. and 2.0 vert. Taking into account the range of 28 000, the average precision is 1:18 000 horizontal and 1:14 000 vertical.

Marker size	12	12	8	8	7	4	4	tv-lines
St. deviation hor.	1.3	1.3	2.6	1.6	2.1	0.7	1.0	Units
St. deviation vert.	1.7	1.4	2.6	1.4	5.4	0.6	1.3	Units

Table 1 Standard deviation of coordinates for various marker sizes

SPATIO-TEMPORAL RESOLVING QUALITY Q

The minimum noise (variance) obtained when calculating the k-th derivative of a signal which is bandlimited with a cut-off angular frequency ω_B, and which is contaminated by additive white noise with standard deviation σ, and sampled at a rate R, is given by

$$\sigma^2_k \geq \sigma^2_{k,min} = (\sigma^2/R)\omega_B^{2k+1}/\pi(2K+1) \text{ units}^2 \text{ (Hz)}^{2k}$$

(Andrews and Jones, 1976; Lanshammar, 1982)

The first bracketed quantity, relating measurement noise and sampling rate, is considered a figure of merit of the measurement system. What the formula shows is how measurement noise may be amplified when derivatives are taken, even with the suitable filter for a bandlimited signal, such as is common practice for obtaining velocities and accelerations in biomechanics. Taking account of the measurement range, and expressing system noise in terms of precision p, we introduce the relative spatio-temporal resolving quality

$$Q = \sqrt{R/P} \qquad\qquad (\sqrt{Hz}), \text{ with } p = \sigma/\text{range}$$

The following formula then obtains: $p_{k,min} = (1/Q)\omega_B^{k+0.5}/\sqrt{\pi(2k+1)}$ which shows the importance of high Q values for a high precision of the result, corresponding to a low value of $p_{k,min}$.

The expression for Q indicates that decreasing the measurement system noise (increasing the precision) is more profitable than increasing the sampling rate. For proper sampling, the sampling rate should obey the constraint $2\pi R \geq 2\omega_B$.

For a number of contemporary motion analysis systems the figure of merit Q was calculated from specifications of sampling rate R and precision P. For systems where precision, as a measure of noise performance, failed to be specified, the quoted figures for resolution were tentatively substituted. Table 2 shows the data and allows a comparison to be made.

System	Sample rate R Hz	Resolution	Precision P	Remarks	Quality Q √HZ
PRIMAS	100		1 : 18 000	X	180 000
			1 : 14 000	Y	140 000
VICON	50		1 : 1 000	3-D	7 000
	200	1 : 400	same?	NAC camera	5 500
ELITE	50	1 : 2 500	same?		17 500
Expert Vision	50		1 : 3 000		21 000
	200		1 : 1 000	NAC, recorded	14 000
CODA-3	300	1 : 6 000	same?	3-D: X, Y only	104 000
		1 : 1 000	same?	3-D: Z	17 000
Selspot	330	1 : 4 000	same?	30 markers	73 000

Table 2 Spatio-temporal resolving quality Q of current motion analysers.

However it does not take into account the adverse effects of non-simultaneous sampling of all markers, which is inherent of non-tv systems like CODA-3 and Selspot. At the 330 Hz sampling rate example with the Selspot system the number of 30 markers is a maximum, and maximum offsets between sampling instants of first and last marker are thereby incurred. Even disregarding the advantage of equidistant simultaneous sampling of all marker positions with tv-systems, all other systems are outperformed by the PRIMAS Q-factor.

APPLICATION: 2-D ESTIMATION OF THE POLAR LOCUS OF PLANAR MANDIBULAR MOTION

Mandibular motion with large amplitudes of the opening/closing activity is investigated in the gnathology department of the University of Groningen in collaboration with mechanical engineering of the Twente University. For the experiments, a lightweight (45 g) bow with retro-reflective markers A, B, C and D is fixed to a protruding mouth bit, which is attached by dental glue to the lower jaw (figures 2 and 4). A spectacle frame carries the reference markers E and F (figure 4). Figure 3 shows a coarse plot of the stationary markers and the four moving marker trajectories of one opening cycle.

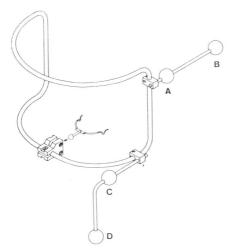

Fig. 2   Marker bow with mouth bit.

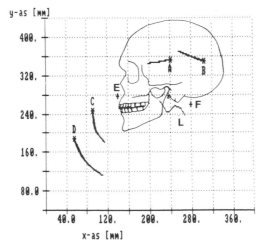

Fig.3   Marker trajectories A, B, C, D reference markers. E, F polar locus L.

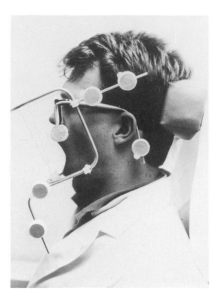

Fig. 4   Bow and spectacles in place.

From these data the polar locus L is estimated and plotted in the same figure. As the locus of the instantaneous centre of rotation, L describes the combined rotational and translational movement of the mandible. Note how the polar locus of the wide jaw opening movement relates to curvature of the tuberculum articulare contact surface, sketched with the skull in an approximate position and scale in figure 3.

The experimental set-up is further illustrated with figure 5. Attention is drawn to the bright ambient light conditions, which by PRIMAS' dedicated electronically-shuttered cameras do not detract from the rated performance.

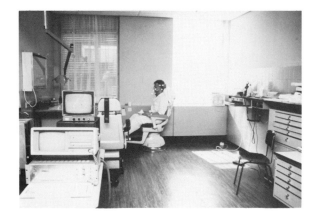

Fig. 5   Experimental set-up.

REFERENCES

Andrews B J and Jones D (1976). A note on the differentiation of human kinematic data. *Dis. 11th int. Conf Med Biol. Engng.*, Ottawa, 88-89.

Furnée E H (1967). Hybrid instrumentation in prosthetics design. *Proc. 5th int. conf. Med. Biol. Engng.* , Stockholm, 446.

Furnée E H, Halbertsma J M, Klunder G, Miller S, Nieukerke K J, van der Burg J and van der Meché F (1974). Automatic analysis of stepping movements in cats by means of a television system and a digital computer. *J. Physiol.*, **240**, 3-4.

Furnée E H (1984). Intelligent movement measuring devices. *Proc. 8th int. Symp. External control of human extremities*, ETAN, Belgrade, Suppl 40-53.

Furnée E H (1986). High resolution real-time movement analysis at 100 Hz. *Proc. North Am. Conf. Biomechanics*, Montreal, 273-274.

Furnée E H (1988 a). On camera speeds for human movement analysis. Preprint *5th int. Conf. Biomed. Engng.*, Singapore, 4.

Furnée E H (1988 b). Principles and performance of some opto-electronic motion analyzers. Preprint *5th int. Conf. Biomed. Engng.*, Singapore, 4.

Lanshammar H (1982). On precision limits for derivatives numerically calculated from noisy data. *J. Biomechanics*, **15**, 459-470.

Paul J P, Jarrett M O and Andrews B J (1974). Quantitative analysis of locomotion using television. *Abstr. World Congr. ISPO*, Montreux, 43.

25

ON THE LOADING OF THE METATARSAL BONES AND THE ROLE PLAYED BY THE PLANTAR APONEUROSIS WITHOUT MUSCLE ACTIVITY

H A C Jacob

INTRODUCTION

Even since the French anatomist Bertin in 1754 realised that the structure of the foot was something more specific than just flat terminal supports (Abramson, 1927), and for the first time described the 'vault' of the foot (la voûte du pied), the arch of the foot has continued to remain a matter of contention. No matter whether the 'key stone' of the arch some investigators were looking for was to be found in the talus or in the navicular, or whether there were in fact five longitudinal arches running side-by-side rather than a dome-like structure, it was unanimously decided, already a century ago, that the ligamentous ties on the concave side of the foot must play an important role in maintaining the structural form under load (Weber and Weber, 1836; Fick, 1911). Not only were the calcaneocuboid, calcaneometatarsal and calcaneonavicular ligaments mentioned, but also the plantar aponeurosis was considered as being of paramount importance in supporting the longitudinal arch. How exactly the bones of the normal foot are loaded is still not known since controversial opinions exist regarding the manner in which the longitudinal arch of the foot is supported by the muscles and ligamentous structures that span the arch on its plantar side. Whereas some, like Meyer (1873), Flick (1911), Jones (1941), Manter (1946), Hicks (1954, 1955) and Wright and Rennels (1964) believe that the plantar ligaments and aponeurosis play a major role in supporting the arch, others like Abramson (1927), and Preuschoft (1970) are of the opinion that the muscles are the main agents. Others, again, are convinced that both ligaments and muscles play a complementary role, depending on the momentary situation of the foot, i.e. whether standing, raised on the toes, etc. Such opinions have been forwarded by Basmajian and Stecko (1963), Mann and Inman (1964), and Kummer (1984).

Generally, it is assumed that bones are subjected primarily to compressive forces. How far the metatarsal bones are also exposed to some degree of bending is uncertain. Most commonly, the ties that support the arch are assumed to act at the two extreme ends of the latter - neglecting the possible effect of the short plantar ligaments and the muscles that insert at, or near the bases of the metatarsals - thereby forming a simple three bar truss with purely compressive forces that act along the axis of the metatarsal bones. Preuschoft (1970) has shown that the action of the plantar flexors of the metatarsophalangeal and interdigital joints could have a powerful influence on the share of load taken by the plantar aponeurosis in supporting the arch and this deserves closer attention in all future attempts at analysing the forces in the structure of the foot.

To investigate this matter further, strain measurements were carried out on a fresh cadaver foot, as follows:

MATERIAL AND METHOD

A complete foot with about 300 mm of the shank had been deep-frozen following amputation because of a tumor just below the knee. The foot belonged to a 37 year old man who had been otherwise healthy and had used his feet until shortly before the operation.

The skin with subcutaneous fat was completely removed to facilitate access to the exposed bones and also to observe interosseus movements that might otherwise be masked by the covering skin. The abductor muscles of the great and little toes were removed and the tuberosity of the calcaneous then resected by cutting along two planes while preserving the attachment of

146

the achilles tendon and the plantar aponeurosis to it. In this manner the tuberosity with the achilles tendon and aponeurosis could be lifted off (figure 1) after detaching the flexor brevis muscles from the latter. It was now possible to remove the flex.hall.brev.muscle, the adductors of the great toe, and such interossei muscles that covered the shafts of the first three metatarsal bones.

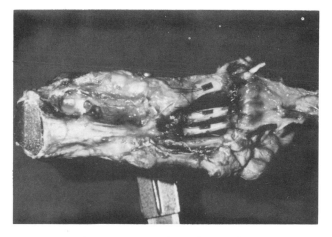

Fig.1   The plantar aponeurosis elevated anteriorly for application of strain gauges.

After exposing the middle portion of the first three metatarsal bones, electrical resistance strain gauges were applied (figure 1) after stripping off the periosteum and degreasing the bone with a mixture of ether and pure ethyl alcohol. The strain gauges of 3 mm gauge length were then glued on, directly opposite each other, one on the dorsal and the other on the plantar side of each bone, using a quick-setting cyano-acrylic glue. Flexible lead wires with polyvinyl insulation were soldered onto the strain gauge tags and the whole assembly protected against moisture and mechanical damage by a coating of fast-setting silicone rubber. All six strain gauges (from the three first rays) were connected to an automatic scanning and measuring unit. Each strain gauge was connected in 1/4 bridge fashion, the Wheatstone bridge being supplied from a source of 0.5 V at a carrier frequency of 225 Hz.

The resected tuberosity of the calcaneus was reattached in its original position with the use of acrylic bone cement after drilling a few shallow holes of about 10 mm diameter to anchor the cement within the two parts to be reunited. Care was taken to bring the tuberosity into its original position (including compensation for the width of the gap made by the saw during the resection procedure). The foot was now held in a jig (figure 2) after anchoring the free ends of the tibia and fibula in a steel tube with acrylic bone cement. The steel tube of 100 mm length, 60 mm diameter and 3 mm wall thickness was rigidly attached to the end of an arm. By adjusting the position of the arm, the foot could be brought into the foot-flat position. A horizontal platform under the foot was forced up against it by means of a special balance arrangement (figure 2) thus exerting a vertical ground reaction force on the foot, the magnitude of which remained independent of the position adopted by the foot on the platform. The platform was free to move in a horizontal plane on ball bearings so that no horizontal component of force greater than about 0.006 x the vertical load could inadvertently come into play.

A strain-gauged force transducer was placed between the tubercle of the calcaneus and the platform while the heads of the metatarsals all made contact with a flat bar that lay on the same platform (figure 2). The achilles tendon was not fixed and therefore the foot was free to move in the ankle and talocalcaneonavicular joints permitting force distribution across the foot without any external restraint whatever. The strains in the first three metatarsal bones were measured in three steps up to a maximum total load of 324 N under the foot. Following this, the plantar aponeurosis was cut through and the strain measurements repeated.

Finally, the whole plantar aponeurosis was removed for testing its elastic properties. The central portion of the aponeurosis of about 50 mm length proved to be of essentially constant width and thickness and therefore this section of the specimen was kept free while on either side of this length grips were attached to facilitate tensile loading of the same in a universal materials testing machine. A "C-clamp" extensometer (furnished with strain gauges) was attached to the aponeurosis over a length of about 40 mm by means of loose sutures into which the ends of the extensometer were hooked.

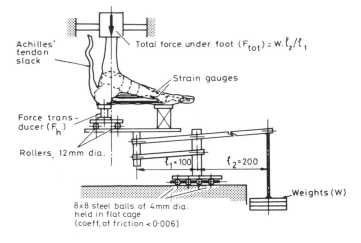

Fig. 2   The jig used to load the foot.

OBSERVATIONS AND RESULTS

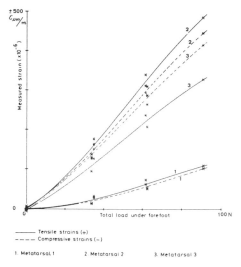

Fig 3   Strains measured on the dorsal and plantar aspects of the first three metatarsal bones at mid-length, versus load under the whole forefoot.

Figure 3 shows the results obtained with the foot acted upon by a total load of up to 324 N. The zero-readings of the strain gauges were taken at a total pre-load of 30 N so that the maximum strains shown correspond to a total load difference of 294 N. In the following report only the load differences (taken from the pre-load level) will be discussed. The maximum load on the heel measured 203 N, so that the total load under the heads of the metatarsals amounted to 91 N or 31% of the total load on the foot. The strain readings which were taken with progressively increasing and decreasing loads were observed to definitely follow an hysteresis loop (the measured points have been entered in figure 3) but the graphs plotted represent the best fitting polynomials of third degree, this being considered sufficient for the present purpose. Repetition of the tests several times confirmed the readings.

A calibration test performed by individually loading the metatarsal heads with known forces and simultaneously measuring the strains that developed, showed that the maximum forefoot load of 91 N was shared by the first, second and third metatarsals and the rest of the forefoot in the ratio of about 1.3:2:1:4, respectively.

The signs, and relation of magnitude of the strains measured show that the ground reaction forces under the metatarsal heads lead primarily to bending stresses in the diaphysis of the metatarsal bones, even with the plantar aponeurosis intact. On severing the attachment of the plantar aponeurosis to the forefoot, however, the bending stresses were observed to increase by about 14% in the first metatarsal and by about 8% in the second. The third metatarsal did not show any appreciable difference in strains when loaded with or without the aponeurosis. Young's modulus of the plantar aponeurosis was found to be 1236 N/mm^2.

DISCUSSION AND CONCLUDING REMARKS

The strain measurements performed show that in the absence of muscle activity, the first, second and third metatarsal bones are mainly loaded in bending (like cantilevers), and as such are stressed to about ± 1.6, ± 7.0 and ± 5.5 N/mm^2, respectively (assuming a modulus of elasticity of 15 kN/mm^2 for cortical bone) when the forefoot is loaded with a total force of 91 N, that corresponds to a total load of 294 N on the whole foot. Any doubt as to whether the elastic properties of the plantar aponeurosis had suffered any change and could therefore have been responsible for inadequate bracing of the arch during the test, might be dismissed on the grounds that the Young's modulus determined ($E = 1236$ N/mm^2) proved to be even 46% greater than the values given by Wright and Rennels (1964). The higher modulus of elasticity of the plantar aponeurosis obtained in the present study precludes the objection that a loss of stiffness might have been responsible for the conclusion arrived at above.

The present investigation has therefore shown that the action of the flexor muscles, and not so much the plantar aponeurosis, might be expected to have a major influence on the nature of loading within the structure of the foot, making it only then behave like a truss. Without the supporting action of these muscles, the metatarsal bones act more like cantilever beams, as suggested by Preuschoft (1970). On severing the attachment of the plantar aponeurosis to the forefoot, the bending stresses were observed to increase by about 14% in the first metatarsal and by about 8% in the second. The third metatarsal was observed to show no appreciable difference in strains whether loaded with or without the plantar aponeurosis.

Since the flexor muscles of the toes would play a decisive role in determining the mode of loading of the metatarsal bones and adjoining articulations, the results of this investigation might be of value in understanding the pathomechanisms of such disorders of the feet that involve fatigue fractures of bones or the unphysiological loading of joints and ligaments of the forefoot as a consequence of fatigued or otherwise functionally inadequate muscles.

Acknowledgements

This investigation is part of a Ph.D. research programme being currently carried out jointly by the Bioengineering Unit of the University of Strathclyde, Glasgow, and the Orthopaedic Department of the University of Zurich, Balgrist. The author wishes to thank Professor J P Paul and Dr A C Nicol of the Bioengineering Unit, and Professor A Schreiber of the Orthopaedic Department for their support.

REFERENCES

Abramson E (1927). Zur Kenntnis der Mechanik des Mittelfusses. *Skand Arch Physiol.*, **51**, 175-234.
Basmajian J V and Stecko G (1963). The role of muscles in arch support of the foot. *J. Bone Jt Surg.*, **31 B**, 1184-1190.
Fick R (1911). *Handbuch der Anatomie und Mechanik der Gelenke.* Part III, Gustav Fisher, Jena,

pp 648-649.

Hicks J H (1954). The mechanics of the foot. *J. Anat.*, **88 (1)**, 25-31.

Hicks J H (1955). The foot as a support. *Acta Anat.*, **25**, 34-45.

Jones R L (1941). The human foot. An experimental study of its mechanics, and the role of its muscles and ligaments in the support of the arch. *Am. J. Anat.*, **68 (1)**, 1-39.

Kummer B (1984). Biomechanik des Vorfusses. *Orthop Praxis*, **7**, 521-527.

Mann R and Inman V T (1964). Phasic activity of intrinsic muscles of the foot. *J Bone. Jt. Surg.*, **46 A**, 469-481.

Manter J T (1946). Distribution of compression forces in the joints of the human foot. *Anat. Rec.*, **96 (1)**, 313-321.

Meyer H (1873). *Die Statik und Mechanik des menschlichen Knochengerüstes.* Wilhelm Engelmann Verlag, Leipzig, 1973, pp 386-387.

Preuschoft H (1970). Statische Untersuchungen am Fuss der Primaten. *Z. Anat. Entwickl.-Gesch.*, **131**, 156-192.

Weber W and Weber E (1836). *Mechanik der menschlichen Gehwerkzeuge.* Dietrichsche Buchhandlung, Göttingen, pp 386-387.

Wright D G and Rennels D C (1964). A study of the elastic properties of plantar fascia. *J. Bone. Jt. Surg.*, **46 A**, 482-492.

26

KINESIOLOGIC AND BIOMECHANICAL ANALYSIS OF THE KNEE JOINT AFTER TOTAL KNEE REPLACEMENT

T Mitsui, S Niwa, H Honjo, T Hattori and H Hasegawa

INTRODUCITON

It has been well known that short term results of total knee replacement become better due to improvement of the design biomechanically and to establishing of correct surgical technique. Kinesiologic and biomechanical factors are considered to have an effect on the long term success of total knee replacement. This report relates to the moment and force on the knee joint calculated using biomechanical parameters in order to investigate the relationship between clinical results and the load on the knee joint during walking after Total Knee Replacement.

MATERIALS AND METHODS

23 joints and 17 cases (14 female, 3 male) were investigated. 14 joints in 9 cases were diagnosed as rheumatoid arthritis and 9 joints in 8 cases osteoarthritis. These knees were replaced using a surface type prothesis, the Kinematic knee system.

Patients' ages ranged from 29 to 78 years (mean age 59 years). The mean period between operation and follow up was 2.6 years. 10 normal subjects with no symptoms or history of knee joint disorders were also tested as controls. These subjects' ages ranged from 36 to 51 years with a mean age of 44.5 years.

The evaluation items are:

a clinical evaluation using our knee score system,

the alignment of lower limb on standing using X-ray and photography during walking,

the moment acting on the knee joint in the frontal plane calculated from the alignment of the lower limb and the load on the knee measured by floor reaction force during walking when the sagittal component changes from foreward to backward; that is corresponding to mid stance of walking cycle,

joint force using the calculation of Dr R J Minns.

RESULT

Clinical Evaluation

In evaluation of knee joint function the average of total points was 42.6 points before, and 71.9 points after surgery; 13 cases out of 17 had pain during walking and at rest preoperatively; 11 cases out of 17 cases showed complete relief of pain postoperatively. The mean range of motion after surgery was 85.9^0, and 14 cases had more than 90^0 of flexion. Walking ability as well as activity in daily life was improved. Independence was gained in activities of indoor daily life even though Japanese daily life style activity is difficult.

Alignment

Regarding the tilting of pelvis, a tendency for the contralateral side of the pelvis to be significantly raised was recognised in the total knee replacement (TKR) group, in comparison to normal subjects during walking. The functional axis of the lower limb and axis of the tibia showed a tendency for reduced inclination laterally during walking as compared with one legged standing.   This indicates that there is an important difference between static and dynamic situations.

It is considered that TKR cases have reduced load sharing between lateral and medial compartments during walking as compared to normal subjects. (Figure 1).

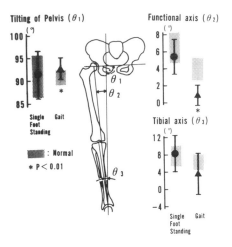

Fig. 1   Alignment of lower extremity.

Moment

When the moment acting on the knee joint after TKR in the frontal plane is anticlockwise, that is positive, an adduction moment of the knee is indicated. These knee moments in the TKR group, with a mean value 3.315 (Body weight). cm were similar value to those of the normal group with a mean value 3.405 (Body weight). cm. The standard deviation was a little greater in the TKR group, than in the normal group. The relationship between moment acting on the knee joint and the angle between the functional axis of the lower limb and the line of gravity was noted.

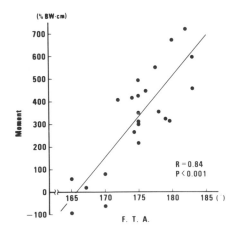

Fig. 2   The correlation between the adduction moment at the knee joint and FTA.

The smaller the lateral inclination of the functional axis and the lower leg axis became, the smaller the adduction moment became. It is considered that the adduction moment is reduced by putting the line of load nearer to the centre of the knee, and the tilting of the pelvis is also reduced. The close relationship between the adduction moment and the femoro-tibial angle (FTA) on one legged standing was recognized in the TKR group. The mean adduction moment of 4 joints with FTA greater than 180^0 was 6.083 (Body weight). cm (4.592 to 7.029), a high value. (Figure 2).

According to the investigation of the relationship between the setting angle of the prosthesis and the moment, the mean setting angle of the tibial component was 91.5^0 (84.5^0 to 98.0^0), while the mean setting angle of the femoral component was 83.2^0 (78^0 to 88^0). The setting angle of the tibial component showed a higher correlation with moment than the femoral component. A positive correlation was recognised between adduction moment and the setting angle of the tibial component as shown in figure 3. It is considered that the setting of components in various positions makes the adduction moment larger. In particular, it is considered very important to achieve proper setting position of the tibial component at surgery. The femoro-tibial angle (FTA) of the lower limb showed a mean of 189.5^0 (183^0 to 195^0) preoperatively, a mean of 177.8^0 (174^0 to 179^0) postoperatively. The moment acting on the knee joint was markedly reduced to approximately 40% of the preoperative value. Reduction of adduction moment is considered due to improvement of dynamic alignment.

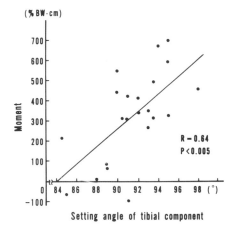

Fig. 3   The correlation between the adduction at the knee joint and the setting angle of the tibial component.

Joint Force

The average total load, including medial and lateral load on knees of TKR cases and normal subjects is as follows:

Lateral load for TKR:	64.7%BW
Lateral load for normal subjects:	27.7%BW
Medial Load for TKR:	224.1%BW
Medial load for normal subjects:	235.7%BW
Total load for TKR:	288.8%BW
Total load for normal subjects:	263.4%BW

A significant difference between normal subject and TKR cases was not recognised. It was considered important to know the proportion of the total load transmitted on the medial compartment of the knee.

To know the joint force, as well as load on the medial and lateral component of the knee joint is important in finding the setting angle of the prothesis, so as to make it smaller than before surgery, and restore the normal value. An increase in FTA indicates greater adduction loading on the knee. (Figure 4).

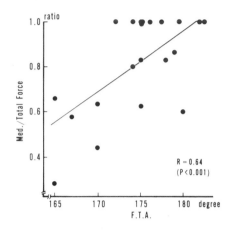

Fig. 4   The correlation between FTA and the ratio of the force in the medial compartment to total knee joint load.

A large amount of lateral inclination of the tibia indicates that a high percentage of the total joint load is transmitted by the medial compartment.

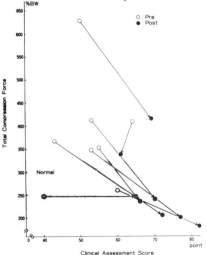

Fig. 5   Figure showed that clinical evaluation using our knee score system was improved and the total compression force on the knee joint was decreased.

Thus, it is suggested that assessment of the alignment of the lower limb and gait pattern can provide an approximate estimate of the load, which seems to determine the long term results of surgery. Total compression force and clinical score before and after TKR is shown in figure 5.

DISCUSSION

This investigation attempted to estimate the moment and the joint force on the knee at the phase when the direction of the sagittal component of the floor reaction force changed from forward to backward, which corresponds to the midstance of the walking cycle. An attempt to estimate joint force during walking was made by Morrison (1968), Johnson and Waugh (1979) and Röhrle et al., (1984).

The value we calculated for the joint force is similar to previously reported results. However, we added the element of dynamic floor reaction force during walking, based on the calculations of R J Minns et al., (1982), without investigation of muscle strength around the knee. Investigation using different walking speeds is necessary in addition to comfortable speed only. TKR has become widely indicated with developments in design and surgical technique, the younger patients are able to undergo TKR, and higher social goals are achieved.

Emphasis tends to be placed on short term results at the time of exercise and clinical results after surgery. However, prevention of contracture, weakness of muscle strength, and loosening of prosthesis are problems to be overcome in order to obtain favourable long term results.

Thus, it is important clinically to know how muscle strength and walking pattern influence the load on the knee. We consider that in the future, the relationship between kinds of artificial joints, surgical techniques and improvement of muscle strength should be investigated, expanding clinical data and establishing a closer model to the living body.

CONCLUSION

The results of TKR for RA and OA were investigated from a biomechanical viewpoint, and the following conclusions were made:

TKR groups showed a tendency of less lateral inclination of the lower limb functional axis as well as of the tibial axis, because the contralateral side of the pelvis is raised during walking as compared with one legged standing.

Adduction moment acting on the knee joint has a close relationship with lower limb alignment during walking, and FTA.

Adduction moment of the knee joint has a close relationship with the setting angle of the tibial component.

Load on the knee joint with TKR shows no significant difference to normal knees.

Load on the knee joint with TKR shows a close relationship with alignment during walking, and FTA.

REFERENCES

Harrington I J (1976). A bioengineering analysis of force actions at the knee in normal and pathological gait. *Biomed. Engng.*, **11**, 167-172.

Johnson F and Waugh W (1979). Method for routine clinical assessment of knee-joint force. *Med. Biol. Engng. Comput.*, **17**, 145-154.

Minns R J, Day J B and Hardinge K (1982). Kinesiologic and biomechanical assessment of the Charnley 'load angle inlay' knee prosthesis. *Engng. Med.*, **11**, 25-32.

Morrison J B (1968). Bioengineering analysis of force actions transmitted by the knee joint. *Biomed. Engng.*, **3**, 164-168.

Rohrle H, Scholten R, Sigolotto C and Sollbach W (1984). Joint force in the human pelvis-leg skeleton during walking. *J. Biomechanics*, **17**, 409-429.

27
RIGID PLASTER DRESSING - SIMPLE PYLON TECHNIQUE FOLLOWING BELOW-KNEE AMPUTATION

I J Harrington, M McPolin and G James

INTRODUCITON

LeMesurier (1926) reviewed the experience with Canadian amputees of World War I and concluded that temporary artificial limbs hastened stump shrinkage and conditioning. Burgess and Romano (1968) recommended rigid plaster dressings to control oedema and immediate post-surgical fitting of prostheses to facilitate early ambulation. Others have advocated using air-splints to promote wound healing and allow early ambulation following amputation. The main disadvantage with all of these methods is that they required skilled application by trained personnel. They are also expensive. Since 1977, we have experimented with a rigid plaster dressing and a simple pylon designed for early ambulation in the peri-operative period following below knee amputation. It allows full weight bearing, its application does not require special training and it is inexpensive. Over 50 patients have been treated to date.

METHODS

Rigid Plaster Technique

Snug fitting stockinette is applied over the stump dressing and extended to the groin. A plaster dressing is applied in the operating room and during its application, the amputated limb is lifted from the O.R. table using the free end of the distal stockinette, figure 1. This simple manoeuvre provides support to the posterior tissue flap and takes tension off the anterior suture line.

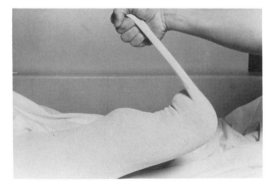

Fig. 1   Stockinette applied to amputated limb, elevating it from the O.R. table prior to plaster application. The cast is extended to the groin.

This is an important technical detail necessary for stump healing since wound breakdown can often be attributed to tension, shear stress or direct bone pressure from the tibia at the front of the stump. A 1" (2.5 cm) rubber Penrose drain is inserted prior to stump closure. The drain is removed through a window cut in the cast at 48 hours. The stump is otherwise undisturbed for 5 - 7 days. If there are no early complications, the original plaster dressing is removed at 1 week for wound inspection and dressing change. A new well-moulded cast is then applied and a pylon attached.

Application of Pylon

The pylon is constructed by splitting a length of 2" (5 cm) copper tubing at one end to create a tripod base, figure 2. Each metal flange is 2" - 5" (5 - 7.5 cm) long. The base is secured to the moulded stump cast which extends to the groin. Plaster is used to attach the pylon to the cast. A rubber crutch tip is placed at the end of the pylon for non-slip floor contact.

Fig. 2   Application of tri-flanged copper pylon, temporarily fixed with adhesive tape prior to application of a few rolls of plaster of Paris.

Physiotherapy

At this stage, the physiotherapist teaches the patient muscle strengthening exercises, transferring from bed to chair and walking - first utilising a walker and then independently, figure 3. The patient is allowed partial to full weight bearing as tolerated and is encouraged to be ambulatory throughout his hospital stay. The cast-pylon is changed weekly thereafter for wound inspection and to accommodate for stump shrinkage.

Since 1977 the cast-pylon technique has been used by the senior author for all below knee amputees. It has been possible to apply the pylon one week after surgery in most cases. Patients have reacted well to it, becoming ambulatory usually within 48 hours of application.

Delayed healing did occur with some patients, requiring special wound care. This often could be provided using the same plaster technique - cutting a window in the cast to allow wound inspection and daily dressing changes, figure 4.

Those patients who were unable to walk independently with the pylon found it useful as an aid to transfer from bed to chair. Seven patients have undergone bilateral below knee amputation. All of these individuals were able to walk on bilateral pylons with the aid of crutches or a walker. We have used the rigid dressing pylon technique with above knee amputees but have not found it to be particularly useful. Many of these patients are in the geriatric age group and for medical reasons often do not progress to a definitive prosthesis.

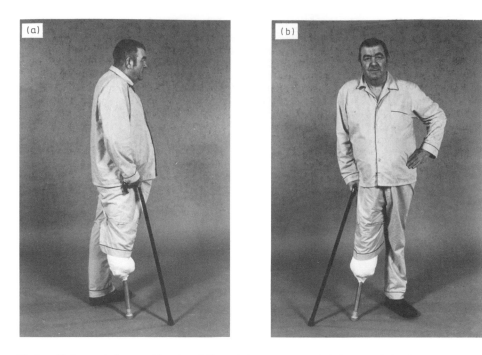

Fig. 3   Patient ambulatory with pylon, full weight bearing at one week following below knee amputation.

RESULTS

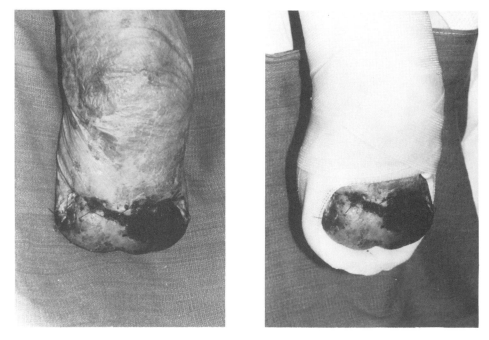

Fig 4   Delayed healing of below knee amputation stump. The stump cast with an anteriorly placed window allows the application of dressings to the tissues while supporting the stump, taking tension off the anterior suture line.

DISCUSSION

Immediate prosthetic fitting in the operating room is a proven technique but it frequently requires the skills of a professional prosthetist: the method is expensive and there is hesitancy to remove the prosthesis for wound inspection. Inflatable splints and other such devices manufactured commercially are also expensive and can be difficult to apply. The technique described in this chapter offers all of the advantages of an immediate fit prosthesis and/or inflatable splint device with the added benefit that it is easy to learn, easy to apply and does not require special training. The materials used for its application are readily available in any hospital setting so that it can be applied, removed and re-applied as required with ease. Clinical results to date have been gratifying.

REFERENCES

Burgess E M and Romano R C (1968). The management of lower extremity amputees using immediate post-surgical prostheses: *Clin. Orthop.*, **57**, 137-156.
LeMesurier A B (1926). Artificial limbs, their relation to the different types of amputation stumps, *J. Bone. Jt. Surg.*, **8A**, 292-324.

28

BLOOD, MARROW AND BONE CEMENT

E R Gardner, R Wilkinson and I G Stother

INTRODUCTION

Fixation of joint prostheses using polymethylmethacrylate bone cement (PMMA) has become common practice in orthopaedic surgery since its introduction by Charnley in 1960. Since then this technique has been widely used for many types of total joint arthroplasty, providing stable initial implant fixation which allows early mobilisation (Charnley, 1965).

Ideally, the cement acts as a means of load transfer from the prosthesis to the bone across the prosthesis cement and the cement bone interfaces. The main clinical problem following joint replacement is component loosening (Carlsson and Gentz, 1980; Moreland, 1988). Loosening can be defined as gross movement between prosthesis and bone which occurs by cement fracture or interface breakdown between either the bone and cement or the implant and cement (Gruen *et al.*, 1979).

To study the aetiology of loosening, the mechanical properties of bone, cement and implants must be known. We have studied the mechanical properties of PMMA using material mixed in the laboratory or obtained during arthroplasty. In addition, the effect of PMMA-precoating of metals on the PMMA-metal interface shear strength has been investigated.

MATERIALS AND METHOD

Cement Compression Studies

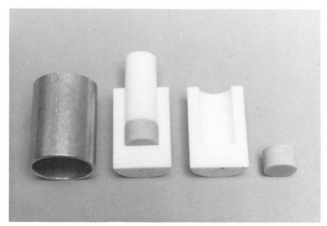

Fig. 1   Mould for production of cement test specimens.

PMMA specimens were prepared using four types of proprietary bone cements; Palacos R, Palacos E, Simplex and CMW 1. The cement was mixed by adding the liquid monomer to the powder in a ceramic bowl and beaten at 1 Hz with a stainless steel spoon for 90 seconds. The mixture was allowed to stand for a further 150 seconds. From each complete mix, approximately 3 ml of cement was inserted into PTFE split moulds 15.1 mm internal diameter as shown in figure 1. A central PTFE piston was inserted and a 1 kg load applied for 15 minutes while the cement cured. The samples produced were 7-13 mm in length. All specimens were stored in air at room temperature for between 4 and 11 days before testing. Similar specimens were made using the cement displaced from the proximal femur during insertion of the femoral component of a total hip replacement (THR) collected in theatre and placed in moulds. Similarly, the displaced cement from total knee replacements (TKRs) was collected and samples produced. The cement used for all THRs and TKRs was Palacos R.

Compression tests were performed using an Instron mechanical testing machine type TTCM. The samples were placed on a steel base plate with a centring mark and covered with a loading plate. All tests were performed with a cross-head speed of 0.5 mm/min.. The Young's modulus and yield stress were calculated from records of loads and cross-head displacement.

Rod-PMMA Push-Out Tests

Titanium rods 6.4 mm diameter were supplied with either a standard shot blasted finish or a thin layer (0.01 mm) PMMA pre-coating as shown in figure 2. The rods were placed in PTFE split moulds and approximately 4 ml of cement inserted around the rods as illustrated in figure 3.

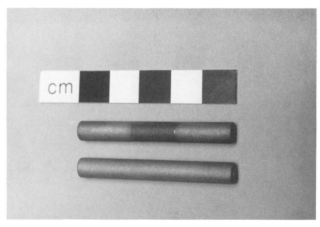

Fig. 2   Titanium rods pre-coated and shot-blasted finishes

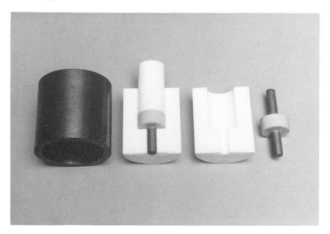

Fig. 3   Mould for production of interface shear strength test specimens.

Samples were made using either clean cement or contaminated cement (Palacos R) displaced from the bone surfaces during the insertion of the components of TKRs at operation. Three types of clean cement were used and all contaminated specimens were Palacos R. A PFTE piston with a central 6.4 mm diameter machined channel was placed over the rod and inserted to compress the cement. A load of 10 N was applied while the cement hardened. All samples were stored in air at room temperature for 4-11 days. The length of the cement surrounding the rod was measured with vernier calipers and the area of the metal-cement interface calculated. Push-out tests were performed on Instron mechanical testing machine, type TTCM with a cross-head speed of 0.5 mm per minute. The loads applied to the samples were recorded until the metal-cement interface failed. The shear stress at the time of interface failure was calculated.

RESULTS

Table 1 gives the results of the compression tests performed on cement samples. The number of specimens tested of each cement type, with the mean and standard deviation for Young's modulus and yield stress are given.

Cement type	n	Young's modulus N/mm^2	Yield stress $N/mm2$
Palacos R	37	2675 ± 233	48.49 ± 5.20
Palacos E	32	2505 ± 207	46.14 ± 4.23
Simplex	23	2515 ± 191	49.31 ± 8.00
CMW 1	12	2354 ± 196	42.90 ± 1.69
Hip	10	2349 ± 204	41.19 ± 2.91
Knee	12	1879 ± 229	36.50 ± 4.60

Table 1 Cement compression tests.

The results were analysed using the Student's t-test. Palacos R has a modulus of elasticity significantly higher than the other "clean" cements ($p < 0.01$). There is no significant difference in yield stress between Palacos R, Palacos E and Simplex. Both Young's modulus and yield stress for CMW 1 are significantly lower than the other cements ($p < 0.05$).

The cement from THR, Palacos R in contact with blood, has a reduction of 12% in Young's modulus ($p < 0.001$) and 15% for yield stress ($p < 0.001$). Contact with marrow fat during TKR results in a 30% drop in modulus ($p < 0.001$) and a 24.7% reduction in yield stress ($p < 0.001$).

Sample type	n	Interface shear strength N/mm^2
Titanium/Palacos R	10	8.81 ± 1.44
Pre-coat/Palacos R	12	11.76 ± 1.70
Titanium/Palacos E	10	9.61 ± 1.28
Pre-coat/Palacos E	10	11.66 ± 2.63
Titanium/Simplex	12	10.89 ± 1.17
Pre-coat/Simplex	12	12.15 ± 1.73
Titanium/knee	10	5.89 ± 1.45
Pre-coat/knee	10	10.27 ± 3.15

Table 2 Push-out test - interface shear strength.

The results of the push-out tests for rod-cement samples are shown in table 2. The number of samples in each group with the mean interface shear strength and standard deviation are given. For all types of cement tested, the PMMA pre-coated titanium has a significant increase in interface shear strength (p < 0.05). The type of cement used does not significantly alter the interface shear stress when the rods are pre-coated. Contact of bone cement with marrow fat during TKR results in a 33% reduction with shot blasted titanium (p < 0.001); and a 12.6% reduction if the rod is pre-coated is not significant at p < 0.05.

DISCUSSION

Fixation of implants by bone cement relies on an intact cement mantle between the prosthesis and the bone. Failure to achieve a tight cement-prosthesis interface and poor trabecular cement penetration have been associated with increased mechanical loosening rates (Amstutz *et al.*, 1976). In theory, if the load applied to a joint prosthesis is transferred smoothly across the prosthesis cement interface and distributed to the underlying bone in a uniform manner, good fixation will be achieved (Matthews *et al.*, 1985).

A strong metal cement interface is produced if the number of cement voids at the interface is minimised. It has been shown that pre-coating metals with a thin layer of PMMA will reduce cement voids and increase the interface shear strength (Barb *et al*, 1982). Our work has confirmed this increased strength for all three types of cement tested and when cement has been in contact with marrow fat.

PMMA pre-coating of knee prostheses could be important in improving fixation. However, reported series of knee arthroplasties using cement suggest that loosening of the femoral component is not a clinical problem, unlike the tibial component (Insall *et al.*, 1985). Load transfer in tibial components occurs through the horizontal condylar component as well as the central peg (Walker *et al.*, 1981). Although, in an ideal knee replacement, when load is evenly distributed between medial and lateral condyles, tensile stresses around the tibial component are low or absent, this is not true for asymmetrical loading (Hayes, 1978). Preoperatively many arthritic joints are malaligned and although efforts at realignment are made during surgery it is unlikely that the 'ideal' loading will be obtained. Thus tensile forces will be experienced across some of the interfaces (Hayes, 1978). Askew and Lewis (1981) estimated in their finite element model that some 30% of load is transmitted as shear across the stem interfaces of the tibial component. Therefore, the reduction in interface shear strength when using cement obtained from knee arthroplasty may influence the expected behaviour of the tibial component. These changes will be reduced if a metal stem is pre-coated with PMMA.

Efforts at improving the stability of the cement-bone interface have included the use of low viscosity cement to improve trabecular penetration (Miller *et al.*, 1976). We have found similar mechanical properties for standard viscosity Palacos R and lower viscosity Simplex and Palacos E. However, CMW 1, another standard viscosity cement has a lower Young's modulus and yield stress than the other cements.

When used in theatre for total hip replacement, the cement comes into contact with bleeding trabecular bone. Cement extruded from the proximal femur and acetabulum after prosthesis insertion is mixed with blood. Capillary bleeding has been shown to reduce cement trabecular penetration (Benjamin *et al.*, 1987).

Most knee replacements, however, are performed under tourniquet and almost no bleeding occurs at the cut bone surfaces. Marrow fat is then the main body fluid covering the cancellous bone. Extruded cement from knee replacement is therefore mixed with marrow fat. We believe this extruded cement from hip and knee arthroplasty is representative of the material used for implant fixation. However, the admixture with blood and marrow will not be uniform throughout the cement.

The effect of additions of known quantities of blood on cement properties has been studied elsewhere (Gruen *et al.*, 1975). However, this may not accurately represent the conditions which exist in theatre. Other workers have looked at blood applied to the upper tibia and its effect on cement penetration (Krause *et al.*, 1980).

In total knee replacement, we found a reduction in Young's modulus of 30% and a 24.7% reduction in yield stress compared with 12% and 15% respectively for hip replacement. This suggests that fat has a greater effect on cement than blood. Ferracane *et al.* (1984) suggested this but in their tests rat abdominal fat was added to PMMA in the laboratory rather than intramedullary contents.

Any alteration in the elastic modulus of cement has important implications for stress transfer in cemented joint replacements. A reduction in elastic modulus will result in increased deformation of the cement for any given stress. A low yield stress would make permanent plastic deformation within the cement more likely. A lowering of the yield stress is likely to be associated with a lower ultimate strength. PMMA is reported to be relatively strong when tested in compression compared with tension and shear (Haas *et al.*, 1975). The reduction in mechanical properties after contact with fat or blood is likely to be more pronounced if tested in tension and shear.

The potentially harmful effect of blood and more importantly fat on the mechanical properties of bone cement requires further investigation. A study is at present underway to examine the effect of bone preparation on extruded cement properties.

ACKNOWLEDGEMENTS

We would like to thank Zimmer for provision of titanium rods, Kirby-Warrick for the Palacos cements, Mr D J A Smith for technical assistance and the theatre staff at Glasgow Royal Infirmary for their cooperation.

REFERENCES

Amstutz H C, Markolf K L, McNeice G M and Gruen T A (1976). Loosening of total hip components: cause and prevention. *Proc. Fourth Open Scientific Meeting of the Hip Society.*, C V Mosby, St Louis, 102-116.
Askew M J and Lewis J L (1981). Analysis of model variables and fixation post length effects on stresses around a prosthesis in the proximal tibia. *J. Biomech. Engng.*, 103, 239-245.
Barb W, Park J B, Kenner G H and von Recum A F (1982). Intramedullary fixation of artificial hip joints with bone cement-precoated implants 1. Interfacial strengths. *J. Biomed. Mater. Res.*, 16, 447-469.
Benjamin J B, Gie G A, Lee A J C, Ling R S M and Volz R G (1987). Cementing techniques and the effects of bleeding. *J. Bone Jt. Surg.*, 69 B, 620-624.
Carlsson A S and Gentz C F (1980). Mechanical loosening of the femoral head prosthesis in the Charnley total hip arthroplasty. *Clin. Orthop.*, 147, 262-270.
Charnley J (1960). Anchorage of the femoral head prosthesis to the shaft of the femur. *J. Bone Jt. Surg.*, 42 B, 28-30.
Charnley J (1965). A biomechanical analysis of the use of cement to anchor the femoral head prosthesis. *J. Bone. Jt. Surg.*, 47 B, 354-363.
Ferracane J L, Wixson R L and Lautenschlager E P (1984). Effects of fat admixture on the strengths of conventional and low-viscosity bone cements. *J. Orthop. Res.*, 1, 450-453.
Gruen T A, McNeice G M and Amstutz H C (1979). Modes of failure of cemented stem-type femoral components, a radiographic analysis of loosening. *Clin. Orthop.*, 141, 17-27.
Gruen T A, Markolf K L and Amstutz H C (1975). Effects of laminations and blood entrapment on the strength of acrylic bone cement. *Clin. Orthop.*, 119, 250-255.
Haas S S, Brauer G M and Dickson G (1975). A characterisation of polymethylmethacrylate bone cement. *J Bone Jt. Surg.*, 57 A, 380-391.
Hayes W C (1978). Theoretical modeling and desing of implant systems, in A H Burnstein (ed.), *Workshop Mechanical failure of total joint replacement*, American Academy of Orthopaedic Surgeons, Atlanta, Georgia, pp 159-175.
Insall J N, Binazzi R, Soudry M and Mestriner L A (1985). Total knee arthroplasty. *Clin. Orthop.*, 192, 13-22.
Krause W R, Krug W and Miller J (1980). Strength of the cement-bone interface. *Clin. Orthop.*, 163, 290-299.
Matthews L S, Goldstein S A and Kaufer H (1985). Experiences with three distinct types of total knee joint arthroplasty. *Clin. Orthop.*, 192, 97-107.
Miller J E, Burke D C, Stachiewicz J W and Kelebay L (1976). A study of the interface between polymethylmethacrylate and living cortical bone under conditions of load bearing. *Trans. Orthop. Res. Soc.*, 1, 191.
Moreland J R (1988). Mechanisms of failure in total knee arthroplasty. *Clin. Orthop.*, 226, 49-64.

Walker P S, Green D, Reilly D, Thatcher J, Ben-Dov M and Ewald F C (1981). Fixation of tibial components of knee prostheses. *J. Bone. Jt. Surg.*, **63 A**, 258-267.

29

ELBOW PERFORMANCE FOLLOWING ELBOW ARTHROPLASTY

A C Nicol, F S Macmillan, W A Souter and F Waddell

INTRODUCTION

A review of literature has shown that there have been many clinical reviews of currently available elbow arthroplasties. Within these reviews the assessment of function has mainly been inferred from two parameters, namely; range of movement and pain relief. The review of literature yielded good evidence from nine studies (433 elbows) that elbow joint replacement does produce improvement in the two areas examined [Kudo et al., (1980), Morrey et al., (1981); Ewalds and Jacobs (1984); Johnson et al., (1984); Rosenburg and Turner (1984); Soni and Cavendish (1984); Souter (1985); Steiger et al., (1985); Trancil et al., (1987)]. However, these clinical assessments provide information on the elbow's performance during single, isolated movements performed in a clinical setting with the associated psychological influences of such an environment. Firstly, this type of testing can give no indication of the "home performance" of the elbow. Secondly, pertinent questioning of the patient upon their level of activity does not constitute reliable evidence of actual change in usage from the pre-operative state. Although the clinical evidence is well founded and it would seem to be a tenable inference that such patients are more active postoperatively, this had not hitherto been scientifically investigated. A study was initiated at the University of Strathclyde, Glasgow, and the Princess Margaret Rose Orthopaedic Hospital, Edinburgh, to define the pre- and postoperative elbow activity of these patients.

Patients were admitted to the study from the Princess Margaret Rose Orthopaedic Hospital, Edinburgh, which is the regional orthopaedic unit for South-East Scotland. All patients suffered from Rheumatoid Arthritis and had been selected to undergo Souter-Strathclyde elbow joint replacement by Mr W A Souter who performed the surgery in each case. One patient was referred to the study from the Western Infirmary, Glasgow. This patient also suffered from Rheumatoid Arthritis and underwent Souter-Strathclyde arthroplasty which was performed by Mr I G Kelly, Consultant Orthopaedic Surgeon.

EXPERIMENTAL METHOD

Strain-gauged electrogoniometers (Nicol 1987) were utilised to measure flexion and extension movement of the elbow joint over periods of 9 to 24 hrs. The electrogoniometer signals were amplified using a custom-built amplifier system (70 x 25 x 67 mm, mass = 0.13 kg) and were recorded on a miniaturised magnetic tape-recorder (Oxford Medical Systems Medilog Recorder 4 - 24, 112 x 86 x 36 mm, mass = 0.4 kg). The mass and dimensions of the data acquisition system were such as to permit the subjects to pursue their daily activities in their accustomed manner (see figure 1).

A Research Nurse was employed as a field worker to collect the patient data. The continuous monitoring equipment was fitted to the patient at approximately 9 a.m. on the test day and removed nine hours later in the case of local patients, and twenty-four hours later for those patients living more than 60 km from Edinburgh. Data were collected from both elbows on three consecutive days one month pre-operatively and at one, three, six and twelve months post-operatively. In addition, nine-hour recordings were made of both elbows for a group of healthy volunteer subjects of similar age and occupation in order to provide a range of normal values of elbow movement. Upon completion of data collection, each tape was despatched to the University of Strathclyde for analysis.

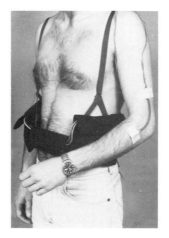

Fig. 1   Volunter subject wearing elbow electrogoniometers and lightweight harness with tape recorder and amplifiers.

ANALYSIS OF DATA

The recorded tape was replayed on a Medilog High Speed Replay Unit (Oxford Medical Systems PB-2 Replay) to a PDP11 computer for analogue to digital conversion. From exhaustive tests an overall system accuracy of $\pm 4^0$ was determined. A typical string of data is shown in figure 2 where a volunteer subject was asked to perform well defined activities during a special recording session. Interpretation of this format of data was very difficult therefore a suite of computer programs was written to represent the electrogoniometric output in terms of features of elbow joint motion.

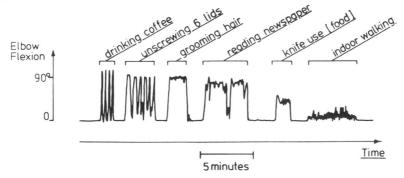

Fig. 2   Elbow motion data for a volunteer subject performing specific tasks.

The features identified can be grouped into two categories: aspects giving information on certain characteristics of movement and aspects giving information on the amount of movement.

Figure 3 shows a record from a nine hour recording for an active normal volunteer. The total number of movements of the elbow is very high but the interesting feature is the low number of movements which commence at small angles of flexion. The elbow is used for few activities which start from a flexion angle between 0^0 and 30^0 and this feature should alter the concern of many clinicians that flexion deformity is a severe limitation for patients.

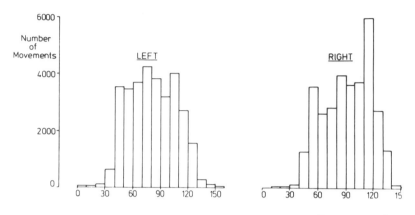

Fig. 3   The number of elbow movements (greater than 3^0) performed by a volunteer subject over a nine hour period in their own home shown to a base of elbow position at the start of each movement counted.

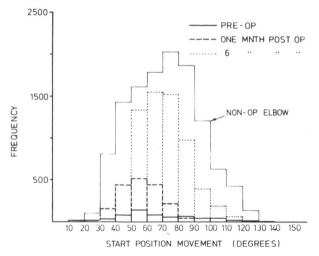

Fig. 4   The number of elbow movements performed by a patient preoperatively and at one and six months after receiving a Souter-Strathclyde elbow arthroplasty shown to a base of elbow position at the start of each movement. Also presented is the number of movements for the non-operated elbow at the six month period.

It is also possible to analyse elbow motion using several other factors such as velocity of flexion or extension, amplitude of motion, duration of stationary periods and results of these are presented in Macmillan (1988).
For a rheumatoid patient the amount of activity in the affected elbow preoperatively is extremely small as shown in figure 4. Following elbow replacement the actual range of possible motion is not substantially improved but the number of movements is dramatically increased. This improvement then continues to approach the activity level of the non-operated and unaffected elbow which is most encouraging for the verification of the surgical procedure in terms of functional improvement. Over the whole patient group, the mean activity level

increased by 159% from pre-operation to 12 months postoperation. The improvement in flexion/extension range of movement was only 28% and the maximum amplitude of movement rose by 53%. These figures highlight the need for careful interpretation of range of motion data in terms of clinical assessment of patient's "functional ability".

The procedure presented in this chapter shows that the measurement of joint motion can be used as an indicator of activity and that the post operative results for elbow replacement show significant benefits to the functional activity of the patient.

ACKNOWLEDGEMENT

This project was financed by a research grant form the Arthritis and Rheumatism Council.

REFERENCES

Ewalds F C and Jacobs M A (1984). Total Elbow Arthroplasty. *Clin Orthop. Related Res.*, **182**, 137-142.

Johnson J R, Getty C J M, Lettin A W F and Glasgow M M S (1984). The Stanmore Total Elbow Replacement for Rheumatoid Arthritis, *J. Bone. Jt. Surg.*, **66 B**, 732-736.

Kudo K, Iwano K and Watanabi S (1980). Total Replacement of the Rheumatoid Elbow with a Hingeless Prosthesis. *J. Bone. Jt. Surg.*, **62 A**, 277-285.

Morrey B F, Bryan R, Dobyns J H and Linscheid R L (1981). Total Elbow Arthroplasty. *J. Bone. Jt. Surg.*, **63 A**, 1050-1063.

Macmillan F S (1988). Aspects of biomechanics of the upper limb following elbow athroplasty and following phalangeal fracture. Ph.D. Thesis, University of Strathclyde, Glasgow.

Nicol A C (1987). A New Flexible Electrogoniometer with Widespread Applications, in B Jonsson (ed.), *Biomechanics XB*, Human Kinetics Publishers Inc., Illinois, pp 1029-1033.

Rosenberg G M and Turner R H (1984). Non-constrained Total Elbow Arthroplasty, *Clin. Orthop. Related Res.*, **187**, 154-162.

Soni R K and Cavendish M E (1984). Review of the Liverpool Elbow Prosthesis from 1974-1982. *J. Bone. Jt. Surg.*, **66 B**, 248-255.

Souter W A (1985). Anatomical Trochlear Stirrup Arthroplasty of the Rheumatoid Elbow, in D Kashiwagi (ed.), *Elbow Joint*, Elsevier Science Publishers, Amsterdam, pp 305-311.

Steiger J U, Gschwend N and Bell S (1985). G.S.B. Arthroplasty: a New Concept and Six Years Experience, in D. Kashiwagi (ed.), *Elbow Joint*, Elsevier Science Publishers, Amsterdam.

Trancik T, Wilde A and Borden L (1987). Capiteloloocondylar Total Elbow Arthroplasty, *Clin. Orthop. Related Res.*, **223**, 175-180.

30

AN ADJUSTABLE EXTERNAL FIXATOR TO PERFORM A

GLENOHUMERAL ARTHRODESIS

F J M Nieuwenhuis and G M Pronk

INTRODUCTION

At the Delft University of Technology in cooperation with the rehabilitation centre "De Hoogstraat" in Utrecht, different aspects of the treatment of patients with a brachial plexus injury have been studied. In the worst case this injury results in a permanently paralysed arm. Then, to regain control of the paralysed arm some patients undergo a glenohumeral arthrodesis, fusing the humerus and the scapula. The arm can thus be controlled by moving the scapula.

The result of this operation in terms of functional benefit, cosmesis and general comfort depends partly on the fusion angles between humerus and scapula. It appears to be very difficult, however, to ensure the optimal result concerning the fusion angles for a patient.

The recommendations about the extent of the optimal fusion angles vary considerably (Barr *et al.*, 1942; May, 1962; Charnley and Houston, 1964; Fontijne, 1965; Rowe, 1974; Cofield and Briggs, 1979). A comparison of a number of these studies yielded the following extents of the optimal fusion angles:

> A variation between 20 and 90 degrees of abduction.
> A variation between 15 and 50 degrees of forward flexion.
> A variation between 45 degrees internal and 40 degrees external rotation.

These variations can be explained by the following arguments:

> The endposition of the hand does not only depend on the extent of the rotations but also on the rotation sequence, which is usually not mentioned.
> The reference position of the arm from which the rotation angles are given is not standardised.
> In follow-up studies it is almost impossible to measure the adjusted fusion angles, because the preoperative reference position of the scapula is lost after the shoulder arthrodesis.

In fact, the optimal set of fusion angles for an individual depends on the functional wishes of the patient, on the morphology of the thorax and the shoulder girdle bones as well as on the severity of the paralysis. Due to the individual variations in these factors, it can be concluded that it is in principle impossible to define an optimal set of fusion angles which is suitable for everybody. (Pronk, 1985).

On the other hand, during an operation the fusion angles are commonly adjusted with the help of simple goniometers. Under normal circumstances this method of defining the threedimensional position of the humerus is itself very unreliable, let alone during an operation. Practically this means that even if an individual optimal set of fusion angles were known, there is every chance that they will differ from the desired ones.

In the current standard orthopaedic text books of the shoulder (Mosely, 1972; Post, 1978; Depalma, 1983) several operation techniques have been described to achieve fusion of the glenohumeral joint. The most successful techniques are mainly based on the principle in which the bones are fused by means of bone grafts and/or screws near the contact area between the humerus and scapula. In order to unload these bonegrafts and/or screws of the torque applied by the weight of the elevated arm, these methods are used in combination with a long arm plaster cast joined to a thorax plaster cast. In this way the average time to fusion of the shoulder joint should be 12-16 weeks. (Cofield and Briggs, 1979; Depalma, 1983).

In the following a new operation technique using an external fixator will be discussed, which solves most of these problems. By means of this fixator it is possible to preoperatively fix the bones and to determine the individual optimal fusion angles experimentally.

METHOD AND TECHNIQUE

The method is based on the external fixation principle which allows the arthrodesis to be performed in three stages. In the first stage only the fixator is placed, in the second one the optimal fusion angles are determined and in the third stage, one or two weeks later, the actual arthrodesis operation is performed.

In the first stage three pins are inserted percutaneously in the scapula and humerus, respectively. Because of the three dimensional composition of the shoulder mechanism, the construction will always be loaded by the weight of the arm in three directions. Therefore, in order to minimise stresses in the pin-bone interface, the pins in the scapula as well as in the humerus must be inserted as much as possible in a triangle. The shape of the scapula and the need for sleeping comfort, however, would be in favour of three pins more or less in one line in the dorsal plane. Hence, the exact placement of the pins will be based on a compromise. The best solution is to place one pin in the scapular spine near the medial border, one in the tip of the acromion and one in the intersection between acromion and the scapular spine. Within certain limitations of this fixator the pins can be positioned arbitrarily. After all pins have been placed, two T-shaped frames are connected to the pins by means of small clamps (figure 1). Then, two joints with a connecting rod in between are attached to the two T-frames by means of two screws as shown in figure 2. This stage is finished by setting commonly used fusion angles and tightening the joints of the fixator.

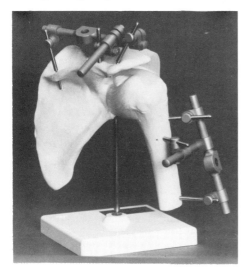

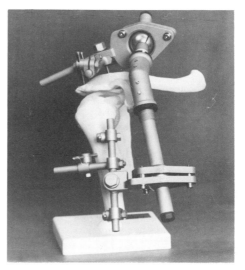

Fig. 1   The T-frames connected to the arbitrarily inserted pins.

Fig. 2   The connection rod with joints attached to the T-frames.

During the second stage the fusion angles can be optimised experimentally in the rehabilitation centre. This is done by successively untightening the two joints, repositioning the humerus and tightening the joints again. The fusion angles can be tested using an elbow orthosis (Cool, 1976) with the elbow stretched as well as in the 90 degrees flexed position. The connection rod is bent 40 degrees in the middle so that the rod can be positioned more or less parallel to both the scapular spine and the humerus. A wide sweater can therefore be worn over the entire construction in order to minimise the negative cosmetic effect.

For the third stage, once the optimal angles are found, the patient returns to the hospital for the actual arthrodesis. The connecting rod is removed by untightening the screws between the joints and the T-frames. The operation is then performed while the T-frames remain attached to the pins. Now, the arthrodesis is reduced to removing the cartilage from the articular surfaces and denudation of the inferior surface of the acromion. If necessary, the space between humeral head and inferior surface of the acromion can be filled up with a bone graft taken from the humeral head. After closing the wound, the connecting rod can be replaced on the T-frames. Since neither the position nor the orientation of the two joints of the fixator have been changed, the position of the bones and the extent of the fusion angles are the same as adjusted before the operation.

As a result of removing the cartilage, some space exists between the humeral head and the glenoid. To cope with this problem the connecting rod of the fixator consists of three parts: two telescopic tubes and a lug. On each side of the lug one tube is connected with a long bolt located inside the tube. The extent to which the tubes slide inside the lug, can be limited with a nut on the tube. By adjusting these nuts and bolts the connecting rod can be shortened in the directions of both telescopic tubes (figure 3).

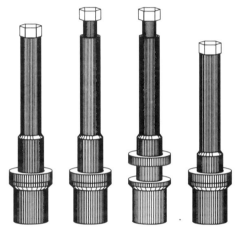

Fig. 3   Shortening of the connection to diminish the room between the two bones.

Once the two bones stick together, it is also possible to apply a compression force in either direction by further shortening of the connecting rod. The compression force can be monitored by means of strain gauges which are placed on the lug.

DISCUSSION

After four applications of this external fixation method it can be stated that the results of this technique are very promising. Experiences from each application have learned the advantages and limitations of this technique, from the surgical as well as from the patient's view.

From the surgical view, there is a number of advantages with regard to other methods described in the literature. The most important ones are:
    The operation is divided into two rather simple parts:
        The insertion of three pins in the scapula and three in the humerus. The pins can be placed arbitrarily.
        The denudation of both articular surfaces as well as the inferior surface of the acromion and placement of some bonesplints from the humerus between the humeral head and the acromion.
    The adjustement of the fusion angles which is most timeconsuming, takes place outside the operation theatre.
    By turning the long bolts in the telescopic tubes in the fixator the humeral head can easily be compressed against the glenoid and if desired against the acromion.
The advantages concerning the patient are:
    The fusion angles can be optimised experimentally. The patient has therefore influence on the setting of these fusion angles. Since these angles play an important role in the

final result in terms of cosmesis, functional benefit and general comfort, this aspect is very essential.

Compared with methods which use a thorax-arm plaster cast the external fixator gives much more comfort. The patient is able to live quite normally during the 12-16 weeks fusion process, as long as some care is taken considering the loading of the pins.

The shoulder girdle will not stiffen because the patient can move normally during the growing process. Therefore, the period of physical therapy needed after the fixator is removed is a couple of weeks shorter than when a thorax plaster cast is used.

During the weeks in which the optimal fusion angles are determined, the patient can evaluate the advantages of an arthrodesis. Since the actual arthrodesis operation has not been performed at this stage, the patient can still decide to refrain from the arthrodesis.

In the four applications so far, patients felt quite comfortable with this system: the first patient returned to work with the fixator, the third went riding on a bicycle and the last drove a tractor. The last patient used the possibility of this system to weigh the advantages and disadvantages of the arthrodesis: he was very uncertain about the choice of whether or not to have an arthrodesis. The fixator was placed and after wearing the fixator for two weeks, he decided not to go on with the arthrodesis.

The weakest part of this technique is, as in external fixation techniques in general, the bone-pin interface. There is a risk of loosening of pins and/or occurence of infections. This risk is higher when the pins are heavily loaded. Therefore, the pins must be positioned as far as possible in a triangle. Additionally, with this external fixator this risk can be minimised by giving a compression force. Should a loosened pin occur, compression can be increased which stabilises a loose pin against one side of the pinhole. In one of the four cases all three pins in the scapula were loosened probably because the patient, in order to test the adjusted fusion angles, went bicycle riding. However, when after the operation the compression force was increased and the patient advised to refrain from further bicycling the pins fastened again and the arthrodesis was consolidated 14 weeks later. The second patient had a pin infection shortly after the fixator was placed. In order not to take any unnecessary risks, the fixator was removed.

In cooperation with three hospitals in the Netherlands, a pilot study using this system will be undertaken in the near future. Presently, a measurement technique to monitor the consolidation during the growing process is being developed similar to the measurement system as described by Cunningham *et al.*, 1987.

REFERENCES

Barr J S, Friedberg J A, Colonna P C and Pemberton P A (1942). A survey of endresults on stabilization of the paralytic shoulder. Report on the Research Committee of the American Orthopaedic Association. *J. Bone. Jt. Surg.*, **24**, 699-707.

Charnley J and Houston J K (1964). Compression Arthrodesis of the Shoulder. *J. Bone .Jt. Surg.*, **46 B**, 614-620.

Cofield R H and Briggs B T (1979). Glenohumeral Arthrodesis. Operative and long-term functional results. *J. Bone. Jt. Surg.*, **61 A**, 668-677.

Cool J C (1976). An elbow orthosis. *Biomed. Engng.*, **11**, 344-347.

Cunningham J L, Evans M, Harris J D and Kenwright J (1987). The measurement of stiffness of fractures treated with external fixation. *Engng. Med.*, **16**, 229-231.

Depalma A F (1983). *Surgery of the shoulder*. ed. 3, J B Lippincott, Philadelphia.

Fontijne W (1965). Shoulder arthrodesis (in Dutch). Dr Thesis, Groningen.

May V R (1962). Shoulder fusion. A review of fourteen cases. *J. Bone. Jt. Surg.*, **44 A**, 65-76.

Moseley H F (1972). *Shoulder lesion*. ed. 3, Churchill Livingstone, London.

Post M (1978). *The Shoulder; Surgical and nonsurgical management*. Lea and Febiger, Philadelphia.

Pronk G M (1985). A reflection on the consequences of the shoulder arthrodesis in case of a brachial plexus injury (in Durch). Delft University of Techn., lab WMR & CE. Report N-248, 28p.

Rowe C R (1974). Re-elevation of the position of the arm in arthrodesis of the shoulder in the adult. *J. Bone. Jt. Surg.*, **56 A**, 913-922.

31

A METHOD OF NON-INVASIVE FRACTURE STIFFNESS

MEASUREMENT

K M Shah, A C Nicol and J B Richardson

SUMMARY

Fracture healing remains unpredictable. The development of methods of treating fractures is hindered by a lack of an objective measurement of bone union. This pilot study was undertaken to develop a non invasive method of assessing union in tibial fractures by measuring fracture stiffness.

INTRODUCTION

Fracture Stiffness

This is an objective measure of the mechanical properties of the fracture callus. The chief function of the skeleton is to provide stability, and this function is lost when a long bone fails by fracture. Fracture healing is directed to a return of this function, and therefore a mechanical measure of healing is appropriate. The gradual change in tissue during healing can be characterised by its stiffness, changing from a Young's modulus of 0.05 N/mm^2 for granulation tissue to 20000 N/mm^2 for mature bone, representing a 400000 fold change (Perren, 1979). The clinician requires to know when healing has advanced sufficiently to enable splintage to be dispensed with.

Methods of Assessing Union

Many different methods have been used by surgeons to assist in the evaluation of the strength and rigidity of fractures. The most commonly utilised method is the clinical assessment. Watson Jones (1943) advised "Union is sound when tenderness has disappeared, when no pain is elicited by straining the fracture, and when there is no longer elasticity or springing of the fragments".

Since their introduction to medical science, X-rays have been utilised to evaluate the stage of healing by the observation of the callus around the fractured bone fragments.

Researchers have worked on methods which would give a more objective assessment of fracture healing. Johannsen (1973) studied the process of healing fractures after intravenous injection of 87 m Sr. Hughes (1980) utilised 99 m Tc-MDP systemically and compared profiles of the radionuclide in the fractured and normal legs. Abendschein and Hyatt (1972) described the use of ultrasonics for calculating the modulus of elasticity of bone in experiments on guinea pigs. Puranen and Kaski (1974) presented their method of Osteomedullography for demonstration of the veins in the bone as a means of monitoring bone healing. Sekiguchi and Hirayama (1979) suggested a method of clinical application of bone percussion (percussion note) producing a wave signal which was used to evaluate the extent of bony consolidation after fracture.

Many researchers have studied the physical and mechanical properties of the callus to utilise them in the assessment of its strength. Burny (1979) described the deformation of a fixed beam method on fractures treated by external fixators. Jorgensen (1979) presented a method of measuring fracture stiffness using a simple dial micrometer attached to the pins of an Hoffman external fixator from which the steel bar had been removed. Although these methods are simple and can be used in the clinical environment, they cannot be utilised in situations where fractures are being treated by conservative means.

Fracture Stiffness as a Measure of Fracture Healing

By using fracture stiffness as a functional measure of healing the progress of patients can be monitored and compared. The system used must be accurate enough to permit these comparisons and precise enough to be used in the monitoring of progress.

Prediction of the Fracture Stiffness that Equates with "Healing"

The stiffness of an intact tibia has been measured as 63.5 Nm/deg for men and 41.3 Nm/deg for women (Jernberger, 1970). Our results show values of the stiffness as 65 Nm/deg and 72 Nm/deg respectively in two tibiae.

From their original data it is possible to calculate the stiffness at which both Jernberger and Edholm (1984) allowed patients to walk free of support (figure 1). Edholm's criteria was 12 Nm/deg while Jernberger allowed his patient to walk free of any support between 5.5 and 9.4 Nm/deg. Burny's group used the figure of 15 Nm/deg, which seems reasonable as no cases of refracture occurred in patients allowed to walk free of support above this level. This represents 23% of the value for an intact tibia, if 63Nm/deg is taken as stiffness of the normal tibia.

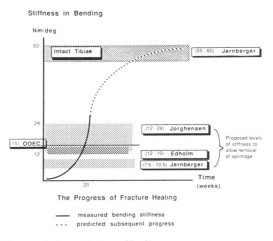

Fig. 1   Bending stiffness variation with time of healing.

Pilot Study

A pilot study was carried out to evaluate the proposed method. The method (figure 2) works on the principle that if the load (F) applied at a certain known distance (Y) from the fracture is measured, giving the moment (FY) at the fracture site, then measuring the angular/deflection (θ) occurring at the fracture site provides the data to calculate fracture stiffness (FY/θ).

Two patients have been examined at different stages of fracture healing. The results from a patient with a distal third tibial shaft fracture are shown (figure 3). Though radiographic criteria for union were not fulfilled at 17 weeks, fracture stiffness was measured to be 20% of the intact contralateral tibia. The decision to graft the bone was deferred and splintage continued. Stiffness progressed with no complications.

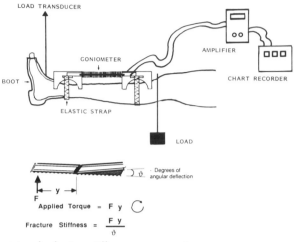

Fig. 2   Loading system for fracture stiffness measurement.

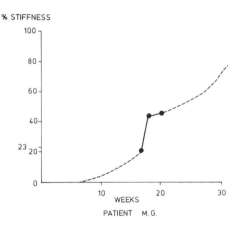

Fig. 3   Changes in fracture stiffness for patient M.G.

METHOD

The leg to be tested is covered in an elasticated cast sock to decrease the soft tissue movement.   "Orthoplast bridges" are applied on either side of the fracture with elasticated straps, and the ends of the "flexible electrogoniometer" (Nicol, 1987) are then attached to these bridges. The ends of the electrogoniometer are placed 55 mm on either side of the centre of the fracture as determined radiographically. Load is applied at the level of the tibial tuberosity, and a load transducer measures the load at the ankle joint. The distance "Y" is measured from the ankle joint to the centre of the fracture. The mean value of repeated readings is taken. The outputs from the electrogoniometer and load transducer are channelled through the amplifier to a chart recorder, thus obtaining simultaneous plots of force and angulation of the fracture.

DISCUSSION

Sources of Errors

The pilot study has revealed certain sources of error. In this system the total error can be represented as follows:

Total Error = Instrument Error + Observer Error + Soft Tissue Error

Instrument Error

With suitable amplification the error of the electrogoniometer is +/- 1%.

Soft Tissue Error

Soft tissue interposition between the electrogoniometer and the tibia provides the major source of error. This can be decreased by improved attachment of the contact bridges, and by comparing the normal with abnormal limb to express stiffness as a percentage. This will avoid population differences in pre-tibial soft tissues.

Computerisation will allow immediate calculation of the fracture stiffness from the data collected. This will allow efficient use of the information in the clinical setting with immediate decision on fracture management. Storage and analysis of all the data by microcomputer will allow assessment of the accuracy of the system and assess the effects of the proposed changes in the methodology.

A reliable portable system for the non invasive measurement of fracture stiffness would allow clinical assessment of fractures with immediate decision on fracture management. This would decrease the use of X-rays for assessment of fracture union, thus saving on expense and radiation exposure to the patients.

CONCLUSION

The strain gauge goniometer is capable of measuring fracture stiffness in patients. Results are encouraging and it is suggested that such a system should be tested on a wide clinical basis.

REFERENCES

Abendschein W F and Hyatt G W (1972). Ultrasonics and physical properties of healing bones. *J. Trauma*, **12**, 297-301.

Burny F L (1979). Strain gauge measurement of fracture healing in, A F Brooker and C C Edwards (eds.), *External Fixation: The current state of the Art*, Williams and Wilkins, Baltimore, pp 371-382.

Edholm P, Hammer R, Hammerby S and Lindholm B (1984). The stability of union in tibial shaft fracture: its measurement by a non-invasive method. *Arch. Orthop. Traumat. Surg.*, **102**, 242-247.

Hughes S (1980). Radionuclides in Orthopaedic Surgery. *J. Bone. Jt. Surg.*, **62 B**, 141-150.

Jernberger A (1970). Measurement of stability of tibial fractures. A mechanical method. *Acta. Orthop. Scand.*, Suppl. 135.

Johannsen A (1973). Fracture healing controlled by 87 m Sr uptake. *Acta. Orthop. Scand.*, **44**, 628-639.

Jorgensen T E (1979). Simple mechanical method of assessing fracture healing in A F Brooker and C C Edwards (eds.), *External Fixation: The current state of the art*, Williams and Wilkins, Baltimore, pp 383-390.

Nicol A C (1987). A new flexible electrogoniometer with widespread applications, in B Jonsson (ed.), *Biomechanics* X-B, Human kinetics Publishers Inc., Illinois, pp 1029-1033.

Perren S M (1979). Physical and biological aspects of fracture healing with special reference to internal fixation. *Clin. Orthop. Related Res.*, **138**, 175-196.

Puranen J and Kaski P (1974). The clinical significance of Osteomedullography in fractures of the tibial shaft. *J. Bone. Jt. Surg.*, **56 A**, 759-776.

Sekiguchi T and Hirayama T (1979). Assessment of fracture healing by vibration. *Acta. Orthop. Scand.*, **50**, 391-398.

Watson-Jones R (1943). *Fractures and joint injuries.* 3rd edition Vol II, pp 752. E & S Livingstone, Edinburgh.

32

EXPERIMENTAL AND THEORETICAL DENTAL

BIOMECHANICS

P M Calderale and M Rossetto

INTRODUCTION

The Department of Mechanics, Polytechnic of Turin, and the School of Dentistry have been involved in collaborative research over the last ten years.

The research has concerned the remodelling of the mandible due to tooth loss, its relationship to stress and strain patterns, and the biomechanical behaviour of prostheses used to restore bilateral distal tooth loss which is the most widely encountered clinical situation. Recently we have extended the work to the biomechanical behaviour of dental implants.

Studies have been made using reflection photoelasticity and strain gauges on dried mandibles. These techniques show only the surface behaviour. For the investigations of dental implants transmission photoelasticity has been used. The study of the whole structure, and the approach to *in vivo* behaviour requires the use of a 3 D finite element model (FEM) of the mandible. Our model has been validated, reproducing the experimental conditions for a dentate mandible.

In this chapter we describe our methodologies and the main results obtained. We also present the validation results of the 3 D FEM.

EXPERIMENTAL METHODS

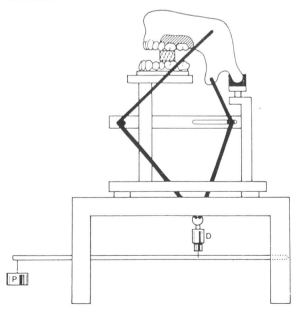

Fig. 1   Device constructed to apply physiological forces on mandible.

179

The strain patterns on the surface of dried mandibles were studied qualitatively using reflection photoelasticity. The reinforcement effect produced by the presence of the photoelastic resin was avoided by using a material (PL-1, Photolastic Inc.) with a low Young's Modulus (about one quarter of the dried bone modulus) and by limiting the layer thickness to 1.5-3 mm. Quantitative information was obtained using strain gauges.

Physiological loads and constraints are applied to the mandible by means of the device shown in figure 1. It consists of a brass replica of the upper arch and glenoid fossae, which matched the dried mandible. Temporal and masseter muscles are simulated by wires or rope bundles bonded to the muscular insertion areas of the mandible by cyanoacrylate cement. The direction of loading can be varied by means of pulleys. The total load applied to the mandible under test could be increased continuously, and measured by dynamometer.

The main limitations of this technique is that dried bones are investigated, and these have mechanical properties which are different from those *in vivo*. We overcome this problem by using comparative criteria.

REMODELLING ANALYSIS

The bony tissue undergoes a continual remodelling process from the time growth ceases until death. This leads to the gradual internal remodelling of the bone by replacement of primary bone with a system of secondary osteons and osteon remnants. If the process proceeds to the outer surface of the mandible external remodelling is produced, which can lead to changes in shape of the bone involved. In the human mandible this process is particularly marked especially in the condyles. The degree of remodelling and the new shape of the bone are closely related to changes in the dental arches, such as tooth loss and dental abrasion. Therefore, remodelling can, to a certain extent, be considered as a functional adaptation to the new occlusal situation. The changes are presumably due to alterations in the direction and intensity of forces applied to the mandible during function, and in the intercuspal position.

Mongini *et al.*, (1977) and Calderale *et al.*, (1980) found a relationship between structure, shape and strain patterns in different mandibles. The isostatic flow on the horizontal ramus of the mandible in centric occlusion was comparable with that of a beam subjected to bending and shear, and the orientation of the isostatic lines on the condyles corresponded to the different condylar inclination. Mongini *et al.*, (1979) showed there was a significant similarity between the isostatic lines and the direction of the main bony mandibular trabeculae as observed radiographically. Mongini *et al.*, (1981) used strain gauges to analyse the surface strains in four mandibles under various loading conditions. The data confirmed that the temporomandibular joint was load bearing and that strains were greater on the balancing side than on the working side.

ANALYSIS OF REMOVABLE PARTIAL DENTURES

Reflection photoelasticity has been used to analyse the biomechanical behaviour of different removable partial dentures (RPD) (Calderale and Rossetto, 1985; Calderale *et al.*, 1986; Pezzoli *et al.*, 1986).

The partial dentures analysed were those which replaced missing teeth distal to the standing dentition. The dentures are supported both by the teeth and edentulous mucosa. Because of such mixed support these dentures can apply torque and harmful stresses to the abutment teeth, and compress the edentulous ridge causing resorption. Clinical experience, biomechanical concepts and the literature (Kratochvil, 1963; Cecconi *et al.*, 1971; Kratochvil and Caputo, 1974) suggest that the ideal prothesis should present the following characteristics; even load distribution between abutment teeth and edentulous ridge, axial loading of abutment teeth and uniform loading of the edentulous mucosa.

Six different removeable partial dentures were tested, having the following characteristics:
Distal rest and circumferential clasps (Akers).
Mesial rest and circumferential clasps (Nally Martinet).
With continuous clasp and resilient connectors (Koller).
With precision attachments and splinted abutments.
With stress breakers and splinted abutments.

With telescopic crowns, distal rest and splinted abutments.

The dentures were constructed on the same dried covered mandible which was bilaterally edentulous distal to the first premolar. A rubber bolus was fixed between the first and second premolar of the two arches on the right side. Prostheses were first classified by determining the minimum load which caused the first isochromatic fringe. As all tests were carried out on the same mandible, the difference in minimum load depended upon the mode of loading and therefore on the RPD being tested. Clearly the larger the load required to generate the first isochromatic fringe, the better the biomechanical behaviour of the prosthesis.

Each denture was then loaded to a constant value of 130 N and the isoclinic lines on the buccal surface of the mandible recorded. The isostatic lines were then plotted.

As each prosthesis transferred the load to the mandible in a different way, different strain patterns resulted. The previous work on dried mandibles showed that the isostatic lines on the buccal surface of the dentate mandible had a regular, smooth distribution and we assume that a regular isostatic pattern represents a good denture behaviour. This allows the biomechanical behaviour of each prosthesis to be estimated qualitatively from the isostatic flow disturbance. Results of the analysis and the prosthesis are shown in figure 2.

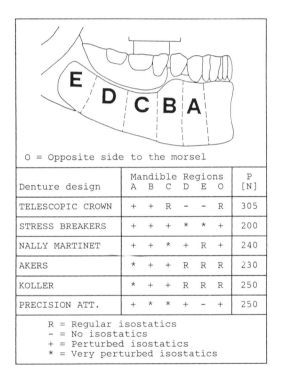

O = Opposite side to the morsel

Denture design	Mandible Regions A B C D E O						P [N]
TELESCOPIC CROWN	+	+	R	-	-	R	305
STRESS BREAKERS	+	+	+	*	*	+	200
NALLY MARTINET	+	+	*	+	R	+	240
AKERS	*	+	+	R	R	R	230
KOLLER	*	+	+	R	R	R	250
PRECISION ATT.	+	*	*	+	-	+	250

R = Regular isostatics
- = No isostatics
+ = Perturbed isostatics
* = Very perturbed isostatics

Fig. 2   Qualitative evaluation of isostatic disturbance and load causing first isochromatic fringe (P).

The prosthesis incorporating telescopic crowns presented an isotropic zone under the abutment teeth, similar to a beam subjected to a constant moment, which suggested that this prosthesis loaded the abutment teeth equally.

The telescopic crown and Koller removable partial dentures showed the best biomechanical behaviour, and these were tested under increased loads. Strain patterns were recorded under loads of 350, 400, 450 and 500 newtons. Increasing load further caused debonding of the photoelastic coating which had a lower adhesion to bone than to metals because of the presence of aluminium powder in the adhesive. Comparing the isochromatic fringes in these prostheses (Figure 3) suggests the following conclusions:

At lower loads the prostheses incorporating the crowns produced less strain in the edentulous area. Both the dentures presented a black area under the abutment. On the unloaded side a greater strain was generated by the continuous clasp denture.

At the higher loads the Koller denture changed its mode of loading the mandible and the magnitude and distribution of strains became practically the same for both types of dentures. However the isotropic zone remained under the denture incorporating the telescopic crowns but the black area under the abutment disappeared with the denture containing the continuous clasps.

The denture incorporating telescopic crowns showed the best biomechanical behaviour because it caused the first isochromatic fringe at the highest loads and the smallest disturbance to the isostatic flow. Futhermore it loaded abutment teeth equally. The prosthesis incorporating the stress breakers showed the worst biomechanical behaviour, causing the first isochromatic fringe at the lowest load and the largest disturbance in the isostatic flow. The remaining prostheses showed a behaviour intermediate between these extremes.

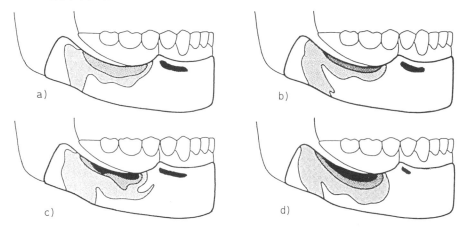

a)   b)

c)   d)

Fig. 3   Isochromatic fringes: a) Telescopic crowns RPD at 350 N, b) Koller RPD at 350 N, c) Telescopic crowns RPD at 500 N, d) Koller RPD at 500 N.

THEORETICAL ANALYSIS

In order to incorporate the internal structure of the mandible it is desirable to use a suitable mathematical model employing, for example, the finite element method (FEM). We have developed such a model of the mandible and have validated it by reproducing the experimental conditions outlined above (Calderale *et al.*, 1987).

FEM Model of the Mandible

The external geometry of the mandible was characterised using a digital electronic measuring machine (IOTA-P1, DEA Turin) and the finite element mesh generated using a preprocessor program. The thickness of the cortical bone was obtained by dissecting a mandible at the points corresponding to nodal lines of mesh. Internal elements were modelled as spongy bone. The mandible was assumed to be symmetrical with respect to the sagittal plane.

Teeth were simulated by three elements. A microanalysis of the periodontal ligament showed that models without ligaments were effective except close to the teeth.

The finite element program used is the NASTRAN 63. The model has 1442 6 and 8 node solid elements, 260 3 and 4 plate elements and gap elements (see Figure 4). The plate elements are thin and do not increase the stiffness of the mandible, but are used to determine the surface strains and stresses. The gap elements are used to simulate the unilateral upper arch constraints in centric occlusion, or to simulate a bolus during biting. A first version of the model without gap elements and with vertical fixed constraints on the teeth gave non-physiological results. The presence of gap elements requires a non linear solution.

Only static conditions are simulated, and the temporomandibular joints are modelled by clamping constraints on the upper nodes of the condyles.

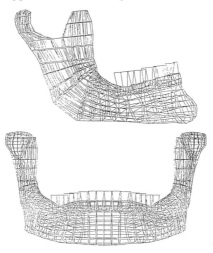

Fig. 4   Mesh of a mandible

Model Validation

To validate the model the numerical predictions it produced were compared with experimental results.

The experimental results were obtained with a dried mandible and appropriate mechanical characteristics were obtained from Yamada (1970): Cortical bone Young's Modulus, E = 86 GPa, spongy bone, E = 350 MPa; and dentine E = 13 GPa. The Poisson's ratio was assumed to be 0.3; because of the doubt about the value of this parameter a parametric analysis was carried out, which showed that results do not change substantially by varying Poisson's ratio within the range 0.2 - 0.4.

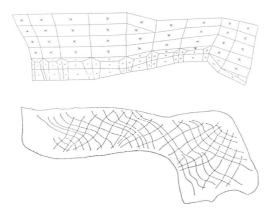

Fig. 5   Experimental isostatic flow of a dentate mandible (above) and finite element principal strain directions (below).

The forces applied simulate exactly the experimental conditions. To simulate the centric occlusion only one half of the model was used, utilising the symmetry of the mandible. The principal strains were predicted on the surface of the condyle coincident with the position of the strain gauge in the experimental study, allowing the comparison between prediction and experimental values. The numerical value, (-1.2 microstrain) was the range of experimental strains obtained by Mongini *et al.*, (1981) under the simulated load conditions.

The strain directions obtained with the model are also in agreement with the isostatic lines obtained experimentally, and this is shown in figure 5. To simulate the insertion of a rubber bolus between the teeth in the experimental study gap elements were incorporated between the molars and premolars. Predicted strains in the working side were in the range of experimental values. In the balancing side the strains were greater than on the working side, confirming the experimental results. The magnitude of the strain on the balancing side was, however, lower than that found experimentally. This is presumably due to the non-symmetry of the mandible used in the experimental tests and perhaps the experimental constraints were more complicated than the simulated one. However, the present model can be utilised as a first approximation in these conditions because it reproduces the experimental behaviour qualitatively.

CONCLUSION

The experimental methods employing dried mandibular bone have given information about the relationship between remodelling and strain patterns in the mandible and can be successfully used to study the biomechanical behaviour of dental prostheses.

The finite element model simulates satisfactorily the mandibular behaviour in centric occlusion and gives acceptable qualitative results for simulated unilateral biting.

REFERENCES

Calderale P M, Fasolio G and Mongini F (1980). Experimental analysis of stresses influencing mandibular remodelling. *Acta Orthop. Belg.*, **46 (5)**, 601-610.

Calderale P M and Rossetto M (1985). Analysis of load transfer between removable partial dentures and mandible by reflection photoelastic technique. *Int. Conf. expl. Mech.*, Beijing, 1015-1020.

Calderale P M, Rossetto M and Pezzoli M (1986). Biomechanical analysis of coupling between mandible and removable partial dentures, in G Bergmann, R Köbel and A Rohlmann (eds.), *Biomechanics: Basic and Applied Research*, Martinus Nijhoff Publishers, Dordrecht, pp 745-750.

Calderale P M, Rossetto M, Trittoni L and Garro A (1987). Studio sperimentale della mandibola umana mediante tecniche numeriche e sperimentali. *XV AIAS Conf.*, Pisa, 265-278.

Cecconi B T, Asgar T and Dotz E (1971). The effect of partial denture clasp design on abutment tooth movement. *J. Prosthet. Dent.*, **25**, 44-51.

Kratochvil F J (1963). Influence of occlusal rest position and clasp design on movement of abutment teeth. *J. Prosthet. Dent.*, **25**, 44-51.

Kratochvil F J and Caputo A A (1974). Photoelastic analysis of pressure on teeth and bone supporting removable partial dentures. *J. Prosthet. Dent.*, **32**, 52-60.

Mongini F, Calderale P M and Barberi G (1979). Relationship between structure and stress pattern in the humane mandible. *J. Dent. Res.*, **58**, 2334-2337.

Mongini F, Calderale P M and Fasolio G (1977). Biomechanical analysis of remodelling in the human mandible and mandibular condyle. *Proc. 1st Mediterranean Conf. Med. Biol. Engng.*, Sorrento, Italy, **1**, 41-43.

Mongini F, Preti G, Calderale P M and Barberi G (1981). Experimental strain analysis on the mandibular condyle under various conditions. *Med Bio Engng Comp*, **19**, 521-523.

Pezzoli M, Rossetto M and Calderale P M (1986). Evaluation of load transmission by distal-extension removable partial dentures using reflection photoelasticity. *J. Prosthet. Dent.*, **56**, 329-337.

Yamada H (1970). Strength of biological materials. F Gaynor Evans (ed.), Williams & Wilkins, Baltimore.

33

DISCUSSION: BIOMECHANICS

Fisher asked Clarke is there was evidence that particles such as barium sulphide and zirconium were introducing more damage to the joints than PMMA debris and whether ceramic heads were likely to be more resistant to damage of this kind. *Clarke* indicated that while he knew of cases of wear of titanium, they had no experience of measured wear in ceramics, but in every joint replacement all parties must be concerned at the possibility of invasion by any particulate matter, especially for patients aged 45 years who may well be subjecting the joint to loading for another forty years. He had no evidence or experimental data relating to wear of ceramics in that circumstance. *Jacob* followed on asking if Clarke was aware of tests with titanium nitride as a hard coating on to titanium balls. *Clarke* indicated that he had no experience of this, but he doubted whether such treatment would have any effect in circumstances where wire mesh titanium beads or chrome beads penetrated the joint. Responding to a question from *Harrington*, *Clarke* indicated that in USA surgeons were having bad experiences with screw-in rings for acetabular cups manifested by lucent lines around the screw thread producing a deficient fixation. Their preference was for hemispherical cups with screw or peg fixation, since this was technically less demanding. *Jacob* commented that in Switzerland this type had been abandoned because of difficulties in cutting the thread due to the top breaking through the ventral part of the acetabulum very close to major nerves and vessels.

Davidson asked Clarke about the black debris which he had referred to and its physical/chemical composition and the source of the material. *Clarke* indicated that he was not certain whether it was titanium oxide or particulate titanium. It was not clear whether the staining came from the joint itself or from fretting corrosion or against the titanium stem. He felt on balance that the staining originated from wear on the ball head of the femoral component. *Pearcy* asked for information about the lubricating mechanism in artificial joints and *Lawes* suggested that there could be a range of possibilities from synovial fluid if there was sufficient capsule left: more generally thin and watery fluid is found on aspiration and some joints appeared to be run in the dry condition. *Hamblen* indicated that arthrograms of successful joint replacements showed very small false cavity development indicating that the amount of lubricating volume was very small. *Fisher* suggested that the presence of wear particles indicated contact between the surfaces with very little fluid for lubrication, but *Scales* responded that serum reduced friction and is definitely necessary for the function of the joint as had been shown in simulator tests. *Cameron* indicated that he had been involved in the manufacture of the balls of femoral prostheses and suggested that the micro finish generally adopted for these heads was not necessarily required. *Clarke* however contended that the micro finishing was not purely aesthetic but was necessary for reduction of the wear rate of the polyethylene.

Hamblen raised the practical point of the surgeon removing a loosened acetabular cup in a patient whose femoral component was firmly fixed. Would minor scratches on the femoral head so prejudice the wear of a replacement cup that the femoral component should be replaced also? *Clarke's* view was that a fixed femoral component should not be removed and he pointed out to Malcolm's analysis of long term Charnley prostheses which had not failed after lifetimes of nine to twenty-two years. In each case the femoral component was firmly fixed despite a considerable amount of polyethylene and cement debris but almost every acetabulum that he looked at was surrounded by a fibrous sheath with large amounts of histeocytic processes ongoing. The concept of wedging of femoral stems due to Ling was raised and *Harrington* suggested to Lawes that when cement yield or fracture resulted in a subsiding stem, it was likely to produce wear particles in the cement and consequently fretting of the cement mantle. *Lawes* suggested that in the tapered wedging situation a minor fracture of the cement would produce a further locking and this would probably be a once and for all occurrence, not a repetitive fretting movement which would almost certainly produce loosening in the long term.

The relationship between calculated predictions of joint force and actual force measurement was raised by *Barton*. *Procter* agreed that in this field there is incomplete knowledge since he knew of no experimenters at that time who had predicted joint forces and had the opportunity also of measuring from an implanted prosthesis. Procter also referred to the consequences of trauma in respect of joint replacements, a field in which there is very little information. *Clarke* returned to the comparison of analysed and actual loads, pointing out that cadaver femurs could not survive loads higher than 7 kN in general and not more than 4 kN in fatigue, whereas the University of Strathclyde had published predictions of joint forces of ten times body weight for stair climbing, using joint forces of 7 kN and higher. How could this paradox be resolved? *Paul* indicated that while there might be some doubt about these very high values of forces, one must be aware that the fatigue tests are conducted on non-living bone without the self-repair mechanism, and that the joint forces are not transmitted along the shaft of the femur but are, in fact, largely transmitted back to the pelvis by the muscles inserted in the proximal femur.

Shepherd asked Clarke if the current profusion of orthopaedic companies producing implants with porous coatings for bone ingrowth was surprising when Clarke's results showed severe problems with prostheses of these types. *Clarke* indicated that his group had started porous ingrowth studies in 1975, and until 1988 he had been very enthusiastic, although he was now somewhat ambivalent because evidence from animal studies and clinical reports suggested that there was a body of scientific data indicating that these implants might be dangerous. He hoped that porous systems would work and indeed porous hemispherical sockets seemed to be working very well with short term follow ups of four to six years. He was not so sure of the success of the stems and consequently his preference would be for cemented stems until the technology had improved. The critical question was whether cemented press-fit or porous ingrowth techniques, all of which had potential advantages and disadvantages would give consistently good results in younger patients.

Harrington asked if Clarke or Lawes had considered coating the femoral stem of an implant with a deformable polymer to make a press-fit type of prosthesis to eliminate the stress relief phenomenon at the upper end of the femur and consequent bone lysis. His group had recently undertaken animal experiments indicating less bone resorption around this type of implant and it has been used clinically in Italy by Bombelli with good results. *Clarke* felt that where you have highly stressed components and the possibility of micromotion polyethylene was definitely contra-indicated. *Jacob* indicated that as far as he knew only one surgeon was now using this type of prosthesis and this was Bombelli.

Davidson asked Scales his comments about using the polyethylene sleeve on the tibial post in the knee joint which he had shown, and in particular whether it was cemented or whether the polyethylene sleeve was a press-fit along with the tibial post into the tibia. *Scales* indicated that the sleeve was threaded and the tibia tapped. Depending on the diameter of the marrow cavity the smaller of the two sizes of tap might be used, but in no case was bone cement used. No incidence of loosening had been found. *Davidson* continued asking if any indication of micromotion or wear had been found between the bone and the sleeve. *Scales* indicated that none had been observed. *Clarke* asked Scales what staining due to different alloys has been found in tissues during reoperation. *Scales* indicated that a lot of staining had been seen with titanium, a minor amount with cobolt chrome. In femoral components of hip replacements, cobolt chrome heads were used together with titanium 318 stems, and commercially pure titanium for the shaft, although steel had never been used with any of these components. In some of the anodised implants no macroscopic staining had been seen but in some it had been enormous. It was unclear whether this was due to faulty anodising during different coatings. The tissue response seemed to be very variable. In some cases titanium oxide could be seen lying in between fibrous tissues. In other areas some giant cells could be observed and lymphocytes, but no dramatic tissue response except around the knee joint where a lot of fibrous tissue may be seen, probably due to variability of the capsule from patient to patient. With regard to cobolt chrome Scales used a knee joint axle in this material running in polyethylene bearings. In some cases the polyethylene bearings wore and the result was a wobble on the knee joint. This happened only occasionally, possibly only in six cases and the bushes were replaced, the knee having been designed specifically to allow this. *Rowe* asked Scales the manufacturing time for custom implants and whether he had any experience of modular systems. *Scales* did not agree with modular systems since he felt that this compelled the surgeon to fit the patient by and large to the implant, and he felt that the implant should be fitted to the patient. As far as manufacturing time was concerned, the growing prosthesis was the most difficult one to make, but that could be made within 24 hours if several people were working round the clock. This was rarely necessary however since these patients were treated with cytotoxic drugs and the replacement could be delayed six weeks after diagnosis. *Hamblen* asked Scales for his views on the attachment of muscle to these prostheses, and in reply *Scales* indicated that at the outset it was thought that everything should be attached to the implant to get movement and in some cases the implants were completely covered in terylene net giving a magnificent movement just after operation. Unfortunately in a gradual process over three months, everything went solid and the patient developed a flexion deformity of the hip and since she now uses an artificial limb, this is producing considerable low back pain. The limb was also adducted and this technique has now been abandoned. It is now believed that the muscles get attached to the capsule which always forms around the implant and then the capsule can move differentially to the implant. *Scales* did not believe that one could ever restore the muscles in exactly the correct position. Another important point was that such implants should always be highly polished, and the technicians are trained to feel every aspect of the implant to identify sharp corners and remove them so that the tissues could slide over the implant. In the same way with intramedullary stems in bone cement, it is desirable to have polished stems so that the cement locks into the bone. The macromotion should be between the stem and the bone cement.

Childress asked Nicol if the goniometer which he had described could be used in motion analysis. *Nicol* indicated that it could be used for gait analysis, although it could not obviously be maintained in the laboratory axis system and thus was potentially less useful than a displacement analysis system. A scheme for gait analysis using these goniometers had been set up by *Rowe* which measured the kinematics of gait of patients after joint replacement surgery.

Small asked if Nicol had made any attempt to measure the load bearing capacity and the stability of the elbow joint replacement which he had described. *Nicol* agreed that the absence of a radial head in the elbow joint replacement was likely to result in reduced load bearing capacity, although the patients seemed to have adequate hand function by virtue of the intra-osseous membrane between the radius and the ulna. A similar type of replacement had been developed by Amis and Miller using a small radial head articulating against the humeral component giving a procedure which is surgically more demanding and which has not yet numbers of patients comparable to the Souter-Strathclyde device so that comparisons were difficult. *Goh* asked Scales about the possibility about applying CAD/CAM technology to custom prostheses. *Scales* indicated that the group at Stanmore had looked at this some time previously and Walker the new Director of Biomedical Engineering was trying to develop it. It was found to be costly and the number being produced did not justify introduction of the system. Similarly, it is necessary to have NMR and CT scanning to utilise such a system effectively.

Murdoch asked Niwa for his opinion of the data of Andriacchi and his co-workers which showed in patients having osteotomy of the tibia a correlation between good clinical results and the development of low adduction moments in patients' gait. This group found all other parameters not to be significant except the unequal pace. *Murdoch's* interpretation of this was that patients who limped most protected their knees best and therefore ensured better results. *Niwa* agreed that small adduction resulted in a good clinical result. *Murdoch* suggested therefore that clinicians who did not have gait analysis laboratories could do a simple measurement of stride length and conclude that the more unequal the stride length, the better the clinical result ought to be.

34

CLINICAL SERVICES: COST AND EFFECTIVENESS IN

PROSTHETICS AND ORTHOTICS

G Murdoch

In addressing the task viz., clinical services in prosthetics and orthodics, note that I have been asked to write about "cost" **and** "effectiveness". I have not been asked to write about "cost effectiveness". Associated with the latter are notions such as "value for money" and "getting it done as cheaply as possible". These notions encompass very complex concepts including political aspects and accordingly the base from which I must start is "to do the best we can". I propose, therefore, to outline the best that we have tried to do in Dundee.

PROSTHETICS

In the field of prosthetics the system we have established has been based first on the concept of the clinic team and second on a study of the circumstances of amputations performed in the Tayside Region prior to the establishment of Dundee Limb Fitting Centre. That study demonstrated that the average number of amputations done per annum per amputating surgeon was 1.8 per annum. Moreover, the person who was alleged to have done most had performed none.

I do not propose to go into these circumstances in detail but what became clear was a need to concentrate experience. It was eventually possible to establish that all amputation surgery in the Tayside Region be performed by one of two teams of amputating surgeons each comprising one consultant orthopaedic surgeon and one trainee.

That in itself, of course, was not enough. The overwhelming majority of amputations are for vascular disturbances of the lower limbs in the elderly. The disruption of a body image of long standing is extremely difficult to compensate. The patient is handicapped additionally by diminishing physical and emotional resources. For the ageing patient it may seem to be the beginning of the end and physical rehabilitation is limited further by concomitant disease of the heart, lungs, bones, joints, brain and blood vessels. Yet the situation need not be as bad as described by Little *et al.*, (1974) in a paper entitled "Vascular amputees: a study in disappointment" in which they stated that only 7% of 67 patients felt that amputation had resulted in significant improvement. Cameron *et al.*, (1979) surveyed the post operative situation in Tayside and found a very different picture - "far from being a depressive exercise, conducting this survey so often lifted the heart when again and again one came across elderly amputees enjoying an "Indian summer" of mobility and social involvement". Little and colleagues completed their contribution by stating that the group responsible for the surgical and rehabilitation management must retain contact with the amputee and not rely on other resources of community support.

Clearly the first necessity was to establish a system whereby we were alerted to the existence of a candidate for amputation and to promote rapid assessment and the institution of integrated patient care. Assessment consisted of an appreciation of the patient as a whole, of the specific pathology and of the desirable level of amputation, if indeed amputation was necessary at all.

The concentration of expertise of medical, nursing, physical and occupational therapy and prosthetics has ensured realistic operation of the clinic team. In the case of the amputee, it ensures the recognition of concomitant disease and its treatment; the control of stump volume and the concentrated experience of surgery has produced a very high level of surgical technique. Equally there are in benefits in wound healing, reduction of infection and control of pain. The development of integrated rehabilitation in terms of everyday activities such as donning and doffing of clothes, and the performance of a planned exercise programme has ensured increasing independence and the involvement of relatives.

In Dundee we have carefully monitored our own progress from the inception of the unit (Murdoch *et al.*, 1988). We noted that from 1966 to 1983 an improvement in amputation levels from a ratio of 1 below knee (BK) : 2 above knee (AK) to one of 3BK:1AK had been achieved. In addition a success rate in terms of wound healing in the region of 90% for BK amputations was attained. In view of these results we set up a prospective study **externally monitored throughout** by Peter McCollum of Dublin to confirm these figures, and, in so doing highlight the value of modern vascular investigation techniques and the concentration of clinical expertise.

Prospective Study of Amputation Surgery Performance

100 consecutive patients with major peripheral vascular disease were admitted prospectively into the study. The mean age was 73 and the male/female ratio was 59:41. It was considered that no vascular surgery or no further vascular surgery was possible in all cases. Diabetes was present in 45. Using the thermoscan and skin blood flow results 10 were recommended for an AK amputation.

Nine had a BK recommended but in fact had an AK performed for other reasons e.g. flexion deformity. This left a total of 81 patients on whom vascular laboratory analysis had suggested that a BK amputation could be carried out and was indeed carried out.

Of the 100 patients the operative mortality was zero. All AK amputations healed, one after a revision. Of the 81 BK amputations there were 6 failures. Of the 75 patients left seven BK stumps 'failed' in that they required a 'wedge resection' all of which were successful. Thus 100 patients entered into the trial, 19 were selected for AK amputation with a successful result. Of the 81 recommended for BK and who had BK surgery, 75 were successful. This gives a ratio of 3BK:1AK and 93% successful healing rate for BK amputation.

Rehabilitation Outcome: 1232 primary amputees: 19 years

Discharge from DLFC occurs on average on the 41st day after admission and 81.8% are sufficiently independent to return home or to be placed in sheltered accommodation. In each case a definitive prosthesis was supplied and the first out-patient review took place at 6 weeks after discharge. Progress is reviewed, training reinforced, the prosthesis checked and changes made if required. Review is thereafter at 3 or 6 month intervals until a stable situation is reached and thereafter yearly contact is maintained. The mean age of the patients is 70 years, 86.8% have vascular disease including diabetes as the causal pathology and mean survival is 3 years 1 month against 10 years for their age peers. 87.1% of the patients are supplied with prostheses and walked an average of 2496 steps a day in contrast to their age peers who walked an average of 8366 steps per day.

The control of stump volume permits fitting with a definitive Patellar Tendon Bearing prosthesis at about 25 to 28 days post-operatively: the socket lasts 180 days before requiring renewal.

With regard to the prosthetic fitting of the above-knee amputee, the in-patient rehabilitation in DLFC permits of day to day changes in alignment, sometimes new sockets and occasionally different components leading towards a final prescription.

Aspects of Costing

To cost this exercise would involve attributing money values to each element of the process of rehabilitation and numbers are issued by Health Boards, for example, with respect to the cost per week per patient.

Equally, there are some numbers which are alleged to relate to the price paid by the government health authorities to the commercial companies for the prostheses they purchase, albeit, that these prostheses up until July 1988 have been prescribed, cast, measured and fitted by Health Service employees.

Such experience that we have had in Dundee and which has been experienced in the University of Strathclyde in their clinical outlet at the Southern General Hospital (SGH), Glasgow, suggests that these costs can be considerably reduced.

The cost of a SGH conventional PTB including a Sach foot and laminating material but excluding labour costs was £70 in May 1988. In contrast the price for a similar but heavier prosthesis from the commercial company was £460. To renew a socket costs £765 !

The cost of a modular BK prosthesis at SGH ranges from £115 to £195 in contrast to a cost of about £700 commercially. At above-knee level and not including laminating material the SGH cost ranges from £256 to £444.35 as opposed to commercial prices of the order of £1000+. Major repairs ranged from £275 to £650 in 1986.

No doubt it would be possible to estimate and cost the proportion of time spent by the secretary who runs the system. Equally a price could be put against the Vascular Laboratory assessment of tissue nutrition and a price too, could be put upon the two or three days hospital stay prior to surgery, the one or two days immediately following surgery and the stay in Dundee Limb Fitting Centre and the total sum on a best care basis is of the order of £4500-£5000. The cost, however, of not doing the best you can is, I submit, very much greater.

The Randomly Assessed Patient

Where no system exists and the patient is eventually recognised as a possible candidate for amputation the medical decision as to this status will be delayed if there is no direct route of referral. Once the patient is seen as possibly requiring an amputation he/she may be referred to any one of a number of specialist departments, but usually internal medicine, general or vascular surgery. Many of these Units are ill-equipped to assess the level of amputation and may indeed proceed forthwith to perform an above the knee amputation in order to obtain early primary wound healing. It should be noted that the surgeon in these circumstances is likely to have done no more than three or four amputations in a year, is unaware of the possibilities for prosthetic fitting and the different levels of functional activity that can be achieved.

Aside from wrong level selection we are likely to find poor skin abutment, haematomas and delayed healing. Moreover we know from a study performed by our Senior Nurse some years ago that bed sores were encountered in 38% of patients entering our service. The limited experience of those involved in random amputation surgery means that the management of pain and stump environment is likely to be less than adequate leading to further delay in referral to a Limb Fitting Centre and an increased stay in hospital, usually in an acute bed. Very often, as the days stretch on, the patient is discharged home with an unhealed stump, discomfort and pain. At home there is likely to be inadequate support and almost total dependence on the visits of a District Nurse. Even if the stump is healed the patient is likely to languish there because of the problems of bandaging, lack of mobility and developing flexion deformities. In these circumstances the patient is ultimately referred for fitting and is given an appointment. If the patient is deemed unfit for fitting and this will be so if the wound is not healed or the stump swollen then the patient will be referred back home and suggestions made as to dressings by the District Nurse and a plea for increased bandaging of the stump.

Bandaging is difficult to do, must be performed regularly, requires training and should not be employed until the wound is healed. Isherwood *et al.*, (1975) demonstrated the very high pressures that can occur inadvertently even when bandaged by an experienced 'bandager' viz. as high as 130 mmHg.

When the patient is ultimately deemed suitable for fitting then the stump may be cast, the patient measured and a date for fitting established. Unless stump volume has been controlled it will be found at the fitting stage that the prosthesis will not fit and the same situation may well exist on the occasion of delivery of the prosthesis. Thereafter follow-up will depend very much on the fashion established in the individual limb fitting centre, but I suspect the initial fitting would be for a stump of increased volume and require an early return for a new socket, despite the wearing of three or four socks.

If you want confirmation of this scenario then you will find it in the "McColl" Report (1986).

Further Aspects of Cost

How to cost this sorry tale - it may be that from the time of the patient's first serious complaint and concern for himself or herself, there are no additional costs in the sense that little money has been spent to alleviate the patient's suffering and pain. When the patient is eventually admitted to hospital and surgery is carried out under this random system then costs are in my view, increased - very often, markedly. The patient is likely to spend more time in acute beds or at home where Social Security and District Nurse costs have to be accommodated.

Malone *et al.*, (1979) examined the economic effects of the introduction of a system of vascular investigation, below-knee posterior flap surgery, immediate post surgical fitting and an accelerated rehabilitation and compared the results with that obtained before in a Veterans Administration (V.A.) hospital. The reduction in rehabilitation time from 125 to 32 days if applied throughout the V.A. system would have saved $17M. out of an actual cost of $24M.

ORTHOTICS

In the case of orthotics the problem nationwide is even worse as it is virtually impossible to compare the best practice with the widespread bad practice in monetary terms. In prosthetics from one patient to another there are common factors involved. In the field of orthotics, however, with a widespread pathology and the presentation of that pathology individual to each patient, it is not possible to apply solutions by rote. This is not to say that it is not practised or attempted; for example, the provision of identical (aside from size) plastic ankle/foot orthoses for all degrees of all conditions affecting the ankle/foot complex is widely practiced; all the devices have been prefabricated and supplied off the shelf. To do the best you can in orthotics requires at least as careful an assessment as is required in prosthetics. The solutions are not as they are in prosthetics where there are norms of an almost absolute nature with fine variations in terms of volume, shape and alignment. Patient/device matching in orthotics has to be much more precise not only in the prescription but also in the design and fabrication of the device. The variations of performance that can be achieved in the use of a polypropylene ankle/foot orthosis simply by altering the ankle/foot angle slightly or applying small differences in the thickness of the material or the trim are considerable. The quality of fit will depend on the ability to assess, analyse, design and fabricate the appropriate device and that can only be done by professionals who have been properly trained and educated. There is an enormous dearth of both prosthetists and orthotists in the United Kingdom. The result is that, certainly in orthotics, many of the people purporting to be orthotists are no more than salesmen and many of the devices supplied are discarded by the patients. An amputee may go on grumbling until a solution of some kind leading to satisfaction is obtained. A patient requiring an orthosis, but failing to get satisfaction, often discards the device.

CONCLUSION

The important thing is to provide effective treatment. If the treatment is effective it may on the face of it be more costly than treatment which is ineffective. Then you have to ask the question "how do you cost bad treatment or the total lack of treatment"? In my belief the best value for money will be obtained by continuing investment in education and training.

To be effective in this field the cost must be found to educate and train all clinic team members including the prescribers.

REFERENCES

Cameron J U, Murdoch G, Troup I M, Wood M A, Napier A and Lloyd R (1979). Amputee Rehabilitation: A Scottish Survey, in R M Kenedi, J P Paul and J Hughes (eds.) *Disability*, Macmillan, London, pp 210-215.

Isherwood P A, Robertson J C and Rossi A (1975). Pressure measurements beneath below-knee amputation stump bandages: elastic bandaging, the Puddifoot dressing and a pneumatic bandaging technique compared. *Br. J. Surg.*, **62**, 982-986.

Little J M, Petritsi-Jones D and Kerr C (1974). Vascular Amputees: a study in disappointment. *The Lancet*, **1**, 793-795.

"McColl" report (1986). Review of Artificial Limb and Appliance Centre Services - The Report of an independent working party under the chairmanship of Professor Ian McColl Vol. I and II. The Department of Health and Social Security, London.

Malone J M, Moore W S, Goldstone J and Malone S J (1979). Therapeutic and Economic Impact of a Modern Amputation program *Ann. Surg.*, **189**, 798-802.

Murdoch G, Condie D N, Gardner D, Ramsay E, Smith A, Stewart C P U, Swanson A J G and Troup I M (1988). The Dundee Experience, in G Murdoch (ed.) *Amputation Surgery and Lower Limb Prosthetics*, Blackwell, Edinburgh, pp 440-457.

HISTORY (STATE) OF p 93

CAD/CAM in 1989 (time of publication)

p 93

Future in CAD/CAM p 93

35

EVALUATION OF CAD/CAM IN PROSTHETICS

G R Fernie and A Topper

INTRODUCTION

Computer-aided design (CAD) systems in prosthetics begin by the input of measurements from the residual limb. These measurements can be taken by means of tapes and calipers or by means of more sophisticated shape sensing technology. The CAD system then generates a modified shape appropriate for the socket by means of one of five processes:

The most appropriate socket shape is selected from a library of predetermined shapes and scaled to fit the measurements (Saunders et al., 1985).

Standardised modifications are selected from a library and applied to the measured shape to simulate the modification process done on plaster replicas (Davies, 1986).

A mathematical stress analysis technique such as finite elements is applied to calculate an appropriate socket shape based upon an assumed pattern of interface stress and tissue mechanical properties that are either assumed or measured (Krouskop et al., 1987).

The shape is modified freely on the computer by the operator. This fourth option is often applied following the use of one of the previous options in order to allow for local custom modifications to refine the shape.

The shape of the residual limb and of the muscles and bones which comprise that limb are represented graphically by digitising X-ray tomographic scans. The operator then uses his judgement to design a socket to fit the anatomy appropriately (Vanderlinden and Warzee, 1986).

There is little actual computer-aided manufacturing (CAM) in prosthetics at this time. It is true that the output from the CAD is generally directed to a computer controlled routing machine or a numerically controlled mill in order to carve a mould for the socket, but the socket is still formed over the mould in the same way as in the conventional process. In effect, the CAM has only made it possible for the CAD to replace one stage in the prosthetic socket fitting process and has not, by itself, contributed to the socket production process. Automated or semi-automatic vacuum forming and heat shrinkable sockets have been used in the CAD/CAM process but these techniques are also sometimes used in the conventional socket production process (Foort et al., 1984; Davies et al., 1985).

The step from CAD directly to a socket has not been implemented, to our knowledge, although several possible technologies such as computer-controlled variable molds, extrusion through a variable profile mandril, and three-dimensional shape reproduction by laser-activated polymerisation in a fluid have been suggested. Very preliminary experimentation in our laboratory has indicated other possible approaches, e.g. extruding a bead of polyethylene, distortion of a wire net.

There are potential advantages of CAD/CAM that may be realised as the technology evolves in this application. For example, the quantification of shape has been shown to be of value in the education of prosthetists and provides hard data needed for the scientific investigation of the fitting process (Fernie et al., 1984). In addition, the technology may lead to more uniform standards of fitting and may provide the means of accessing prosthetic services from a broader geographical base. The ability to recall a previous shape at a subsequent fitting may be found to be the most valuable feature.

The question remains however, how "good" are these systems at producing sockets that fit? The purpose of our evaluation of CAD technology in prosthetics at this time is to determine whether this technique is capable of producing the same or better fitting sockets than the conventional method in a reasonable time frame, so that we may be able to decide whether it is financially viable to introduce it into service at this time and whether the result will be acceptable for patients. The secondary purpose of our evaluations is to attempt to predict the future role of CAD/CAM in prosthetics and the possibility of expanding capabilities to include prosthesis assembly, alignment and cosmetic finishing.

EVALUATION METHODOLOGY

The evaluation was designed to give clear and unbiased answers to two questions:

Can CAD fit a patient as well or better than the conventional method within a similar time frame?

What is the potential for CAD/CAM in prosthetics?

The first question indicated that a comparative study is required. Comparing the new method with a standard also provides a control for variability in the subjects. There are three possible ways of determining which of two prosthetic sockets is the better fit:

The amputee's opinion,

Professional opinions,

Objective measurement.

Objective measurement would be the preferred method but there are not yet reliable quantitative indicators of quality of fit. Parameters such as pressure, shear stress and tissue oxygen tension may be measured with varying levels of confidence but the scientific basis for interpretation of these measurements is lacking. The protocol relies upon the opinion of the amputee since his opinion is arguably the most relevant and since he can be blinded as to which socket is produced by which method. The prosthetist's evaluation of the sockets is also recorded because these opinions are useful in determining the cause of patient preference and so may be helpful in determining the nature of improvements required to the system.

We chose to have two prosthetists fitting each patient with one responsible for the conventional and one for the experimental method. This protocol was chosen, rather than having the same prosthetist do both fittings, in order to reduce the possibilities of bias and to eliminate the possibility of contamination by knowledge gained during the fitting of one method being applied subsequently to the other.

Since each prosthetist in the pair fits the same number of experimental and control sockets, the evaluation design also becomes balanced with respect to the prosthetist's abilities. However, the conventional technique is not an invariable "gold standard", as the quality of that fit is likely to be influenced by the experience and ability of the prosthetist. Since the possibility exists that a prosthetist who is less experienced at conventional fitting might adapt more quickly or successfully to the CAD/CAM method, the prosthetists were organised into pairs matching, as closely as possible, experience level in terms of years in the field and number of sockets fitted.

The ideal trial would be double-blinded where both the patient and prosthetist would be unaware as to which socket is the control and which is the experimental when performing the comparison. Unfortunately, the prosthetist and other professional observers may easily distinguish between the two sockets on the basis of subtle characteristics that would not be evident to the patient. Consequently, a single-blinded design is the best that can be achieved. To accomplish this, sockets are manufactured from identical materials and one set into identical trial limb hardware. The assignment of subjects to prosthetists, the assignment of techniques to prosthetists and the sequencing of fitting activities are alternated to reduce order effects.

A hard socket with only a single one-ply cotton stump sock and/or a nylon sheath (to reduce friction) has been used throughout our evaluation since this makes the amputee's task simpler. Errors in fit may by masked by thicker socks and flexible sockets. The sockets are transparent to aid in the prosthetist's assessment of changes needed for the next trial.

PILOT STUDY

The Computer Aided Socket Design software (CASD) from the University of British Columbia was selected for our evaluations to date. A pilot study was conducted in 1986 to test the initial protocol and funding was subsequently approved to begin the full scale testing in 1988.

The pilot study involved six prosthetists grouped into three pairs according to experience level. Ten subjects were randomly assigned to the prosthetist pairs. For each subject, the prosthetists were randomly assigned to design a socket using either CASD or the conventional method. The random assignment was intended to reduce the tendency to put more effort into the personally preferred method.

The subject was blind to the method of designing the socket. He walked with each limb in order to determine whether the fit was adequate. He was then asked to indicate the extent of his preference for one socket over the other using a continuous scale. The prosthetists were allowed to ask their subject questions regarding socket fit and were permitted to examine the condition of the residual limb after the subject had walked with each leg. The prosthetists completed a questionnaire with drawings on which problem areas could be marked, and indicated whether they considered the socket to be safe for use at home for one week. The results of this pilot trial were reported by Holden and Fernie (1986).

In the pilot trial, only one attempt was allowed to fit the conventional socket. Three consecutive attempts at fitting the CASD socket were allowed with modifications being made between the attempts. If any CASD socket was preferred by the subject then the trial stopped. The original CASD socket and subsequent modified versions were compared to the conventional socket. The subject was not told whether the subsequent sockets had been modified. Both prosthetists in each pair were present during the fittings of all sockets performed by the pair. Prosthetists purposely showed cooperative interest in both sockets in order to reduce the likelihood that the subject would choose a socket simply because it was perceived to have been built by a particular prosthetist.

A jury comprising one orthopaedic surgeon, one prosthetist, one representative from CAD/CAM industrial technical sales and one senior management consultant to CAD/CAM industries was assembled for a one-day meeting at the end of the pilot study. The results of the pilot trials were presented and discussed and the jurors were given hands-on exposure to the system. Two prosthetists who took part in the pilot trials answered questions privately for the jury. These prosthetists were selected because one appeared to be relatively positive while the other appeared to be relatively negative in opinion.

PROTOCOL CHANGES BASED ON THE PILOT RESULTS

In the pilot trial, three prosthetist pairs were used representing three skill levels. Involving such a large number of prosthetists limited the opportunity for individuals to gain experience with the CAD/CAM technique. In the full study, we have compromised and have two pairs of prosthetists representing two levels of experience.

Our experience has shown that the present state of CASD prosthetic fitting frequently requires several iterations to obtain a successful fitting of an amputee. Three attempts were permitted in the pilot trial but this has now been extended to a maximum of five. This is not to suggest that the system would likely be considered to be commercially practical if five iterations were needed, but is intended to enable a determination of the capacity of the system to achieve an equivalent quality fitting. Of the nine subjects fitted to date in the full study, one subject preferred the CASD socket after the fourth try and one after the fifth. This would seem to indicate that these extra iterations, at least during the learning stages, are useful. If a trend in the number of iterations required is observed, then this may provide data on the learning time required by the prosthetists on the new system and may indicate when maximum productivity has been achieved. It may be found to be acceptable economic practice to perform several iterations routinely, as shapes are more quickly modified with CAD/CAM.

Since the comparison is with a conventional socket, it is important to be certain that the conventional socket is competently fitted. In the pilot study it was found that the conventional socket was not always ideal. Consequently, the present protocol permits the prosthetist to revise the conventional socket if he feels that it could be improved. Modifications have been made to the conventional sockets for seven of the first nine subjects. The revised protocol is summarised in figure 1.

There has been a resistance by both prosthetists and amputees to the decision to use hard sockets with only a one-ply sock. This decision has necessitated clarification on several occasions. We are not advocating hard sockets for all amputees but simply using this method of fitting as an assessment tool.

The pilot test of the jury procedure was valuable and provided useful advice in selecting and structuring the full study. The pilot jury concentrated on comparisons between CAD/CAM and conventional methodologies and on the likely development path of CAD/CAM. The full jury trial will be presented with terms of reference that include an additional and important dimension. New technologies that are alternatives to CAD/CAM will be emphasised since the value of CAD/CAM must be compared with technologies, e.g. new materials that allow sockets to be formed directly on the residual limb, adjustable sockets that may be capable of simulating any desired shape and then acting as the form for production of a socket.

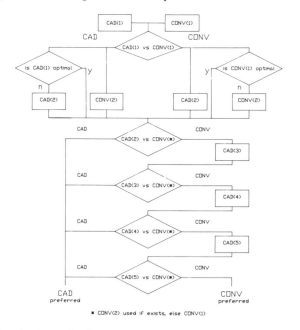

Fig. 1   Modified evaluation protocol.

SUMMARY

A protocol for the evaluation of CAD/CAM technology in prosthetics has been developed and pilot tested. This protocol has been designed to provide an unbiased assessment of the present and potential value of CAD/CAM. Matched pairs of prosthetists alternatively fit using the experimental method and using conventional plaster casting and modification. The comparison is performed by the amputee who is blinded to the method of production of two apparently identical prostheses. The assessment of potential is performed by an independent jury examining the results of the trial and interrogating experts.

REFERENCES

Davies R M (1986). Computer aided socket design - the UCL system. In R M Davies (ed.), *Annual report of the Bioengineering Centre. Roehampton*, pp 9-31, ISSN 0262 8619.

Davies R M, Lawrence R B, Routledge P E and Knox W (1985). The Rapidform process for automated thermoplastic socket production. *Prosthet. Orthot. int.*, **9**, 27-30.

Fernie G R, Halsall A P and Ruder K (1984). Shape sensing as an educational tool for student prosthetists. *Prosthet. Orthot. int.*, **8**, 87-90.

Holden J M and Fernie G R (1986). The results of the pilot phase of a clinical evaluation of computer aided design of trans-tibial prosthesis sockets. *Prosthet. Orthot. int.*, **10**, 142-148.

Foort J, Lawrence R B and Davies R M (1984). Construction methods and materials for external prostheses - present and future. *Int. Rehab. Med*, **6**, 72-78.

Krouskop T A, Muilenberg A L, Dougherty D R and Winingham D J (1987). Computer aided design of a prosthetic socket for an above knee amputee. *J. Rehab. Res. Development*, **24 (2)**, 31-38.

Saunders C G, Foort J, Bannon M, Dean D and Panych L (1985). Computer aided design of prosthetic sockets for below knee amputees. *Prosthet. Orthot. int.*, **9**, 17-22.

Vanderlinden J and Warzee G (1986). Computer-aided design and manufacture applied to modular shoe inserts. *Abstract Fifth Wld Cong. int. Soc. Prosthet. Orthot., Copenhagen,* 311.

36

THE FUTURE IN CAD/CAM

D Jones and J Hughes

INTRODUCTION

Progress in prosthetics has always been achieved with the assistance of advances in many related fields; for example deriving from:

Progress in surgical and clinical techniques

Development of new materials and process for prosthesis manufacture

Improved understanding of the biomechanical characteristics of the situation

Improved attitudes in the health care professions leading to the recognition of new possibilities for rehabilitation and the development of the clinic team concept.

A general desire for greater professionalism and improved standards of education and training.

Some of these factors are recognisable as "hard" technology, some are identifiable as know-how and some are attitudinal changes. All of these are necessary for the successful adoption and implementation of new methods into clinical practice.

Prosthetic CAD/CAM has emerged very rapidly as a field of interest around the world. This emergence has already generated a variety of emotions in the profession; it is viewed with suspicion and even opposition by some, and with excitement and uncontrolled optimism by others. The true potential for prosthetic CAD/CAM can reasonably be expected to follow some middle ground and in the formative years of the field efforts must be made to nurture and actively manage all of the know-how, attitudes and techniques so that any sustainable future is possible.

MANAGING TECHNOLOGICAL TRENDS

Before speculating on future trends in prosthetic CAD/CAM it is useful to identify some of the motivations for the consideration of such technology.

Some, but not all, motivations derive from a recognition of problems with existing methods. This class of problems might include dissatisfaction with the control of quality, speed of supply, or the ability of the contemporary processes to cope effectively with individual patient needs.

These factors could be said to provide a "problem pull"; resulting in an attempt to solve these problems with known and directly applicable solutions. The sum effect is to address the perceived problems until the marginal cost of further improvements outweighs the apparent marginal benefits. Those working within the field compose these solutions by examining their own expertise and knowledge. The danger in depending upon these, essentially introspective, motivations is that a vulnerability to outside events can be established. In this scenario, potentially important developments and innovations in other fields can be overlooked or misunderstood and those improvements which are gained are likely to be in modest increments.

Other motivations derive from an essentially external "technology push" when developments in a totally different field promise to influence the way prosthetics is practiced. Such a technology push ideally requires people with insight into both the problem domain and into how these new technologies could be applied. The danger in this scenario is that changes proposed without a balanced consideration of all the necessary factors can, in the longer term, lead to waste and dissolution. New technology, particularly that which is little understood within a field, can be entrancing and attractive for a while. The honeymoon period can generate significant optimism which is typically impossible to sustain in clinical practice. On the other hand, the attractions of this external technology push is that it can offer radical new solutions which would have been simply impossible in a previous era.

One of the familiar management challenges within Bioengineering is how to, on the one hand, cultivate the new insights made possible by those entering and looking at a field with new eyes (which are not constrained by the beliefs and previously accepted methods of working) whilst on the other hand remain in touch with clinical and economic practicalities; the real problems. This balanced culture is going to be essential for the effective development of the prosthetic CAD/CAM field.

PROBLEM PULL

One reason why CAD/CAM technologies are being considered is that the current methods of creating prostheses do not universally allow the desired control of quality, speed of supply and flexibility of operation. The supporters of a CAD/CAM approach suggest (Jones, 1988; Lord and Jones, 1988) that it should be possible to aim for the advantages of mass production technology whilst retaining the flexibility for custom device manufacture which is one of the strengths of contemporary practice. Such an achievement offers benefits far greater than those made possible by earlier notions of mass production technology.

The application of standardised components for many of the structures of prostheses offered the prospect of greater productivity but the necessity for a customised interface between the stump and socket has limited the overall impact on patient service. The creation of the stump socket interface remains a skills based operation highly dependent for its quality on the poorly understood expertise of the individual prosthetist.

The skills involved in cast taking and rectification are difficult to learn and are in short supply worldwide. Although there are many proposed techniques for cast taking and rectification for various amputation levels, none demonstrate a sound, experimentally supported general theory for the biomechanical "correctness" of the method.

The defence for any technique is frequently that "it seems to work in our hands" or "no patients have ever complained". A problem facing all device development in this field is that we have no reliable benchmarks for evaluation purposes. The fact that different techniques can apparently perform adequately in some situations should not be surprising; there is no universal and quantifiable standard of socket fit. When patients accept a socket as satisfactory it may say more for the natural adaptability and tolerance of the individual, and the nature of their previous prosthetic experiences, than the quality of the prosthetic result.

The factors which serve to provide a pull to generate new technological solutions are therefore those which stem from the nature of the expertise of the prosthetist, the availability and transferability of that expertise and the tools available for rapid production. An introspective view of prosthetics provides no easy solutions to these problems and so it is rational to look for ideas outwith the field.

TECHNOLOGY PUSH

Reseach and development efforts across a broad front could have an impact upon prosthetics in the medium term. The most obvious are those that relate to work in the fields of robotics, artificial intelligence and computer science. Newer trends under the collective label of Desktop Manufacturing are also worthy of note.

Any current CAD/CAM approach to prosthetics must start by performing the equivalent to taking a plaster cast of the stump. Developments in robotics may offer new opportunities for the sensing of shape and other properties of the stump. Current CAD/CAM systems have tended to rely on structured light (Fernie *et al.*, 1985; Jones *et al.*, 1986) to obtain stump shape information but the idea of sensing tissue compliance along with the vision sense is receiving attention in a number of Centres. Such a combined sensor is likely to be difficult to develop.

Having captured stump properties the CAD/CAM approach must allow the visualisation and manipulation of shape. This is to allow the prosthetist to perform the equivalent of stump cast rectification. Advances in computer technology offer the prospect of innovation in the display and manipulation of such shapes. As computer power and graphics performance continues to rise for a given price, the advantages gained in terms of realism and interaction at the human computer interface will have direct application in prosthetic CAD/CAM at the design and manufacturing stages.

Manufacturing aspects of the process have received relatively little attention to date but the interactive simulation of manufacture which becomes more practical with powerful computer systems can become a very useful tool. A particular challenge in prosthetic CAD/CAM derives from the fact that often the manufacturing batch size will be just one unit; perhaps a socket shape for one patient. Clearly any error invalidating a manufacturing cycle will instantly halve the productivity. Whilst small batch sizes are common in precision engineering fields it is normally accepted that several trials will be required to iron out the errors prior to the definitive manufacturing run. This would really be unacceptable in prosthetics and in Glasgow we believe that simulation of manufacture can be a significant aid in reducing the likelihood of errors.

In Glasgow as elsewhere, there is a recognition that the ideas of knowledge-based expert software could have application as a socket design and limb alignment advisor within the CAD/CAM process. A major effort is still needed to overcome the knowledge bottleneck; due to simple lack of knowledge about how expert prosthetists perform the skilled tasks in prosthesis creation.

What can be learned from advances in advanced design and manufacturing technology? Close examination of the practices of Computer Integrated Manufacturing (CIM) show that whilst these ideas are interesting they are focussed on the needs of large factories and large batch size. Flexible Manufacturing Systems (FMS) approaches are interesting because FMS intends to supply special products, customised to individual needs. However, FMS depends on a supply of diverse subcomponents to provide the variety of products and it is not easy to visualise a prosthetic socket constructed in this scenario.

A future concept of Desktop Manufacturing offers a relatively unconstrained generation of a three dimensional object such as socket or even a complete prosthesis. Examples of such systems are already emerging and are worthy of note.

Stereolithography Apparatus by 3D Systems, Inc is an example of stereolithographic printing in which a laser light source solidifies a series of layers of photoinitiated liquid plastic. The design of the object is controlled by a software programme. The object is created by printing successive layers, one on top of the other. The process brings together the technologies of radiation chemistry, computer-aided design, laser light and laser image formation.

Ballistic Particle Manufacturing by Automated Dynamics Corp aims to create three dimensional shapes by directing streams of metals or composite materials at a target surface, building up the shape layer by layer.

Photochemical Machining has been proposed which forms patterns and shapes by photochemically crosslinking a polymer or degrading it, depending on its chemical composition. The material is exposed simultaneously to two intersecting laser beams and the chemical reaction occurs at the intersection of the two beams.

Selective Laser Sintering has been proposed to create a solid object by directing a laser beam at a batch of metal or plastic powder. The controlling software controls the movement of the laser which heats and sinters a thin layer of the powder to create a solid pattern. Successive layers of the powder are similarly sintered until the solid object is built up.

There is a need for new techniques and technologies which may assist in the task to gain understanding of the dynamic characteristics of the stump-socket interface. Improved data gathering through the use of three dimensional diagnostic imaging offers the prospect of improved mapping of the shape and distribution of stump tissues and their behaviour within a prosthetic socket. Advanced tissue modelling can help to provide new knowledge to guide socket design philosophies in the future.

THE FUTURE IN CAD/CAM

Present systems applying CAD/CAM in the prosthetics field fall short of the potential for the concept and many problems remain to be solved. It is important not to take a short term view of the field and neither praise or condemn too strongly the work so far. Current systems can perform no better than the philosophy of their design permits and this will remain limited until greater understanding of the biomechanical facts and the existing expertise of the master practitioner have been gained. So far the developments in CAD/CAM have served to point out the real gaps of knowledge and how little we still know about how prosthetists practice and how sockets should be designed and assessed.

The future success of CAD/CAM in prosthetics and orthotics is not assured because of the application of the new technologies. Multidisciplinary efforts must be made to solve the real problems and these must be allowed to guide the application of appropriate new technology.

REFERENCES

Fernie G R, Griggs G, Bartlett S and Lunau K (1985). Shape sensing for computer aided below-knee prosthetic design and manufacturing techniques. *Prosthet. Orthot. int.*, **8**, 12-16.

Jones D (1988). Impact of advanced manufacturing technology on prosthetic and orthotic practice. *J. Biomed Engng.*, **10**, 179-183.

Jones D, Mackie J C H and Hughes J (1986). An image processing approach to stump shape measurement. *Abstracts 5th Wld. Cong. int. Soc. Prosthet. Orthot*, Copenhagen, 305.

Lord M and Jones D (1988). Issues and themes in computer-aided design for external prosthetics and orthotics. *J. Biomed Engng.*, **10**, 491-498.

37

OPTIMISATION OF ALIGNMENT TECHNIQUES IN LEG

PROSTHESES

S E Solomonidis and W D Spence

INTRODUCTION

The basic requirements for a lower limb prosthesis are function and comfort, structural strength and cosmetic restoration. Various designers have given different orders of priority to these requirements. It has been known for manufacturers to place cosmetic restoration ahead of all other requirements. It is our experience, which is based on extensive surveys carried out at a Limb Fitting Centre, that no amputee will tolerate a prosthesis if it is not comfortable and functional under prolonged use. One would expect however, that a sound design would be based on a compromise of the various factors without unduly neglecting any.
It is evident that the quality of the interface between the amputee and the prosthesis, i.e. the fit of the stump socket, is most important in an artificial limb. Fitting of a socket to an amputee however, to this date, remains an empirical procedure dependent on the skills of the prosthetist. Indeed at present there is much controversy in the USA regarding the best design of socket for above knee amputees with the well established quadrilateral socket being challenged on a number of issues. The main criticism levelled is its apparent failure to provide adequate mediolateral stability (Sabolich, 1985). Until an effective technique which would allow the measurement of direct pressure and shear force at the interface is developed, the subject of socket design is likely to remain subjective and controversial.
Assuming that a satisfactory socket has been obtained and that appropriate components have been selected, the next important requirement, is the provision of a satisfactory alignment of the various parts of the prosthesis. Alignment is also an empirical procedure. During dynamic alignment the prosthetist observes the gait, receives feedback from the amputee and using subjective judgement aims to achieve the most suitable limb geometry for best function and comfort. Alignment may be described as a "biomechanical" operation during which an attempt is made to find optimal load distribution and transfer between the amputee and the prosthesis in order to achieve as normal gait as possible. However, apart from simple diagrams depicting the expected directions of the forces on the amputee's residual limb at various points of the gait cycle (Radcliffe and Foort, 1961; Radcliffe, 1970) there is no objective information in the literature to guide the prosthetist during his deliberations. It is the purpose of this chapter to highlight the parameters that are mostly effected by alignment changes and to present information which will assist the prosthetist in his efforts to determine the "optimum" alignment configurations for a given amputee/prosthesis combination.

MEASUREMENT OF PROSTHETIC ALIGNMENT

During the seventies evaluation programmes of several below knee (BK) and above knee (AK) modular artificial limbs were undertaken in this University on behalf of the Scottish Home and Health Department (Solomonidis, 1975, 1980). The objectives of these programmes were firstly to assess the clinical suitability of the various systems and secondly to produce design information relating to the clinical and constructional requirements of modular prostheses. In order to study the mechanical characteristics of the alignment devices supplied with the systems and assess their suitability with respect to the range of adjustment, measurements of the configuration of the prostheses were made. A prosthetist defines prosthetic alignment subjectively in terms of the approximate flexion and adduction angles of the socket and the relationship of both ischium and greater trochanter and knee to the heel of the shoe when the patient is standing (e.g. Radcliffe and Foort, 1961; Radcliffe, 1955, 1970). Such loose definitions, although they may be adequate for a successful fitting, were far from sufficient for the development of a measurement system for the evaluation process. Before alignment measurements could therefore be made it was necessary to develop a system which would allow "alignment" to be determined as accurately and repeatably as possible. It was also thought desirable that the new definitions would be as close to those used by practicing prosthetists as possible. Special equipment was designed and constructed to facilitate the measurement procedure. Detailed description of the definitions and measuring procedure is given in Zahedi *et al.*, (1986).

The results obtained from the evaluations and from subsequent studies clearly demonstrated that a patient and prosthetist may be satisfied by a range of alignments rather than by a single alignment as was conventionally believed prior to this work. To illustrate this finding figure 1 shows a BK patient who was aligned on 19 different occasions by the same prosthetist using the same prosthesis. In the figure the foot has been maintained fixed to the ground as a reference, and the prosthesis has been drawn to scale to show the position and orientation of the socket axis corresponding to the different alignments. As seen, it is not a single line but a band of lines. These findings prompted the group to carry out a study of alignment in greater depth. The aims of the study were to establish the range of acceptable alignments for below and above knee amputees, to determine the repeatability of achieving "optimum" alignment, to study the variations in the pattern of the loads transmitted by the prosthesis and the contralateral leg and in the gait pattern corresponding to a range of alignments.

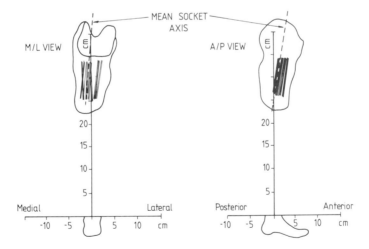

Fig. 1   Socket position and orientation for 19 acceptable alignments by one prosthetist on one below knee patient.

METHODOLOGY

In this study a systematic detailed investigation on below and above knee amputees resulted in the collection of alignment data corresponding to 400 fittings, and over 100 gait analysis tests were carried out during which kinematic and kinetic data of the amputees' locomotion were acquired.

The Patients and Prosthetists

Ten (BK) and fifteen (AK) amputees of mean age 63, SD ± 10.8 and mean activity level 20.8, SD ± 12.3 (Day, 1981) participated in the study. One prosthetist was responsible for casting and fitting all patients with Otto Bock modular limbs. The BK amputees were fitted with patellar-tendon-bearing sockets with supracondylar strap suspension and the AK amputees with quadrilateral suction sockets. Five prosthetists were involved in the alignment of the prostheses.

Data Acquisition Systems

The kinetic and kinematic parameters during locomotion were recorded in the University's Biomechanics Laboratory using the following systems: a six quantity strain gauged shank pylon load transducer and electrogoniometer system for recording prosthetic loading, Kistler force platforms for ground to foot load measurement on the prosthetic and contralateral side, the Strathclyde TV/Computer system for detection of the joint centres of the lower limbs, a three-dimensional cine camera system for measurement of the kinematics of the whole body and foot print patterns for determination of the temporal-distance parameters. The alignment configuration corresponding to each fitting was measured using the custom made equipment described above.

Data Analysis

All the alignment data obtained were processed using various statistical packages in order to determine if any interrelationships among various parameters existed. For the analysis of the biomechanical data use was made of a three-dimensional model which calculated the loading at the ankle, knee and hip joints and the angular and linear displacements of the legs and trunk. Software was developed for signal processing and statistical treatment of the results.

Experimental Procedure

Each patient was dynamically aligned by the prosthetist using conventional clinical routines to determine the best limb configurations. The alignment was measured and recorded and the patient was questioned using a structured questionnaire which discussed comfort and alignment of the prosthesis. The subjective assessment of the prosthetist relating to gait deviations and patient/prosthesis matching was also recorded. Gait analysis testing was then undertaken. This procedure was repeated by the participating prosthetists several times.

In an additional series of tests using six of the AK patients, the alignment of the prosthesis was systematically changed from the clinically "best" configuration to a mal-aligned prosthesis by specific increments in the A/P plane in order to study the biomechanical effects of the changes. The changes included foot plantar/dorsi angles, socket flexion/extension angle and knee forward/backward position.

DISCUSSION OF RESULTS

Alignment Measurement Data

Analysis of the alignment data i.e. relating to the geometrical configuration of the prostheses confirmed the findings made during the evaluation programmes viz. that a prosthetist cannot repeat any given alignment at will, and a number of alignments are acceptable to the patient and prosthetist. Different prosthetists produced different ranges of alignments on any one patient and these ranges varied in the A/P and M/L planes with different prosthetists. The amputee's ability to accept various alignments is undoubtedly related to the degree of control the amputee has over the prosthesis. It was found that the below-knee patient tolerated a greater degree of variability in alignment than the above knee subject, suggesting that he has greater control over his prostheses. From this study the acceptable range of alignments was determined, recommendations for optimum bench alignment settings were made and design criteria for alignment devices for below and above-knee amputees were suggested (Zahedi *et al.*, 1986). One reason for the existence of the variability in the alignment parameters that was found was thought to be interrelationships between certain parameters. It was found that, although certain trends were apparent, in general there was considerable scatter in the results no doubt due to the inability of the prosthetist to accurately adjust the geometry of the limb he was attempting to align. The best relationships found were in the case of the AK amputee, socket flexion versus knee setback (figure 2), and for the BK amputee, foot set-in versus socket M/L tilt.

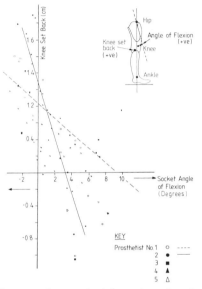

Fig. 2   Socket angle of flexion versus knee set back for various above knee patients and prosthetists.

Figure 2 shows that there is approximately a linear relationship between knee set-back and socket flexion when all the fittings are considered. In this diagram positive knee set-back indicates that the knee is behind the hip ankle (HA) line and is therefore stable. The relationship shows that as the knee is positioned posterior to the HA line socket flexion is reduced and vice-versa. This indicates that the prosthetist has a choice of one of two adjustments for improving knee stability. He could either position the knee posteriorly relative to the HA line or alternatively increase socket flexion, presumably in order to place the hip extensor musculature in a more advantageous position. Although there is scatter in the results it is interesting to note that the gradient of the line of best fit appears to be dependent on the prosthetist rather than on any other variable.

Kinetic and Kinematic Data

The influence of various alignment configurations on the amputee's gait pattern can be quantified by measuring a number of kinematic and kinetic parameters.

The results obtained showed that there is a significant step-to-step variability which must be quantified for each individual subject before comparisons can be undertaken (Zahedi *et al.*, 1987). Due to fluctuations in the stance phase time of successive steps a Fourier analysis technique which "time" normalised the signals before averaging was employed.

The kinetic data of the lower limbs during walking proved to provide the most useful parameters for the selection of the "true" optimum alignment. The ground reaction forces and the loading at shank level for below knee amputees and at the artificial knee for above knee amputees were found to be the most important parameters in the interpretation of the effects of alignment changes on amputee gait. The loading on the prosthetic side was found to be the most affected by alignment changes, whereas the loading pattern on the contralateral side altered in order to compensate for changes in the swing phase duration of the prosthetic side.

The kinematic data of the lower limbs provided the most suitable information for quantifying large variations in limb motion. Differences in the movement patterns caused by changes in alignment, which could be seen by visual observation, were best quantified in terms of joint angles. The temporal-distance parameters of gait provide useful additional information and facilitate the determination of the optimum alignment from several acceptable alignments. On their own however, the temporal-distance parameters do not allow firm conclusions to be drawn.

The Effect of Misalignment

Since the effects of various acceptable alignments on gait were, in some cases, too small to produce significant changes, an extensive series of tests involving deliberate but controlled misalignment of the prosthesis was undertaken. The most important observations made in these tests may be summarised in the following:

Kinematic Parameters

Trunk Movements

Both rotational and translational movements were considered. It was found that each individual patient had his own characteristic movement patterns of the trunk which were not greatly affected by alignment changes.

Kinematics of the Prosthetic Limb

Alignment changes caused compensatory changes in the angular displacement of the prosthetic thigh throughout the gait cycle as shown in figure 3. As the socket flexion/extension angle was altered the amputee's hip joint angle characteristic curve was changed so that the foot would always land on the ground in approximately the same position relative to the body.

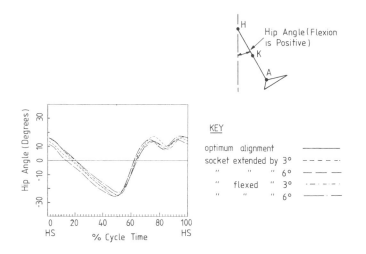

Fig. 3   Effect of changing inclination of socket on hip kinematics on the prosthetic side of an above knee amputee.

Kinematics of the Contralateral Limb

No marked changes were observed in the sagittal plane angular displacements of the sound limb when alignment of the prosthesis was changed.

Ground Reaction Forces

It was found that the ground reaction forces, especially the fore and aft shear force and the progression of the centre of pressure were very sensitive to alignment changes.

Kinetics of the Lower Limbs

Prosthetic Side

Figures 4, 5 and 6 are typical graphs showing the effects of alignment changes on kinetic parameters.
Figure 4 shows the effect of changing the foot dorsi/plantar angle on the anterioposterior (A/P) ankle moment pattern. Although the peak moment value did not appreciably change there were obvious changes in the duration of the plantar and dorsi flexion moments and the transition between them. The effect is similar to that obtained when an amputee is walking up or down a ramp.
Figure 5 shows the effect of changing the position of the knee joint centre on the A/P knee moment pattern. In this figure the increased stability or instability resulting from shifting the prosthetic knee joint backwards or forwards respectively has been quantified.

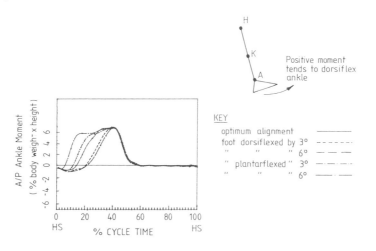

Fig. 4   Effect of changing inclination of foot on the anterioposterior ankle moment pattern on the prosthetic side of an above knee amputee.

Figure 6 shows the effect of changing the foot dorsi/plantar flexion angle on the stability of the knee joint. Although this effect has at times been disputed, it is clearly demonstrated that increasing the dorsi-flexion angle of the foot decreases the stability of the knee joint (and vice-versa) during early stance without appreciably changing the maximum value. It is of interest to note that load monitoring systems employing digital electronics that have been suggested by some investigators that scan the signal in order to display its maximum value would not detect knee instability during early stance.

Contralateral side

Although some noticeable effects of alignment changes on the ground reaction force
and the joint moments at the sound side were observed, no simple interpretation on
the underlying biomechanics could be made. Further detailed study is required.

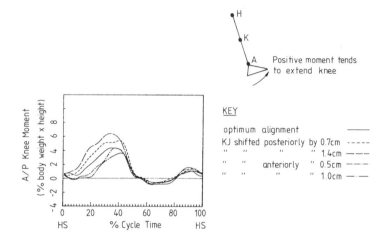

Fig. 5   Effect of changing position of knee joint centre on the anteriorposterior knee moment
pattern on the prosthetic side of an above knee amputee.

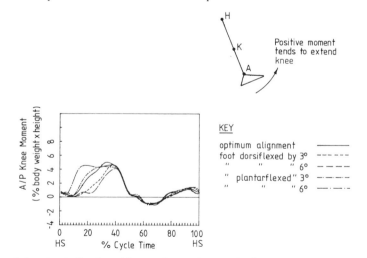

Fig. 6   Effect of changing inclination of foot on the anteriorposterior knee moment pattern on the
prosthetic side of an above knee amputee.

CONCLUSIONS

It is evident that although a prosthetist may be able to satisfy the patient with several
acceptable alignment configurations, he is generally unable to achieve the "optimum"
alignment. Thus, there is a need for a method for assisting the prosthetist in obtaining the
true "optimum" alignment and in selecting the most appropriate limb components. Due to
the large number of variables involved, several kinetic and kinematic parameters need to
be considered in order to provide sufficient information for this optimisation process. Thus,

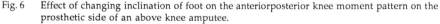

the use of gait analysis facilities in the rehabilitation procedure of the lower limb amputee is visualised.

ACKNOWLEDGEMENTS

The investigators would like to thank the collaborating clinicians Mr K Protheroe, Mr J A Hamilton, Mr J Bingham and Mr G A Whitefield and the staff of the Belvidere Limb Fitting Centre. The dedicated contribution by the patient during the course of this investigation is greatly appreciated. The study was in part funded by the Scottish Home and Health Department.

REFERENCES

Day H J B (1981). The assessment and description of amputee activity. *Prosthet. Orthot. int.*, **5**, 23-29.
Radcliffe C W (1955). Functional considerations in the fitting of above knee prostheses. *Artificial Limbs*, **2** (1), 35-60.
Radcliffe C W (1970). Biomechanics of above-knee prostheses, in G Murdoch (ed.), *Prosthetic and Orthotic Practice*, Edward Arnold, London, pp 191-198.
Radcliffe C W and Foort J (1961). *The patellar tendon bearing below-knee prosthesis*, Berkeley, Biomechanics Laboratory University of California.
Sabolich J (1985). Contoured adducted trochanteric - controlled alignment method (CAT-CAM): introduction and basic principles. *Clin. Prosthet. Orthot.*, **9** (4), 15-26.
Solomonidis S E (1975). Modular Artificial Limbs, First report: "Below-Knee Systems". H.M.S.O., London.
Solomonidis S E (1980). Modular Artificial Limbs, Second report: "Above Knee Systems". S.H.H.D./H.M.S.O. Edinburgh.
Zahedi M S, Spence W D, Solomonidis S E and Paul J P (1986). Alignment of lower limb prostheses. *J. Rehab. Res. Dev.*, **23** (2), 2-19.
Zahedi M S, Spence W D, Solomonidis S E and Paul J P (1987). Repeatability of kinetic and kinematic measurements in gait studies of the lower limb amputee. *Prosthet. Orthot. int.*, **11**, 55-64.

38

CONTROL PHILOSOPHIES FOR LIMB PROSTHESES

D S Childress

INTRODUCTION

An attempt has been made in this paper to articulate some general principles for the design and prescription of artificial limbs. Limb prosthetics is often practiced on the basis of experiences and anecdotal information. The effort here is to develop some general guiding principles that may be helpful in conjunction with experimental and empirical results to assist in the overall development of designs and prescriptions for artificial limbs. Traditional engineering design is a creative endeavour but one that is strongly influenced and directed by principles and theories that enable prediction of design results. Artificial limb design is also a creative endeavour and one that might function more effectively and efficiently if there was more of a theoretical base on which design could be based and upon which creativity could build. Of course, design under stringent constraint conditions (e.g. financial, physical, etc.) - the kind existing in prosthesis design - often results in design compromises that vary from guiding theories and principles. Nevertheless, theories and principles provide references from which designs can proceed.

CONTROL MECHANISMS FOR PROSTHESES

Most, although not all, powered prosthesis fittings for upper-limb amputees use velocity control, either "on-off" or proportional velocity control. Most body-powered prostheses use position control mechanisms in which the position and velocity of the prosthesis part under control is related to a body position and velocity. Forces on the prosthesis part may also be perceived through the control system. Prostheses that relate a body position, velocity and force directly with prosthesis position, velocity, and force seem to perform better than those that do not. The PTB below-knee prosthesis is the classic example of this because knee position and velocity tell the wearer of the position and velocity of the foot, while forces transmitted through the prosthesis to the closely-fitted socket give information about forces. Controls that employ the body's own actuating and sensing systems seem to be incorporated readily and may result in more "subconscious control" than other control schemes. Subconscious control should, in the author's opinion, be a key objective of all prosthesis design.

Simpson (1974) was the first to apply these position control concepts to the control of complex, multi-functional prostheses for high-level upper-limb prostheses. He called the concept "Extended Physiological Proprioception" (EPP), which means control through extension of the body's own physiological control mechanisms into the prosthesis being controlled. He called the position control system he used "unbeatable" because the input and output were directly linked so that a one-to-one relationship always existed (dynamically and statically) between the input and the output. The idea is to use the body's own motor, proprioceptive and sensory system to provide control for the prosthesis. Since the body already uses its own system in subconscious ways the prosthesis control should also be somewhat subconscious because the control is "extended" into the prosthesis.

Simpson's clinical results, although not as widely recognised as they should be, are an existing proof of the validity of this control concept. The effectiveness of PTB prostheses and of body-powered cable-driven upper-limb prostheses also tends to validate the concept. The concept is further validated by the way in which humans incorporate the use of tools (e.g. hammers, racquets, etc.) into their body function in subconscious ways. These tools are extensions of the human body much like a prosthesis is an extension of the amputee's body.

If the human as an "information link" can be considered to be similar to the human as an "information source", then the results of Doubler and Childress (1984a and 1984 b) provide quantitative experimental support of the superiority of the EPP concept when compared with proportional velocity control of the same task. In pursuit tracking studies of band-limited random signals the information transmission rates for EPP position control were roughly twice as high as for the same conditions under velocity control.

SIMPSON'S THEORY OF CONTROL

The idea of D C Simpson has been put into a theoretical construct. It seems to be more than a hypothesis at this stage of development; certainly more than conjecture. It follows, in a form similar to the way it was presented by Childress (1988).

Simpson's Theory: The most natural and most subconscious control of a prosthesis can be achieved through use of the body's own joints as control inputs in which joint position corresponds (always in a one-to-one relationship) to prosthesis position, joint velocity corresponds to prosthesis velocity, and joint force corresponds to prosthesis force.

The kind of system embodied by the theory is shown in figure 1. There are various ways in which this kind of system can be realised, some of which are described in this paper. It should be pointed out that it may be difficult in practical systems to achieve the concept in its entirety.

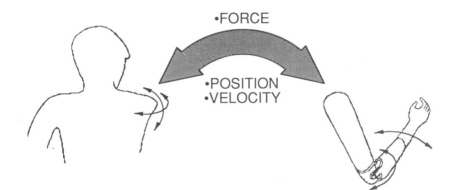

Fig. 1   Diagram showing interaction between human and prosthesis with EPP control of the kind described in the text. The input control channels and the feedback control channels are essentially the same and the scheme results in an intimate human-machine relationship. Feedback is though the control interface that actuates the prosthesis so that the operator gives and receives information in a way that is similar to the interaction found in extension tools used in the human hand (e.g. a hammer).

In our own laboratory, we have developed force-actuated position control systems for electrically powered prosthetic components (Doubler & Childress, 1988) and have begun preliminary clinical testing. Our practical units are only "partially unbeatable" in that, for example, the elbow is unbeatable in flexion but is beatable in extension. This compromise was made in order to simplify the system. Carlson (1985) and Gow *et al.* (1983) also have been developing "unbeatable" position control systems for EPP control of electric powered prostheses.

One of the problems with exploiting EPP control to its fullest advantage is that if it is used as a general control concept in prosthetics, not only for high-level bilateral upper-limb amputees, it may prove to be undesirable to use available body joints as control inputs, even though they may provide excellent control capability. For example, with a partial hand amputation one might not want to use the remaining wrist joint for EPP control of hand prosthesis prehension because that would take away a useful positioner of the hand (i.e. the wrist), a natural subconscious controller of hand position. In like manner, with the below-elbow amputee it may not be desirable to use some proximal joint for control of the prosthesis because that would require harness proximal to the prosthesis and might limit freedom of limb movement on the prosthesis side, as is evident when an amputee uses a cable-actuated, body-powered, below-elbow prosthesis. Therefore, it appears important to have control options other than those requiring joint-position input; options that also might yield subconscious control. Myoelectric control is one method that has worked effectively for control of prehension but it is a velocity control approach that is not considered by the author to be a good control approach for the positioning of artificial joints, even though it may be relatively effective in the control of prehension. One of the reasons myoelectrically controlled below-elbow prostheses are so effective, in the author's opinion, is that in this kind of fitting the human elbow joint controls and senses position and velocity of the electric hand and is aware of forces on the limb. Hence, this fitting is of the EPP type. Of course, with the below-elbow amputee the myoelectric signals for control of the prosthetic hand come from muscles related to hand function and this also tends to lead to somewhat "natural" control. The control of prehension with myoelectric signals appears to be a somewhat generalised notion since prehension is related to general contraction of muscles throughout the body.

An alternate control input option for subconscious control of prosthetic joint position, where it is not convenient to use residual joint position inputs, is to use mechanical inputs directly from muscles. This control of artificial joint position could be achieved through "tunnel cineplasties" in which muscles are attached by direct mechanical link to "unbeatable" control systems of the kind discussed in this chapter. It appears that this kind of human-machine interface may also be able to achieve subconscious control. This is yet to be proved, although evidence from current users of prostheses operated through tunnel cineplasties indicates that this is an excellent control approach.

TUNNEL CINEPLASTY CONTROL

Tunnel cineplasty is a surgical approach that permits muscle forces to be brought outside the body. It was brought to clinical application early in this century by Sauerbruch, a well-known German surgeon. Taylor (1968) presents a thorough description of this approach. It has fallen out of favour since the 1950's, partially because the force and excursion of many muscles could not be easily matched to the mechanical prostheses available and partially

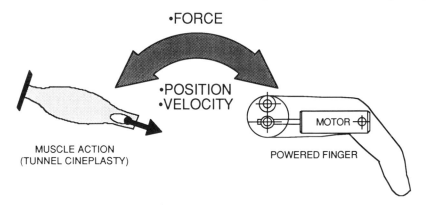

Fig. 2   Diagram showing interaction between a cineplasty muscle and an electric actuated prosthesis component. In this case the component shown is an electric-powered prosthetic finger that is linked to the human muscle through an EPP control element. It is hoped that human-machine control can be achieved in this way that is similar to that achieved with EPP when the input comes from human joints. Again, as described the caption of figure 1, the sending and receiving pathways are coincident.

for other reasons that will not be discussed here. Direct muscle action (concentric and eccentric contraction) through the use of tunnel cineplasties, perhaps miniature tunnel cineplasties, could be used to control powered prostheses in situations where because of insufficient strength and excursion they would not have been successful in the past. Through force-actuated position servomechanisms to control powered prosthetic components, it is possible to operate a powered prosthesis with almost any force/excursion characteristic; hence, it seems the tunnel cineplasty control approach needs to be re-examined in light of new developments of powered limbs and electronic control mechanisms. It is likely that such control may also produce sufficient integration between the user and prosthesis so that subconscious control is possible with this scheme. This leads to the following conjecture:

> Direct Muscle Action Control Conjecture: A natural and subconscious control of a prosthesis can be achieved through the use of the body's own muscles as direct control inputs to position controls in which muscle position corresponds (always in a one-to-one relationship) to prosthesis position, muscle velocity corresponds to prosthesis velocity, and muscle force corresponds to prosthesis force.

The use of "cineplasty" muscle as and input to a EPP control system to control a prosthesis is illustrated in figure 2. The mechanism shown is a "powered finger" but could be any appropriate device.

FEEDBACK THROUGH THE CONTROL INTERFACE

Much has been made of the fact that many control schemes for powered prostheses do not incorporate feedback of information to the user except through vision and through "incidental feedback". Consequently, a number of research efforts have examined the use of supplemental sensory feedback (SSF). So far, the efforts to supplement feedback information have not been very successful. One of the strong points of EPP control and possibly direct muscle control through "unbeatable" control systems is that motor and sensory pathways are both inherent in the control scheme. Childress (1980), in a paper that classifies various sensory feedback approaches and reviews feedback modalities, has called this "control interface feedback". That is, the human operator receives information concerning the state of the prosthesis through the same channel through which the prosthesis is controlled. Hence, it is difficult to separate motor and sensory channels because they are integrated through the design of the control interface. This approach, as exemplified in EPP, has simplicity in that separate motor and sensory feedback channels do not need to be constructed. It also seems to be a rather natural way for a human to interact with an assistive device. The concept is illustrated in figure 1.

FORCE-ACTUATED "UNBEATABLE" POSITION CONTROL MECHANISMS

In the "unbeatable" position control system the input is physically tied to the output so that it is impossible for the input to, so to speak, "beat" the output, as it can in a typical position control mechanism where the output position may be compared with the input position electrically and an error signal created to drive the output in the desired direction to reduce the difference between the input and output signals.

Unbeatable systems can be constructed such that they allow a little "play" (position dead zone) between input and output but this is considered undesirable because it results in a small region of decoupling between the input and output positions. Doubler and Childress (1984b) describe a system that eliminates the position dead zone by comparing force in the actuating mechanism with a reference force. When force deviates from the reference force by a certain amount the motor drives the system so as to reduce the difference between the forces. A dead zone exists in the error system so that the absolute difference between the input force and the reference force must be greater than a specified value before corrective action is taken. This results in what is called a "Force-Actuated Position Control System", where force can vary to some extent in the input signal without consequence in the output, but where positional changes in the input always result in positional changes in the output. The reference force could be made a function of other variables.

PERFORMANCE SPECIFICATIONS AND CONTROLLABILITY OF JOINTS FOR PROSTHETIC LIMBS

Our tracking studies (Doubler and Childress, (1984a) have shown that the shoulder joint in protraction-retraction, and in depression-elevation can pursuit track as well as the elbow joint in flexion-extension and as well as the wrist joint in pronation-supination. This indicates that if we could construct an artificial elbow joint and an artificial supination-pronation joint that had dynamic performance characteristics to match the performance of the corresponding human joints that amputees could control them, through the proper control systems, in the pursuit tracking mode as well as humans naturally control the joints. Of course, normal usage may be somewhat different to pursuit tracking. Nevertheless, the results are encouraging to design engineers because they indicate that there could be merit in building high-performance artificial joints because the indications are that they would be controllable by the human if proper control systems, for example EPP systems, are available. Powered elbows and wrists with angular velocities of between three to five times those currently available (3 to 5 rad/s) should be controllable.

It has been demonstrated that humans can control prehensor velocities of three to four radians per second using myoelectric control (Childress and Grahn, 1985). However, good control of a device with these velocities requires a myoelectric controller with proportional output and with essentially no time delay (Childress and Strysik 1986). Likewise, it is expected that control of a high performance elbow or wrist will require proportional control and a control system without significant delays (less than 1.0 ms). The unbeatable position control system described in this chapter has proportional position control and has essentially no delay.

THE PRINCIPLE OF SYNERGY IN PREHENSION

The author, at a conference in Dubrovnik, introduced an approach to powered grasp (Childress, 1972) that involved the use of two motors to achieve high performance in a prehensor while using low power. The concept is based on the observation that the gripping of reasonably solid objects requires low power because when the fingers are moving freely they have velocity at low torque (low power) and when the object is gripped there is torque but no velocity (low power). Consequently, it is possible to build prehensors that have high pinch force and high velocities but which require low power (Childress and Grahn, 1985). Therefore, it is possible to design a prehensor using several small electric motors so that some of the motors produce the high speed necessary for rapid grasping, while others produce the high forces necessary for firm gripping. We call this general principle the "Principle of Synergy". In the "Synergic Prehensor", a powered gripping unit that we have developed, we use one motor to drive one finger rapidly, at low torque, while we use another motor to drive the high-force finger slowly, with high torque. Since the motors act in opposition to each other, and since a stalled motor is turned "off", it has been possible to achieve closing velocities of over 3.0 rad/s and a pinch force of about 120 Newtons at a power of approximately 1.0 watt. This enables about 1200 operations of the prehensor (full open to full close with a pinch force of about 20 Newtons and full open again) using a 80 mAh, nine volt transistor battery.

POWERED FINGERS

The principle of synergy means that it is theoretically possible to design prehension devices with very small motors. Some motors are now available that are small enough to be put inside hand prosthesis fingers. The finger concept is shown in figure 2 in connection with control using direct muscle attachments. Current technology permits each finger to generate between 30 and 35 Newtons in an adult finger. The vector sum of forces from two powered fingers could therefore be of the order of 60 to 70 Newtons. A grasping speed of 3.0 rad/s in an opposing finger or thumb is desirable. Powered fingers, although still experimental, offer the possibility of fitting partial hand amputees and wrist disarticulation amputees with powered mechanisms, while maintaining appropriate limb length.

SUMMARY

Generalising principles for design and control of powered prosthetic limbs have been presented. While they were originally developed with upper-limb prosthetics in mind, particularly electric powered arms, several of the ideas apply equally well to body-powered upper-limb prostheses, and to lower-limb prostheses. It is hoped that these concepts can be the beginning of a theoretical core for development of a Prosthetics Science that will enable theory, as well as empirical results, to be applied to the design and development of prostheses for amputees.

ACKNOWLEDGEMENT

The author wishes to thank the Veterans Administration Rehabilitation Research and Development Service and the National Institute on Disability and Rehabilitation Research for their sustaining support that made this paper possible.

REFERENCES

Carlson L E (1985). Position Control of Powered Prostheses. *Proc. 38th Annual Conf. Engng. in Med. and Biol.*, Chicago, Illinois, 48.

Childress D S (1972). An Approach to Powered Grasp. in Gavrilovic M and Wilson A B Jr (eds), *Advances in External Control of Human Extremities* (Proc. of 4th Intl. Symp., Dubrovnik), Belgrade, Yugoslavia, (1973), 159-167.

Childress D S (1980). Closed-Loop Control in Prosthetic Systems: Historical Perspective. *Ann. Biomed. Engng.*, 8, 293-303.

Childress D S and Grahn E C (1985). Development of a Powered Prehensor. *Proc. 38th Annual Conf. Engng. in Med. and Biol.*, Chicago, Illinois, 50.

Childress D S (1988). Biological Mechanisms as Potential Sources of Feedback and Control in Prostheses: Possible Applications, in G Murdoch (ed.), *Amputation Surgery and Lower Limb Prosthetics*, Blackwell Scientific Publ. Ltd, Edinburgh, pp 197-203.

Childress D S and Strysik J S (1986). Controller for a High-Performance Prehensor, in A Hahn (ed.), *Biomedical Sciences - Vol 22*, Proc of 23rd Annual Rocky Mountain Bioengng. Symp., Columbia, Mo., Instrument Soc. of America, pp 65-67.

Doubler J A and Childress D S (1984a). An Analysis of Extended Physiological Proprioception as a Prosthesis-Control Technique. *J. Rehab. Res. Dev.*, 21:1, BPR 10-39, 5-18.

Doubler J A and Childress D S (1984b). Design and Evaluation of a Prosthesis Control System Based on the Concept of Extended Physiological Proprioception. *J. Rehab. Res. Dev.*, 21:1, BPR 10-39, 19-31.

Gow D J, Dick T D, Draper E R C and Loudon I R (1983). Physiologically appropriate Control of an Electrically Powered Hand Prosthesis. *Abstracts 4th Wld. Cong. int. Soc. Prosthet. Orthot.*, Imperial College, London, Sept. 5-9, 1983.

Simpson D C (1974). The Choice of Control System for the Multi-Movement Prosthesis: Extended Physiological Proprioception (EPP), in P Herberts, R Kadefors, R Magnusson and I Peterson (eds.), *The Control of Upper-Extremity Prostheses and Orthoses*, Springfield, Illinois, C C Thomas, pp 146-150.

Taylor C L (1968). Control Design and Prosthetic Adaptations to Biceps and Pectoral Cineplasty., in P E Klopsteg and P D Wilson (eds.), *Human Limbs and their Substitutes*, Hafner Publ. Co., New York and London, (reprint of 1954 edition) chapter 12, pp 318-358.

39

MECHANICAL AND CONTROL ASPECTS OF ARTIFICIAL

EXTREMITIES

A Morecki

INTRODUCTION

According to statistical data, about 10 to 12 percent of the population in the developed countries are disabled and half of them need the assistance of special devices for substituting or supporting lost functions in manipulation and locomotion activity.

This discussion focuses on the control aspects of the amputee with upper limb deficiencies and of those patients, who after spinal cord injury on the level C5-C6 are unable to perform some functions of the upper limbs. These limb deficiencies can be an amputation of a part or an entire upper limb or paralysis of parts of the limb.

A block diagram of the information exchange between patients, orthosis and prosthesis and the controlled system is given in figure 1.

In this case we shall define the orthosis/prosthesis of an extremity as an anthropomorphic bionic mechanism or one as near as possible to anthropomorphic, substituting some lost functions of the extremity or replacing its partial function, fixed directly on the stump (Morecki *et al.*, 1984). Correct control of the orthosis/prosthesis, requires at least power to operate the orthosis/prosthesis and also a sufficient feedback system (figure 1). Amputation of part of a limb removes part of the body which means that the body becomes physically incomplete and consequently the control systems are useless and appropriate feedback is no longer generated.

In the case of a unilateral amputation it is usual to perform almost all tasks without use of a prosthesis. Some tasks are simply taken over by the normal hand. Tasks requiring two-handed behaviour are sometimes possible to perform with the normal hand together with the stump. In this case new strategies have to be developed and learned. If the prosthesis is fitted it can replace a physical part of the body but never the destroyed control loops.

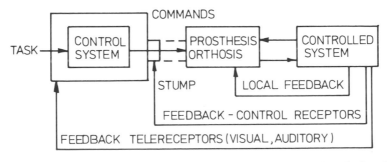

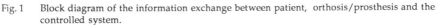

Fig. 1   Block diagram of the information exchange between patient, orthosis/prosthesis and the controlled system.

STRUCTURE AND CONTROL SYSTEM OF ORTHOSIS/PROSTHESIS (GENERAL RULES)

The fitting of a patient with an upper extremity prosthesis, orthosis, implanted stimulator or medical manipulator, should be preceded by a thorough analysis (Morecki *et al.*, 1984). Before a supporting device is applied the following tasks should be completed:

identification of patient's motions and control sites,
determination of the real needs of the patient, adequate to his current possibilities,
determination of the patient's emotional and physical states and his predisposition to work with the apparatus,
consideration of the possibility of the patient resuming his professional life depending on his health, profession, family status, financial situation and the like.

Two examples are given, figure 2a presents a general view of a forearm orthosis with three degrees of freedom, namely flexion and extension in the elbow joint, pronation and supination of the forearm and prehension (opening and closing of the hand). The actuators are of the pneumatic type connected with a set of electropneumatic transducers controlled by EMG signals or microswitches.

Figure 2b shows the orthosis mounted on the wheelchair with a chin control system.

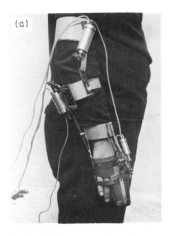

Fig. 2   Upper extremity orthosis with three modes of motion with bioelectric control
2a) Portable orthosis
2b) Orthosis mounted on a wheelchair with chin control.

RESEARCH AND APPLIED PROBLEMS OF PROSTHESIS

The number one problem which is of basic importance, is how to determine the degree of simplification of an engineering model against the archetype and to propose a construction operating on different principles but functionally approximating to the lost part of the extremity.

Experience of the recent years in constructing prostheses of upper extremities indicates that a sensible model of the extremity needs certainly not more than eight degrees of freedom but not less than four degrees.

Problem No. 2 is to choose the type of servomotors and their proper arrangement.

Problem No. 3 involves the choice of method and control system. It is one of the most difficult problems and it is solved by using different principles other than that for living systems (Morecki, 1976).

Problem No. 4 largely associated with the control system, is the choice of suitable sensors and feedback loops. It should be noted that most of the prosthetic systems now used, operate in open loops making use of feedback through sight. Only some systems are equipped with touch (force) and position sensors.

Problem No. 5 concerns the supply system. Electric batteries and compressed or liquid gases (e.g. CO_2) at reasonable cylinder volumes last only for a short time and must often be exchanged or replenished which is in practice very inconvenient to the users.

Problem No. 6 is day-to-day synergy in activities between the patient and the prosthesis, fixed on the stump. Statistics of the practice in countries in which prostheses of upper extremities are in volume production show that only 40% of the patients make full use of the prosthesis.

Figure 3 shows an example of an upper extremity prosthesis with four degrees of freedom with bioelectric control.

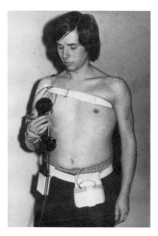

Fig. 3   Upper extremity prosthesis with four degrees of freedom with bioelectric control.

Listed below are some problems which were investigated in connection with the design of efficient devices with several degrees of freedom (Morecki and Kedzior, 1977; Ober, 1977).

nerve signals as a control source for above-elbow prostheses,
development and control of the Warsaw, Boston and Utah arm,
above elbow prostheses with optimised coordination,
cartesian co-ordinate control of a complete arm prosthesis,
methods of bioelectric system control of multifunctional prostheses and information feedback loops,
multifunctional hand prosthesis with myoelectric pattern control system,
artificial tactile perception with computer processing,
multitime approach to coordinated prosthetic control,
end point control using head orientation,
multi-level control of human extremities,
non-numerical control methods.

From this short review it can be seen how much effort was made in the design of efficient multifunctional prosthetic devices, but at present without full success.

At present the main trends of development of the bioelectric upper extremity prosthesis are determined by the necessity to raise the control reliability and to increase the number of performed functions without tiresome concentration of the invalid's attention on the control process.

In the last year we observed the trends connected with the development of multifunctional prosthesis control systems using microprocessors. Numerous groups have attempted to develop control techniques which employ a real time computational operation, including resolved motion rate control adaptive decision aiding and pattern recognition control.

REHABILITATION MANIPULATORS AND ROBOTS

Apart from the prosthetic/orthotic trends, another trend is developing recently, namely boosting of lost functions of the extremities with the use of appliances of the manipulator type. These are either freestanding appliances or manipulators connected with an electric wheelchair. Through their use the patient is in contact with suitably prepared surroundings. The rehabilitation manipulator we shall define as a bionic mechanism, partially anthropomorphic, designed to perform - in a stated range - the functions of man's upper extremity. One should mention here the significant contribution to the development of rehabilitation manipulators made by the investigation in the field of computer control of manipulators conducted in recent years by Sheridan, Whitney, Lyman, Freedy, Roth, Roessler and others.

At this point we confine ourselves to citing three examples, namely the freestanding manipulator, concept of robotic aids for the disabled and the manipulator connected to the patient's head (Ober, 1977).

The study carried on by the interdisciplinary team from the Mechanical Engineering Design Division of Stanford University and the Palo Alto Veterans Administration Hospital spinal cord injury service formulated the idea of utilisation of robotic aids for disabled persons. The robotic aids approach to environmental control was based upon the development of SMART ROBOTIC ARMS, which can be programmed by the patient to perform virtually any manual task. The manipulator system is shown in figure 4.

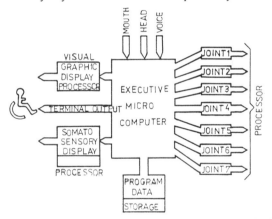

Fig. 4   Proposed multi-processor system for a manipulator for the severely disabled patient.

An example of rehabilitation manipulator designed by J Ober (1977) directly connected to the patient is the manipulator mounted on the head (figure 5). It is most useful in cases of bilateral disarticulation of the shoulder joints. It is an appliance designed for indoor use and on account of its unquestionable merits, such as low weight and a simple control system it is competitive with prostheses and free standing manipulators.

In the last years several projects concerning the applications of robotic aids in rehabilitation engineering were developed. Let us mention some of them:

a microprocessor controlled robotic arm (work-table system with wheelchair chin-controller compatibility).

the Spartacus telethesis, Manipulator Control Studies,

voice command system for motion control,

multi-channel sensory feedback from telemanipulators for the physically handicapped,

a preliminary evaluation of remote medical manipulators,

two wheelchair manipulator systems,

sensor system for automatic grasping and object handling.

This short review shows us the new possibilities in utilisation of robotic aids for substituting of lost upper limb functions.

The last three decades in this very exciting field have brought us quite a lot of data, experience and prototypes, but did we really achieve so much if we consider the patient benefit?

Let us consider why the patient does not like the prosthesis. In the first line because the prosthesis restricts his motion, because of the weight and the discomfort and sometimes because of lack of integration. The patient does not feel that the prosthesis is part of this body. Frequent repairs and bad cosmetic appearance are other reasons. Every patient has different needs and different expectations, different life situations in which he is using the arm prosthesis. In clinical practice all our arm amputees are fitted, or the greatest number of them, are fitted with exactly the same type of prosthesis. The concern relating to rehabilitation manipulators at present is that we have not enough data to make the final decision on their merits.

Fig. 5   Rehabilitation manipulator mounted on the head.

CONCLUSIONS

This short survey shows the ideas, technological possibilities and limitations as well as physical, sociological and economic borders in the design and application of prostheses and robotic aids to the patients with upper limb deficiencies. The analysis is based on experience collected by many interdisciplinary teams with international reputations in the last thirty years.

In our opinion the main problems in the design and application of supporting devices for the handicapped will concentrate on:

relatively simple multifunctional prostheses with microprocessor control,

multifunctional manipulators operated by the patient, with rather simple controls (voice, head, chin, etc.),

rehabilitation systems including a manipulator or manipulators, partly mobile if necessary for very heavy disabilities.

We should also remember that it is not enough to transfer the new technology to the hospitals and clinics, but it is necessary to create conditions for its adaptation and use by the patients. The acceptance by the patient of this kind of supporting device seems to be one of the most important factors in cooperation between medicine and technology.

REFERENCES

Morecki A (1976). *Manipulatory bioniczne* (Bionic Manipulators), PWN, Polish Scientific Publishers, Warsaw.
Morecki A, Ekiel J and Fidelus K (1984). *Cybernetic System of Limb Movements in Man, Animals and Robots*, PWN, Polish Scientific Publishers, Warsaw, Ellis Horwood Ltd, Chichester.

Ober J, Leonhard J and Ogorkiewicz A (1977). On the Head Manipulator for Bilateral Shoulder Disarticulated Children, in A Morecki and K Kedzior (eds.), *Proc. int. Symp. Theory and Practice of Robots and Manipulators,* PWN, Polish Scientific Publishers, Warsaw, Elsevier Scientific Publishing Company, Amsterdam, pp 206-210.

Morecki A and Kedzior K (1977). *Proc. int. Symp. Theory and Practice of Robots and Manipulators,* PWN, Polish Scientific Publishers, Warsaw, Elsevier Scientific Publishing Co., Amsterdam.

40

DEVELOPMENT OF ABOVE-KNEE PROSTHESIS WITH A

FUNCTIONAL KNEE UNIT

A P Kuzhekin, V S Golovin, J S Jackobson and B S Farber

One of the most important problems of normalisation of walking on an above-knee prosthesis is the removal of extra energy expenditures of the whole body and of the intact limb, especially. It is of primary importance for the stance phase of a step during which about 80% of the energy which is required in walking is expended. To normalise the support on the prosthesis two main conditions must be met:

To damp "the heel-strike" by means of knee flexion against spring resistance to an angle of 18^0 (on average) with subsequent extension.

To provide a proper "push-off" from the support during a "toe-off" phase.

Collation of these requirements shows that achievement of the first objective at the current level of technology development is a more realistic task if it can be oriented towards general use in prosthesis construction. Obviously spring-loaded knee flexion during stance phase would decrease asymmetry of the walk and damp impacts at heel-strike. It should be noted that at present minimisation of energy expenditure is considered the dominant criterion of the quality of the walking. Consequently it is of interest of make a preliminary estimate of the influence of a resisted knee flexion device on the energetics of walking. An answer to this question would be convincing proof of the necessity to reproduce this parameter in the prosthetic design.

Based on these arguments a mathematical model of a patient's walking on an above-knee prosthesis has been considered, the prosthesis comprising: a body and two legs with normal inertial characteristics and two weightless feet. One of the legs represents an intact limb and the second one is a prosthetic limb. The intact limb comprises a thigh and a shank with a weightless foot. The prosthetic limb comprises a thigh to the free end of which a console is rigidly attached at the angle. At the end of the console a knee joint of a knee mechanism is located with which a shank socket with a weightless foot is linked.

The length of the prosthetic limb segments has been chosen in such a way that in the vertical position the lengths of both limb thighs are equal and the length of the shank of the prosthetic limb exceeds the length of the shank of the intact limb.

The position of a pivot point of the limbs on the plane is defined by two Cartesian coordinates and the position of the limbs and the body by angular coordinates.

For definiteness the length of a step is fixed and is equal to 0.65 m.

The walk velocity in this case has been chosen as an average one for the patients on above-knee prostheses (0.9 m/s) but the velocity is not uniform. From the movements defined for the hip and ankle joints it is not difficult to calculate the values of the inter-segment angles. Then with the help of the Lagrangian equation of the second order the equations of the motion of the model in question have been formulated from which the values of the resultant forces have been determined. The energy expenditures have been calculated as the summation of the work corresponding to the work done by the joint moments. For this purpose the model had been implemented on a computer. With it the value of the knee angle during the weight-bearing period has been varied. As a result it has been established that spring-resisted knee flexion provides a reduction of energy costs by 13-17% while walking. The results have proved the rationality of a provision of flexion in the design of the above-knee prosthesis.

To accomplish this a number of designs of single-axis units of dual action have been created providing:

Free flexion at angles of up to 130⁰.

Spring-resisted bending in the first half of the weight-bearing period of the step with adjustment of the flexion angle within the limits of 5 to 20⁰, with simultaneous exclusion of the dominant flexion (Jackobson *et al.*, 1987).

The said dual action has been achieved by means of the use of an additional disc (intermediate link) mounted on the axle of the knee joint.

Resistance to bending during the flexure has been provided by an automatic lock controlled by the roll-over of the fore foot. For this purpose the prosthesis has been assembled with the thigh socket in a flexion position of up to about 5⁰. During the final period of rolling over the foot an additional extension has been performed at the knee joint up to 180⁰ due to appropriate resilience of the knee stopper restraining the leg extension, the additional extension being followed by the opening of the knee lock.

To finish bending (to extend the leg completely) a damping spring is provided accumulating a potential energy while bending at the beginning of a stance period of the step. Due to compactness of the design we have succeeded in producing a knee unit of mass 550 g.

The biomechanical studies have shown that while walking on an above-knee prosthesis with an above-mentioned knee unit a goniogram of a natural knee joint has been reproduced completely and not only by the amplitude of the angle of bending but by the temporal distribution of the goniogram components, in particular, by the position of the maximal angle of bending on the time coordinate.

As an example we show a goniogram of the knee unit while an invalid, V-t A. is walking on the given above-knee prosthesis.

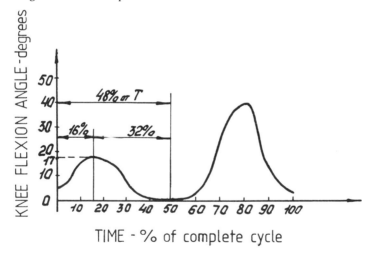

Fig. 1   A goniogram of the knee while walking on an above-knee prosthesis with a knee unit of dual action.

Also known is an above-knee prosthesis with a bouncy knee developed at the clinic of Roehampton (Judge, 1979). Nevertheless this prosthesis has not been widely adopted, as we suppose, because of its sophisticated design. In addition, a disadvantage of the design is the locking of the main flexion under the application of body weight. As a result, before the beginning the flexion of the prosthesis, complete removal of the load from the prosthetic limb must take place and this impedes the natural motor pattern of the gait.

The test of above-knee prostheses with the dual action knee unit in a group of patients have revealed the following:

A normal goniogram of the knee is reproduced that makes rolling over the foot easier and provides smoothness of translational displacement of the body.

Due to the decrease of vertical displacements of the centre of mass of the body the energy expenditure is diminished by 10-12%.

There is no need to apply an extension moment by the thigh stump while weight bearing on the prosthesis and this contributes additionally to the reduction of fatigue during walking.

When stepping with the prosthesis on to the support surface the patient does not feel an impact to the pelvic bones and the spine.

Due to the location of the knee joint axle in front of the line connecting the centres of the hip and ankle joints an unimpeded swing of the prosthesis has been achieved over the support.

An increase of the gait velocity is possible on this prosthesis.

Resistance to flexion is guaranteed when the prosthesis is loaded.

While stepping by the heel on the support surface the whole foot fits closely to it in a more accelerated tempo.

Push-off by the prosthesis from the ground is strengthened to some extent by the additional extension of the knee.

Due to the location of the knee joint axle anteriorly and thus due to the displacement of the centre of mass of the shank and a foot backwards flexion of the leg is accelerated in the swing phase.

The persons being tested have evaluated the prosthesis positively. The simplicity and reliability of the design, its small mass and size guarantee a wide use of the prosthesis in the practice of prosthesis building. At present this design is a prototype one and considered to be a basic model for further development. It has since been improved and advanced.

Concerning the second demand - an improvement of the push-off from support during the toe-off phase - we aim to motorise the prosthesis, i.e. to drive the ankle-joint with an external energy source making the toe-off more normal. This technical solution is also a prospective one but serial production is difficult at present in the prosthetic industry and may be realised in future with further development of a technical level and in particular of power sources.

REFERENCES

Jackobson J S, Kuzhekin A P, Pokatilov A C, Lipovskij V I, Golovin V S, Pauchkov S V, Ivanov B P, Degtjarjov G A, Kiseljov A I and Mazur A P (1987). Protez bedra. *USSR Avtorskoje svidetelstvo No. 1351600.*

Judge G W (1979). Artificial legs. *UK Patent Application GB No 2 014855 A.*

41

DISCUSSION: PROSTHETICS AND ORTHOTICS

In general discussion between Childress, Fernie and McLaurin, it was agreed that CAD/CAM was a method of prosthetic socket production which was still at an early stage and by comparison with industrial design processes, it had a comparatively small amount of man years spent on its development. As a consequence it was not yet at a stage where a realistic comparison could be made between CAD/CAM and other methods of socket construction. *Patel* spoke of his experience with the Ford Motor Company where they have bought in CAD/CAM technologies and have found that the best way of getting data into the system was to build a clay model. Millions of pounds have been spent on development CAD technology but there had to be a time to stop and ask where the technology was leading, and the evaluation of prosthetic CAD/CAM techniques was therefore timely. *Fernie* made the point that it was difficult to decide when to stop and evaluate. He also thought that other construction technology should not be neglected. For instance, an ambition was to spray material on to a stump to form a socket. *Loughran* indicated that there were many CAD/CAM packages in local industry that could be used to develop limb construction systems. He also asked whether any of the developers had stopped to evaluate the shape of the socket in comparison with the shape specified by the data base in the model. *Jones* pointed out the difficulty in milling the mould for socket construction corresponding to the diameter and profile of the tool, the pitch of the helix, the feed rate, the rate of table rotation and tool chatter. He had not however looked at the overall system accuracy.

Murdoch asked Fernie if there were any data from comparisons of sockets manufactured in the conventional way of casts taken and modified by prosthetists. Were the errors the same in both sockets and did this lead to improving the software in some way? *Fernie* responded that software was being improved to eliminate systematic errors, but random errors could only be controlled by tightening up the measuring and forming techniques. *Kirtley* referred to Jones reference to the way that the prosthetist used information about stump compliance and Solomonidis had illustrated the pressure and shear forces acting. It appeared that the whole process of forming a socket is a hand/eye task that embodied the acquisition of medical as well as force data, and the question never arose whether this information was utilised in the CAD/CAM procedures. *Jones* indicated that it was planned to address this problem although he foresaw the difficulty in the building block approach where essentially information is in effect separated from the design information in the way that the prosthetist practises and it was difficult to translate this to the computer for the CAD/CAM process. *Childress* suggested that Jones technique resembled the north-west suspension casting technique which applies pressure to the stump, implying that some shape sensing or shape control was being undertaken in the CAD/CAM procedure. They had been looking at using automated fingers to locally compress the tissue to measure not only the position but the amount of compression that could be obtained under different loading techniques. *Wilson* asked Fernie about the evaluation of the MERU system (University of British Columbia) and whether a diagnostic socket had actually been used to assess the quality of fit of both the CAD socket and the conventional socket. *Fernie* replied that a transparent socket was always used so that the prothetist could obtain the maximum benefit as he progressed from one iteration to the next. Records were kept of the comments of the prosthetist although no quantitative tests had been applied. *Childress* asked Solomonidis whether account had been taken in his alignment investigations of the adjustment that the amputee makes to the alignment of the limb in each case. *Solomonidis* replied that this in fact was what was being measured. The limb itself could not develop load - it was the limb and the amputee using it and indeed it was surprising how quickly the amputees adapted to each different alignment as it was tried. *Kirtley* suggested to Childress that his definition of

Simpson's theory was deficient. There was another aspect that the number of control channels necessary to position the hand in a required position and orientation in space could be reduced by the use of kinematic linkages. *Childress* agreed but said that he did not think that the mechanism should be considered as part of the controller. *Simpson* disagreed. He felt that the total control problem was to position the fingers in a position in space and in orientation, and studies on blind children had convinced him that they related the position of their hands in space according to a polar coordinate system based on the shoulder. Thus the complication of the elbow joint which was only an anatomical method for shortening the shoulder hand distance could be avoided. Certain ergonomists felt that the placing problem was one of tracking the trajectory and ultimate position but studies with blind children had showed that this was not the case. They could manoeuvre articles in the same way as normally sighted people and in any case the normally sighted people did not continuously monitor hand position by eye. *Kralj* agreed with the control principle, but pointed out that the input to the system must be by the function of muscle and using the muscle for control generally interferes with its normal function, and as a consequence there will be a deficit in control sites for, for instance, a multifunction hand. *Kralj* went on to ask Simpson how he would apply his philosophy of control to the lower extremity. *Simpson* had no answer to this but pointed out that they had not had any problems with their patients in finding effective active muscles to control hand position and function without interference with other activities. *Kralj* maintained that to control three coordinates it was necessary to have three control sources. *Simpson* agreed, pointing out that each shoulder offered two control sites - one at the acromium and one at the scapula - so that four control sites were available. *Furnée* referred to a paper by Gibbon reviving the extended physiological proprioception (EPP) philosophy because he was dissatisfied with EMG control. He was therefore surprised that Childress was introducing EMG control since it gave no advantage in proprioception. *Childress* contended that Gibbon's paper was using position control systems which were not unbeatable and so in that sense he was not using EPP. In the future perhaps with tunnel cineplasties a new control site could be developed, to give not only power but also receive proprioceptive signals and this would be an ideal situation.

42

BIOSENSORS - PRINCIPLES AND PRACTICAL PROBLEMS

E A H Hall

THE BIOSENSOR MARKET

Diagnostics as a whole represent a large well established market which is continually expanding. The greatest proportion of this market is directed towards clinical testing with expectations of a European market in the 1990's in excess of 4,000 Million US$. By contrast the entire Biosensor market is very small. Current world estimates stand at 1.5 Billion US$ with an estimated growth rate of around 10%. The growth in Biosensor diagnostics can largely be attributed to the ever increasing need for earlier detection of more and more analytes in widely diversifying applications.

The unique feature of a biosensor is that the device incorporates a biological sensing element; utilising its inherent biospecificity to respond to and transduce the signal due to a target analyte (figure 1). The biorecognition species is immobilised in close proximity or integrated with a suitable transducer to produce a sensor probe capable of reagentless analyte determination.

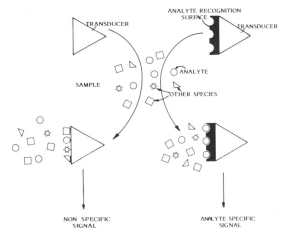

Fig. 1   Schematic representation of a biosensor.

THE ENZYME ELECTRODE

The biosensor was first described in 1962, the term being applied to an 'enzyme electrode', where an oxido-reductase enzyme was retained next to a platinum (Pt) anode in a membrane sandwich. The primary analyte for this electrode was glucose:

$$\text{Glucose} + O_2 \xrightarrow{\text{glucose oxidase}} \text{Gluconic acid} + H_2O_2$$

the Pt anode polarised at + 0.6 V vs SCE responding to the peroxide produced in the enzyme reaction, the technique could be applied to many other oxygen mediated oxidoreductase enzyme/analyte combinations.

The major failings with this system are the dependence of the assay on the presence of oxygen and the potential interference to the electrode signal by other species present in the sample, which are electroactive in the same potential range as peroxide (most notably ascorbic acid). Since the redox enzymes involved in these assays are catalysts of electron transfer, an obvious solution to both the problems cited above, would be to intercept the electron-transfer chain before oxygen and perform the oxidation of the enzyme directly by the electrode itself.

However, electron transfer between electrode and enzyme rarely gives a limiting current and is unsatisfactory as an analytical signal. Various solutions have been sought to solve this paradox and among these the use of low molecular weight redox mediators is frequently favoured to shuttle electrons between electrode and enzyme:

Scheme 1:

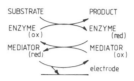

In this scheme, the whole assay procedure is a cyclic process with no net depletion of electroactive species. It can therefore offer significant advantages over the irreversible electrochemistry of peroxide, associated with oxygen mediation.

MEDIATED ELECTRON TRANSFER DEVICES

Many electron acceptors for glucose oxidase have been proposed and to date, those based on the transition metal π-arene complex, ferrocene, remain the most successful. Employment of a mediator in an amperometric biosensor compounds the immobilisation requirement, since not only must the analyte selective enzyme be positioned at the electrode, but the mediator must also be attached.

Technically, the easiest immobilisation procedure is adsorption, and the irreversibility of this technique is increased with the insolubility of the adsorbed species. Multi-step chemical modifications, giving covalent attachment of the enzyme/mediator generally give a more stable configuration but may be deemed less favourable if manufacturing requirements are considered. Ultimately the relevance of a particular immobilisation refinement must be considered with reference to the expected assay environment and the anticipated duration of the analysis. 'One-off' disposable tests are not, for example, concerned with long term operational stability.

IMMOBILISATION IN ELECTROCHEMICALLY DEPOSITED MATRICES

A different line of attack on the problem of co-immobilisation of the mediator redox couple and the enzyme in the vicinity of the electrode surface has been to produce polymeric electrode coatings, which will trap the enzyme within the polymeric matrix and will contain the redox centre as part of the polymer, or retained within the polymer by electrostatic binding.

Of particular interest are polypyrrole and polyaniline since these films can be grown *in situ* by the electrooxidation of the respective monomer (or a derivative) in aqueous solution. The **electrooxidation** of pyrrole for example, is believed to proceed via a reactive π radical cation, which reacts with a neighbouring pyrrole species to produce a chain which is predominantly α, α' coupled. The resulting polymer incorporates anions from the supporting electrolyte and if enzyme is present in the polymerisation solution, it is also trapped in the polymer.

These polymer films exist in both insulating and conducting forms depending on oxidation state and, particularly in the case of polyaniline, pH. While a non-conducting polyaniline film containing the enzyme, glucose oxidase, at pH7 has been prepared the use of a graduated pH gradient for polymer preparation and final entrapment of the enzyme at pH4 gave an enzyme matrix with better electrocatalytic properties.

In fact it is well established that slight modifications in the polymerisation conditions, including the nature of the supporting electrolyte, can drastically alter the characteristics of the resultant polymer. Foulds and Lowe (1986) have optimised the polymerisation conditions required for the entrapment of glucose oxidase in polypyrrole so that the apparent K_M of 31 mmol dm^{-1} (figure 2a) for the enzyme/substrate measured via peroxide oxidation at the electrode, was comparable with the solution enzyme. However, manipulation of the polymerisation recipe to give a less porous film results in a diffusional barrier, which can extend the linear range for the glucose response; this can be of considerable analytical benefit for applications where higher analyte concentrations are required, which would normally be close to saturation levels for the enzyme.

This immobilisation procedure is therefore not only attractive since it allows enzyme to be deposited accurately and easily, directly on the electrode, but it is also possible to manipulate the polymer matrix in favour of diffusional or enzyme kinetics control. The technique can also be extended to give a mediating electrode which circumvents oxygen mediation. Redox mediator derivatives of polypyrrole, e.g. ferrocene amido propyl pyrrole (Foulds and Lowe, 1988) have been co-immobilised with pyrrole in the presence of enzyme to give an enzyme electrode which essentially realises the concept of a reagentless biosensor. Sample glucose concentration in this device can be monitored via the oxidation current, recorded at + 0.3 V vs SCE, due to mediator (see scheme 1 page...) (figure 2).

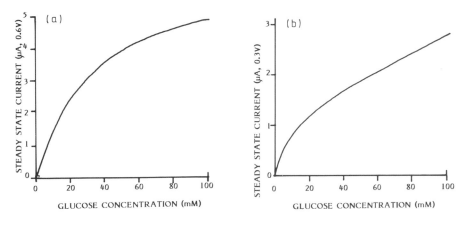

Fig. 2   Current response to varying glucose concentrations at a platinum ink electrode, modified with polypyrrole, containing entrapped enzyme, glucose oxidase.
(a)   assay of glucose via peroxide estimation at + 0.6 V vs SCE.
(b)   assay of glucose via oxidation current due to mediator at 0. 3 V vs SCE.

OPTICAL BIOSENSORS

While *amperometric* biosensors have been the first to be significantly developed, the majority of traditional bioassays which would be suitable for sensor exploitation, are based on photometric methods. The requirement here is a solid phase optical sensor, ultimately where there is no direct contact between sample and light path, causing signal modulation, due to <u>non specific</u> absorption, light scattering, etc. Such an intrinsic optical modulator can be developed using an optical wave guide, where the assay reagents, immobilised at the surface, interact with the light propagating along the waveguide.

Transmission of light along a waveguide (with refractive index n_1) is by internal reflection. When the angle of incidence (θ) exceeds the critical angle there is total internal reflection. Under these conditions the angle of transmission is complex so that the transmitted wave, is an evanescent wave normal to the interface, decaying with a penetration depth (figure 3a),

$$dp = -\frac{\lambda}{2\pi n_1 \left(\sin^2 \theta - n_2^2 / n_1^2\right)^{1/2}}$$

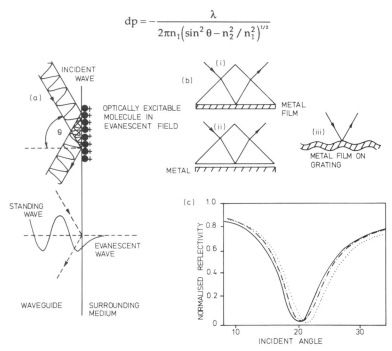

Fig. 3   Optical Transducers

(a)   Propagation of light through a waveguide or optical fibre, by total internal reflection and the development of a standing wave, decaying exponentially into the surrounding medium, with a penetration depth related to the wavelength, and defining the evanescent field.

(b)   Configurations for the excitation of surface plasmon resonance (i) Kretschmann configuration (ii) Otto configuration (iii) sinusoidal metal grating.

(c)   angular shift in resonance excitation for an antibody modified gold metal grating, reacting with antigen.

           (-) gold grating with adsorbed sheep Ig anti-HSA;

           (_ . _) immunocomplex formation with HSA;

           (.....) incubated with second sheep Ig anti-HSA.

In practice measurement would be made at a fixed angle on the steepest part of the curve - e.g. $\theta \sim 15°$ or $\theta \sim 22°$.

INTRINSIC OPTICAL MEASUREMENTS

Optically excitable species immobilised within the evanescent field can interact with the light propagating through the waveguide, thus resulting in a change in the characteristics of the propagated wave. In theory therefore, many of the solution assays that employ optical labels could be converted to an optical biosensor, if both label and biorecognition molecule can react within the evanescent field.

Various enzyme and immunological models have been reported, which rely on this principle. A polystyrene fibre, modified with the pH sensitive dye bromocresol green and with penicillinase adsorbed on the surface, responded to penicillin due to the enzyme catalysed pH change:

$$\text{penicillin} \xrightarrow{\text{penicillinase}} \text{penicilloate} + H^+$$

Sunderland *et al.*(1984) and Badley *et al.* (1987) have reported sandwich and competitive immunoassay systems with antibody immobilised to an optical waveguide. In the former case the 'antibody-sensor' was incubated first with sample antigen and then a second labelled antibody, and in the latter case with a mixture of sample antigen and labelled antigen. In both cases the labelled species involved in the immunocomplex within the evanescent field, could be excited by the light propagating through the guide.

The success of these optical label devices depends largely on the efficiency of interaction between label and transmitted wave, and the elimination of non-specific interactions of the label. For example, pH sensitive dye labels must be operated in buffered samples to prevent an ambient pH response, or non-specific adsorption of labelled antibody or antigen at the surface of the guide must be inhibited with blocking agents.

SURFACE PLASMON RESONANCE

The wave treatment can be developed to consider the excitation of the surface plasmon in a thin metal film. This can be achieved by incident light of the same frequency and wave vector in the plane of the interface. Various configurations exist to achieve these resonance conditions (figure 3b), and since the dispersion relation for the surface plasmon wave is defined by the dielectric constants of the metal film and the adjacent phase, the resonance condition is very sensitive to the binding or interaction of molecules at the metal interface.

Experimentally resonance is observed when coupling takes place between the incident light and the electromagnetic surface wave, as a minimum in the intensity of the reflected light. Surface interactions are observed as a shift in the position of the reflectance minimum.

This technique is particularly applicable to the challenge of unlabelled immunoassay. Cullen *et al.* (1988) have investigated immunocomplex formation at gold diffraction gratings. Antibody or antigen was immobilised on the grating by adsorption and the reaction with the immunocomplement followed as a function of angular shift (figure 3c). A plot of angular shift versus log [protein] was linear in the range 20 μg/ml, and the immunocomplex formation could be followed at a fixed angle of incidence, near the reflectance minimum, by monitoring the intensity of the reflected light.

In this device Cullen *et al.* (1988) were able to show that the response due to a non-specific protein was minimal. Nevertheless, immobilisation by adsorption of the biorecognition species may not be widely suitable or sufficiently stable, and a major research effort can now be seen in the modification of the plasmon interface in order to enable covalent attachment of the recognition species (Liley, Hall and Lowe, unpublished results).

This technique shows a departure from more traditional analytical methods and a demonstration of multidisciplinary expertise that is now a characteristic of the expansion in Biosensor research. Many other optical surface techniques are also finding application in the sensor field, with the information and data available in the various techniques often partially duplicated. The final choice of a transduction system becomes largely dependent on its suitability in a <u>particular application,</u> so that the parallel development of many different general of biosensor devices will continue to provide the best approach to the propagation of these diagnostic devices.

REFERENCES

Bradley R A, Drake R A L, Shanks I A, Smith A M and Stephenson P R (1987). Optical biosensors for immunoassay: the fluorescent capillary device. *Phil. Trans. Roy. Soc. Lond.*, **B 316**, 143-160.

Cullen D C, Brown R G W and Lowe C R (1988). Detection of immunocomplex formation via surface
        plasmon resonance on gold-coated diffraction gratings. *Biosensors,* **3** (4), 211-227.
Foulds N C and Lowe C R (1986). Enzyme entrapment in electrically conducting polymers. *J. Chem.
        Soc. Faraday Trans.* , **82**, 1259-1264.
Foulds N C and Lowe C R (1988). Enzyme entrapment in electrically conducting polymers:
        Immobilisation of glucose oxidase in ferrocene modified pyrrole polymers and construction
        of a reagentless glucose sensor. *Anal. Chem.,* in press.
Sunderland R, Danne C, Place J F and Ringose A S (1984). Optical detection of antibody - antigen
        reactions at a glass-liquid interface. *Clin. Chem.,* **30** (9), 1533-1538.

43

RECENT DEVELOPMENTS IN THE CLINICAL APPLICATION OF

LASERS

J H Evans, H S Gouw and W H Reid

INTRODUCTION

It is interesting to recall that at the time of the birth of the Bioengineering Unit, some twenty five years ago, the laser was a new and exciting tool looking for an application. The search has certainly been rewarded as with so many applications it is difficult to know what we would do without it today. However, we may forget that some of the earliest applications were in the field of surgery and medicine. Some of the results were very good, as in opthalmology, while others were bad or even disastrous. Partly as a result of the adverse publicity attaching to the very bad results and partly due to the highly empirical approach then being adopted, lasers were not readily accepted as therapeutic or medical tools.

However, in the last decade, with the development of many new lasers and a better understanding of the effect of intense light on body tissues, there has been a revival of clinical interest and a major commitment is being made on an international scale.

A new era is dawning as we are now able to predict and encourage tissue reactions in the knowledge that equipment can be made available to develop practical clinical techniques.

PREDICTING THE EFFECTS OF LASER RADIATION ON TISSUE

A laser can produce an intense beam of monochromatic light either continuously or in short pulses. In most clinical applications there are two principal factors which relate to producing the desired reaction in the 'target' tissue while inducing a minimum of disturbance to neighbouring tissues. There are the 'wavelength' and the 'pulse length'.

Pulse Length

At an appropriate wavelength the nature of the reaction will generally be determined by the intensity of the power delivered.

There are basically four different reactions which need to be considered; electromechanical, ablative, thermal and chemical. When the empirically observed reaction zones are mapped on the plane bounded by power intensity and interaction time they appear moderately discrete (Boulnois, 1986; Gouw, 1987). They lie in a band between the constant energy fluences 1 to 10^3 Jcm^{-2}. Short interaction times with correspondingly high peak power densities, produce electromechanical effects; whereas photochemical reactions may require very long exposure. Photothermal reactions are primarily confined within the range 1 ms to 1s although thermal 'side effects' are commonly observed at both longer and shorter pulse lengths.

Wavelength

Many of the above reactions are not tissue specific and thus it is necessary to selectively interact with the 'target' tissue by contrasting it with its surrounding. Each tissue or tissue component has a natural colour which is a reflection of its absorption spectrum. For example blood, which is red, absorbs maximally at a wavelength of 418 nm, which is blue, and has a lesser absorption peak at 577 nm. If a tissue's natural contrast with its surrounding is inadequate then it may be possible to colour it artificially with an added pigment or dye (chromophore).

It is now possible to generate laser energy at many wavelengths in the ultraviolet and infrared as well as across the entire visible spectrum. The problem is to identify the optimal wavelength and to engineer the delivery system for clinical use. Fibre optics are frequently used to guide the light but they have their limitations, particularly in transmitting long wavelengths and high power transients.

Mathematical Simulation

In most applications the target material is not exposed on the surface of the body and thus laser light will have to penetrate overlying tissues in order to reach the target. On entering a tissue and in passing through it, light will be subjected to reflection, refraction, scattering and, eventually, absorption. A good example of an organ commonly treated is skin. This is a multicomponent tissue with very different scattering and absorption coefficients in the overlying epidermis, the dermis and the blood.

Although analytical solutions to the problem of light distribution exist for simpler models it is preferable to use a numerical model. A statistical model, based on the Monte Carlo method, is being used for organs as complex as the skin. This treats the incident light as a large number of small packages, photons, and summates their histories from incidence to absorption. Each history is simulated using pseudo-random numbers to determine those aspects of a photon's behaviour which are governed by probabilities - for example, the scatter path length.

The sites of absorption determine the initial energy distribution and thereafter, if a photothermal reaction is involved, thermal diffusion and convection will redistribute the energy.

RECENT DEVELOPMENTS IN CLINICAL APPLICATION

There is such a wide range of potential and proven applications that it will be useful to illustrate the scope and utility of this relatively new technology.

It is possible to change the focus of the eye by machining the cornea with a laser. The cornea can be recontoured to within a micrometre using an excimer laser emitting in the ultraviolet. The tissue is subjected to photoablation and a negligible amount of heat is generated. A complete, painless reshaping, the equivalent of a permanent contact lens, can be accomplished under computer-control in just a few seconds.

Equally kidney stones can be safely shattered *in situ* and it is possible to remove the plaque which could block coronary blood vessels.

Lasers are now well established as a safe and efficient means of stopping internal bleeding, particularly in the stomach, and they are now finding an application in 'welding' or joining blood vessels and other tissues.

Owing to the specific photochemical reactions that can be elicited, lasers and light delivery systems are now being developed to destroy malignant tumours in organs such as the lungs, liver and brain. The principal technique being utilised is that of 'photodynamic therapy' in which the aberrant cells are labelled using an added reactive chromophore and subsequently irradiated at the appropriate wavelength.

Two further examples are applications related to the skin and will serve to illustrate the utility of the modelling technique in optimising the laser-output characteristics for very different applications in a complex, multicomponent tissue.

Applications to Skin

Using mathematical simulation the distribution of absorbed energy in skin can be computed for any chosen wavelength of incident light. This distribution, which is multidimensional, can best be illustrated as a plot of energy intensity over a two dimensional section of the tissue.

The skin model is ascribed pertinent dimensions and the laser beam is considered normally-incident on the outer, epidermal surface of the skin. The resulting energy distribution is depicted as a profile where the height corresponds to the local energy intensity (figure 1).

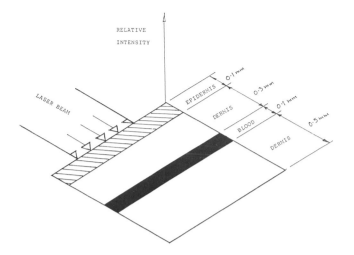

Fig. 1   A plane model of skin incorporating typical values for epidermal thickness and depth and size of peripheral vasculature. The incident light beam is assumed to be both normal and axisymmetric. The magnitude of the local relative density of absorbed energy is displayed on the axis normal to the plane.

Two particular applications illustrate the predictive utility of the model. These relate to the removal of Port Wine Stains and tattoos.

Removal of Port Wine Stain

The Port Wine Stain is of the class haemangioma and comprises aberrant, dilated capillaries in the dermis which give rise to the wine red appearance of patches commonly found on the head and neck. The desired objective is to destroy the aberrant vessels and to reduce the potential for recanalisation by inducing an appropriate degree of perivascular connective tissue damage and contraction (McLeod, 1984; Smart, 1986).

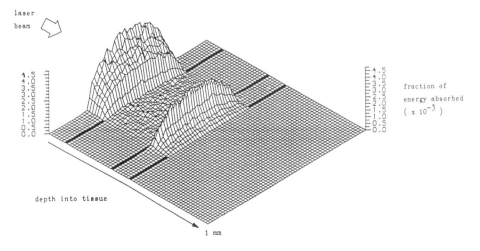

Fig. 2   The relative absorbed energy distribution in skin for incident radiation at a wavelength of 577 nm. Note the relatively high absorption in the vascular bed.

The blood vessel walls are not optically different from their surroundings and thus cannot be targetted directly. The choice is either to label the vessel wall and perivascular tissue with an added chromophore or to utilise the blood as the optical target. Blood has characteristic absorption peaks at 418 nm and 577 nm and it is relatively most absorbent at 418 nm. However, due to the scattering of light in the dermis it transpires that 577 nm is the theoretical wavelength of preference (figure 2).

Clinical investigations at this wavelength suggest an optimum pulse length of the order of 100 µs and an energy fluence of up to 10 Jcm^{-2} (Sobey, 1986; Gouw, 1987; Tan *et al.*, 1989).

A clinic has been established using a flash-lamp pumped dye laser and microprocessor-controlled beam-scanning system for the treatment of port wine stain haemangiomas at Canniesburn Hospital, Bearsden, Glasgow.

Removal of Tattoos

The carbon-based tattoo has obvious potential for phototherapy as the target material is an added chromophore which is not only optically different from its surrounds, by design, but is also a broad-band absorber (Ritchie, 1982; Vance, 1985).

In order to optimise the contrast between the tattoo pigment and the surrounding and overlying tissue a wavelength is chosen to correspond with the minima in the absorption spectra of the naturally occurring chromophores. The effective 'windows' in skin occur at around 700 nm and in the near infrared. The most obvious lasers to use are the ruby, at 694 nm (figure 3), and the neodymium: YAG , in the near infrared. Modelling suggests that in more highly pigmented skin the Nd:YAG has a theoretical advantage but in pale skin the advantage is marginal.

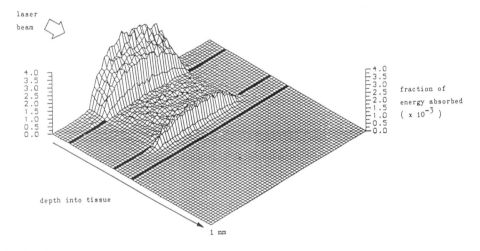

Fig. 3   The relative absorbed energy distribution in skin for incident radiation at a wavelength of 694 nm. Note the relatively low absorption of energy in the vascular bed and good depth penetration.

A ruby laser clinic has been established for eradicating tattoos. Professional tattoos do respond but amateur tattoos in young adults receive most attention. The reaction underlying this therapy is primarily mechanical and requires short pulses of 30 ns duration repeated at monthly intervals in order to reach deeper carbon granules.

SUMMARY

In its relatively short history the laser has made an immense contribution to our well being. Its contribution in the medical field, in general, is expanding rapidly, in part due to the rational, predictive approach which can now be adopted in contrast to the earlier enthusiastic empiricism.

Laser techniques, as currently applied in the practice of medicine and surgery are quicker, cleaner and more precise than those they replace and are generally far less demanding on both the operator and patient.

ACKNOWLEDGEMENTS

Work in this field at the University of Strathclyde was initiated by Dr M W Ferguson-Pell. The mathematical analysis of laser/tissue interaction owes much to Mr A C Veitch of the Department of Mathematics. One author (HSG) was supported by an ORS studentship from the United Kingdom CVCP and a David Livingstone studentship from the University of Strathclyde. The work was supported by grants from the Scottish Home and Health Department.

REFERENCES

Boulnois J L (1986). Photophysical processes in recent medical laser developments: a review. *Lasers med. Sci.*, **1** (1), 47-66.

Gouw H S (1987). Modelling of photothermal interactions in biological tissue with application to laser therapy. Ph.D. Thesis, University of Strathclyde, Glasgow.

McLeod P J (1984). Selective absorption in the laser treatment of tattoos and port wine haemangiomas. Ph.D. Thesis, University of Strathclyde, Glasgow.

Ritchie A (1982). The use of a Q-switched ruby laser to treat blue/black tattoos: an *in vitro* and clinical trial. Ph.D. Thesis, University of Strathclyde, Glasgow.

Smart J (1986). The evaluation of port wine stain haemangiomas before and after treatment by pulsed dye laser. Ph.D. Thesis, University of Strathclyde, Glasgow.

Sobey M S (1986). Flashlamp pumped tunable dye laser treatment of port wine stains. Ph.D. thesis, University of Strathclyde, Glasgow.

Tan O T, Sherwood K and Gilchrest B (1989). Treatment of children with port wine stains using the flashlamp-pulsed tunable dye laser. *New Engl. J. Med.*, **320** (7), 416-421.

Vance C A (1985). An investigation into Q-switched ruby laser treatment of tattoos. M.Sc. Thesis, University of Strathclyde, Glasgow.

44

LASER ANASTOMOSIS AND BONDING OF SMALL

ARTERIES

J Fisher, C A Vance, T J Spyt, S D Gorham, J H Evans, J P Paul and
D J Wheatley

INTRODUCTION

The use of laser energy to thermally denature proteins and produce tissue bonding has applications in both the repair and anastomosis of small vessels. The reported advantages of laser assisted vessel anastomosis over conventional suture techniques include, less foreign body response, less constriction and risk of thrombosis, and greater speed. Several different lasers have been used in studies on a variety of vessels. The carbon dioxide laser has very little penetration in tissue and has been mainly used in animal studies for anastomoses of small arteries having a diameter of less than 1 mm (Quigley et al., 1985; Neblett et al., 1986; Fleming et al., 1988). The argon laser has greater penetration in tissue than the carbon dioxide laser and has been used in the anastomoses of larger diameter vessels with thicker walls. (Gomes et al., 1983; White et al., 1986). The Nd YAG laser has been used for both microvasculature (Jain, 1980) and larger diameter arteries (White et al., 1986) but in the latter study the results were not as good as with the argon laser.

We have investigated the use of the argon laser, to produce tissue bonding and vessel anastomosis in 1 to 2 mm diameter arteries with particular emphasis on its potential application in coronary artery bypass surgery. A novel technique has been used in which a chromophore is applied to the tissue to obtain preferential absorption of the laser energy at the site of anastomosis (Vance et al., 1988). The transmission of the laser energy through the vessel wall, the temperature rises and changes in tissue properties associated with protein denaturation and tissue bonding have been investigated. The mechanical strength of sutureless end to end anastomosis in pigs coronary arteries for different power levels of laser irradiation have been studied *in vitro* in an attempt to optimise bonding conditions. *In vivo*, the laser repair of longitudinal slits in the rat abdominal aorta has been compared to conventional suture repair in 57 rats with survival times up to 10 weeks.

METHODS

Laser and Delivery System

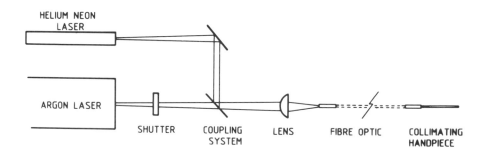

Fig. 1   Diagram of the laser and delivery system.

238

A 5 W argon laser was coupled with a 2 mW helium neon laser to act as a guiding beam and transmitted down a 100 μm diameter optical fibre to a collimating handpiece. A spot size of 1.7 mm diameter was used with a range of powers from 200 to 650 mW. Figure 1 shows the layout of the laser and delivery system.

In Vitro Studies

Tissue and chromophore application:

In vitro studies were carried out using pigs' coronary arteries, nominal diameter 2 mm and wall thickness 400 to 500 μm. Pigs hearts were collected within two hours of sacrifice, vessels dissected from the myocardium and used within 24 hours of collection. Basic fuschin histological dye in alcohol was used as an absorbing chromophore. The chromophore was applied to the tissue using a 100 μm pen nib, which produced a line width of approximately 250 μm on the surface of the tissue. The tissue was surface dried before application of the chromophore and tissue bonding.

Transmission of the laser beam through tissue:

The transmission of the laser beam through the vessel wall was studied using a pinhole and photodiode. Arteries of varying wall thickness were dissected longitudinally and placed on the photodiode. The total laser energy transmitted and profile of the laser beam were measured for varying thickness of tissue. Measurements were made on tissue without chromophore, on tissue with chromophore applied to the cut edges to simulate tissue bonding, and on tissue on which the surface was completely covered with chromophore.

Temperature measurements and identification of a visual end point:

Remote temperature measurements were made using a Heinmann KT infrared thermometer, during irradiation of the tissue. Tissue was irradiated without chromophore at a power of 0.5 W, and with a line of chromophore 250 μm wide applied to the surface at a power of 0.3 W. Temperature measurements were made on both surfaces of the tissue. Photographs of the irradiated surface of the tissue were taken at different temperatures to identify the visual endpoint when tissue denaturation was complete. This corresponded to an increase in intensity of back scattered radiation. This visual endpoint was used as an indication of when to terminate lasing during vessel anastomosis.

Vessel anastomosis:

End to end anastomoses were performed on short lengths of pigs' coronary arteries which had been transversely dissected. Chromophore was applied to the cut edges of the vessel wall, and a single suture used to bring the vessel edges into apposition. Butt alignment of the vessel edges was not achieved in each case and frequently the vessel edges were everted. The quality of such anastomosis was assessed by bursting pressure tests, with the pressure increased at 1.3 kPa (10 mmHg) per second.

In addition to the conventional laser anastomoses, twenty laser anastomosed vessels were reinforced by coagulating a protein solution onto the surface of the vessel using the laser energy. A solution of between 5 and 20 percent bovine serum albumen in phosphate buffered saline, thickened with 5 mg per ml of hyluaronic acid and mixed with 2 μl/ml of chromophore was used. Bursting pressure tests were carried out as described above.

In Vivo Studies

The repair of longitudinal arteriotomies in the rat abdominal aorta using the argon laser were studied in forty two animals and compared to vessel repair by conventional interrupted suture technique in fifteen animals. The rat abdominal aorta has a nominal diameter of 1 to 1.5 mm and wall thickness 100 to 200 μm. In all animals a longitudinal incision 1.5 to 2 mm long was made below the renal artery in a portion of the vessel isolated by microvascular clips. The vessel was washed with saline, dried and chromophore applied to the cut edges. The edges were brought into apposition by applying tension to two stay sutures places beyond the ends of the slit. A 200 mW power level was used for the lasing of these smaller vessels and lasing was terminated when the visual end point was reached. Suture repair was carried out using four or five interrupted 9/0 prolene sutures. Animals were sacrificed acutely and after 1, 3 and 10 weeks. The quality of vessel repair was assessed by macroscopic examination, angiography, bursting pressure tests and histology. In this chapter we report the initial results of these animal studies.

RESULTS

In Vitro Studies

Approximately fifty percent of the laser beam energy was transmitted through 500 µm thick vessel wall, and this was reduced to twenty five percent when a line of chromophore was applied to the cut edges of the tissue. Tissue completely covered with chromophore on the irradiated surface transmitted between five and ten percent of the incident energy. Scattering increased the width of the laser beam by seventy five percent as it passed through the vessel wall, and the peak intensity was reduced to less than thirty percent of the incident value. From measurements of the power needed to produce equivalent increases in temperature with and without chromophore, it was calculated that between fifteen and twenty five percent of the incident energy was absorbed in the normal vessel wall and this was increased between forty and fifty percent when a line of chromophore was applied to the cut edges of the tissue. Application of a line of chromophore to the tissue at the site of anastomosis selectively absorbed a portion of the 1.7 mm diameter laser beam, and allowed the incident power level to be reduced by more than fifty percent. Figure 2 shows examples of increases in temperature in the tissue irradiated with and without chromophore. Tissue denaturation occurred between 65 and 80°C and was indicated by an increase in the intensity of the back scattered or reflected radiation. This could be seen as an intense white spot when viewed through laser safety goggles. This change in appearance of the tissue was used as the end point to terminate lasing.

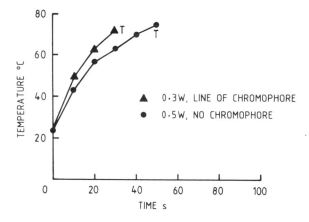

Fig. 2   Increase in temperature in the tissue when irradiated without chromophore at 0.5 W and with chromophore at 0.3 W. T represents the point at which lasing was terminated.

Figure 3 shows the mechanical strength of standard laser anastomoses and the protein reinforced anastomoses. The mean bursting pressure for the standard anastomoses 26.9 kPa (203 mm Hg) was markedly less than the reinforced anastomoses 59.2 kPa (446mmHg). Incident powers of between 0.3 and 0.55 W gave the optimal mechanical strength. At higher powers disruption of the tissue occurred which weakened the anastomoses and at lower powers below 0.3 W excessive drying of the tissue occurred and the denaturation temperature was not consistently achieved.

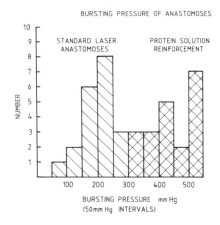

Fig. 3   Bursting pressures of end to end anastomoses *in vitro* for the standard anastomoses and anastomoses reinforced with protein solution.

In Vivo Studies

Forty percent (14 out of 35) of the survival rats with laser repaired vessels died prematurely due to vessel rupture within eight days of operation. There were no failures beyond eight days and there were no failures in animals which underwent suture repair.

Table 1 shows the results of the bursting pressure tests on laser and suture repaired vessels acutely and at one, three and ten weeks. Failures occurred at bursting pressures below 66 kPa (500 mmHg) in laser repaired vessels of animals sacrificed acutely and at one week. The bursting pressures were greater in animals sacrificed at three and ten weeks when the healing processes had increased the strength of the anastomoses. This was consistent with the incidence of vessel rupture causing premature death.

Time Interval	Type of repair	number of rats aortas with a bursting pressure	
		greater than 66 kPa (500 mmHg)	less than 66 kPa (500 mm Hg)
Acute	Laser	1	4
7 Days	Laser	3	1
	Suture	3	0
3 Weeks	Laser	3	0
	Suture	1	0
10 Weeks	Laser	3	0
	Suture	2	0

Table 1   Aorta bursting pressure tests.

Angiography of eight vessels at three and ten weeks showed no constriction of the aorta at the site of repair in either the lased or sutured groups. However, macroscopic observations and histology of the laser repaired vessels at sacrifice showed pseudo aneurysm formation in 29% of the cases. Figure 4 shows a good laser repaired vessel sacrificed at one week and a pseudo aneurysm formed in a vessel sacrificed at three weeks. Histology revealed no pseudo aneurysm in the sutured vessels, but there was some disruption and misalignment of the elastic laminae.

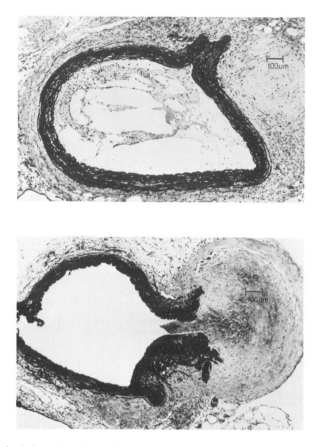

Fig. 4   Histological section of a good laser repaired rat aorta and a laser repaired vessel with a pseudo aneurysm at three weeks, (Scale 100 μm).

DISCUSSION

The use of laser energy to thermally denature proteins and produce tissue bonding and vessel anastomoses has shown encouraging results in some early studies (Fleming *et al.*, 1988), but different laser wavelength, vessel configurations, power densities and spot size have also produced variable results (Vale *et al.*, 1986; Quigley *et al.* 1988). Most studies have been carried out on the microvasculature (vessels less than 1 mm diameter) using a carbon dioxide laser at a wavelength which has very high absorption and little penetration in tissue. In larger diameter arteries with thicker walls such as coronary arteries the radiation from carbon dioxide laser is absorbed at the surface and the use of argon laser with greater penetration is more appropriate. We have used the argon laser along with the application of a chromophore to obtain preferential absorption at the site of tissue bonding. The simple physical studies carried out *in vitro* have provided us with a better understanding of the physical processes which occur during irradiation and bonding of the tissue and has allowed us to optimise the technique before starting animal studies.

     Transmission measurements confirmed that the argon laser was an appropriate choice of wavelength for vessels with approximately 500 μm thick walls, with fifty percent of the energy being transmitted and between fifteen and twenty five percent being absorbed in the vessel wall. Application of a line of chromophore to selectively absorb a small portion of the 1.7 mm diameter beam, increased absorption to approximately forty to fifty percent of the incident beam energy. This allowed a reduction in the incident power level by over fifty percent, and reduced the need for accurate aiming and focusing of the laser beam. As it was not practical to monitor temperature during vessel anastomosis, identification of the change in intensity of the back scattered irradiation following collagen denaturation was an important feature in this study, as it ensured that subsequent anastomoses were performed at the same temperature. This end point has been described as "blanching and contraction of the tissue" by Fleming *et al.*, (1988). Consistent tissue bonding could not be achieved if the collagen denaturation temperature was not reached, but this may have been due to our inability to consistently achieve the same thermal conditions below the collagen denaturation temperature. Therefore, we cannot say conclusively that collagen denaturation temperatures have to be reached in order to produce tissue bonding.

An incident power range of 0.3 to 0.55 W was found to produce optimal mechanical bursting pressure. However, there was considerable variation in the bursting pressure of the anastomoses. These tests were carried out with physiological saline, and many failures were due to small pin holes in the anastomoses which might have been sealed with coagulated blood *in vivo*. The variable mechanical strength was probably caused by variations in technique and alignment of vessel edges. The use of a heat coagulated protein solution in the *in vitro* studies produced a considerable increase in the bursting pressures of anastomoses. Although caution must be exercised when extrapolating from the *in vitro* to *in vivo* situation, such a method may well be of use in strengthening the anastomosis in the acute phase and hence reducing the number of early failures.

The animal model used was not ideal for assessment of this technique as the rat abdominal aorta is only 1 to 1.5 mm in diameter and is a very elastic artery. A longitudinal slit was used as it was not possible to approximate the cut ends of a transected aorta due to the elasticity of the vessel. The main results of this initial animal study were the high number of failures in the laser group within the first eight days and the level of pseudo aneurysm formation found in the survival animals with laser anastomoses. Both of these results may be a consequence of using elastic arteries. Only one other study has reported high failure rates within ten days (Godlewski *et al.*, 1986), but a high incidence of pseudo aneurysm formation has been more widely reported (Quigley *et al.*, 1985; Vale *et al.*, 1986). Pseudo aneurysm occurs when the space between the cut edges of the elastic laminae was filled with connective tissue, and may be caused by initial misalignment of the elastic laminae. The long term healing of the laser repaired vessel was good with no constriction or thrombosis and little foreign body reaction. Nevertheless the short term mechanical integrity of the laser repair and incidence of long term pseudo aneurysm formation requires further investigation in a more muscular, larger diameter artery.

ACKNOWLEDGEMENT

This work was supported by the Medical Research Council and the British Heart Foundation.

REFERENCES

Fleming A F S, Colles M J, Guillianotti R, Brough M D and Bown S G (1988). Laser assisted microvascular anastomosis of arteries and veins: laser tissue welding. *Br. J. Plastic Surg.*, **41**, 378-388.

Godlewski G, Pradal P, Rouy S, Charras A, Dauzat M, Lan O and Lopez F M (1986). Microvascular carotid end to end anastomosis with the argon laser. *Wld. J. Surg.*, **10**, 829-833.

Gomes O M, Macruz R, Armelin E, Ribeiro M P, Brum J M G, Bittencourt D, Verginelli G and Zerbini E J (1983). Vascular anastomosis by argon laser beam. *Texas Heart Inst. J.*, **10**, 145-152.

Jain K K (1980). Sutureless microvascular anastomosis using the Nd Yag Laser. *J. Microsurg.*, **1**, 436-439.

Neblett C R, Morris J R and Thomsen S (1986). Laser assisted microsurgical anastomosis. *Neurosurgery*, **19**, 914-923.

Quigley M R, Bailes J E, Kwaan H C, Cerullo I J, Brown J T, Lastre C and Monma D (1985). Microvascular anastomosis using the milliwatt CO_2 laser. *Lasers Surg. Med.*, **5**, 357-365.

Vance C A, Fisher J, Wheatley D J, Evans J H, Spyt T J, Moseley M and Paul J P (1988). Laser assisted vessel anastomosis of coronary arteries in vitro. *Lasers med. Sci.*, **3**, 219-227.

Vale B H, Frenkel A, Trenka-Benthin S and Matlaga B F (1986). Microsurgical anastomosis of rat carotid arteries with the CO_2 laser. *Plastic and Reconstructive Surgery*, **77**, 759-766.

White R A, Kopchok G, Donayre C, Lyons R, White G, Klein S R, Abergel P and Uitto J (1986). Laser welding of large diameter arteries and veins. *Trans. Am. Soc. artif. intern. Organs.*, **32**, 181-183.

45

UPPER EXTREMITY ASSIST DEVICES FOR CONTROL OF THE

TETRAPLEGIC HAND

P H Peckham, M W Keith, G B Thrope, K C Stroh, J R Buckett, B Smith
and V L Menger

INTRODUCTION

Neuroprostheses for restoration of controlled prehension-release have been in limited outpatient clinical use for ten years. The neuroprosthesis provides control of two types of grasp; palmar prehension (or three jaw chuck pinch) and lateral prehension (or key grip). The subject proportionally controls both the position of his fingers and thumb and the grasp force by a single command control source, generally the position of the shoulder opposite to the limb being stimulated. The clinical system operates as an open loop system, with the human operator using visual feedback to ensure adequate grasp on an object. This system has been supplied to twenty six subjects for outpatient use over the period of 1978 through 1987. This report summarises the selection and implementation process and reviews results of the clinical assessment during that period.

METHODS

Patient Selection

Criteria for involvement in the functional neuromuscular stimulation (FNS) programme are intact C5 or C6 musculature, intact vision, controlled spasticity, stability in a wheelchair, medical stability, positive attitude and motivation with family support, and availability of excitable musculature. Clinical assessment is carried out by a medical and engineering team consisting of the programme director, orthopaedic hand surgeon, rehabilitation engineer, and hand therapist, with consultation as required to other disciplines, primarily urology and neurology. Assessment which is specific to the FNS system consists of sensory evaluation, manual motor test, range of motion, and motor neuron mapping. The former three tests ensure that the extremity has sufficient function in the absence of stimulation.

System Implementation

Candidates were implanted with chronically indwelling percutaneous electrodes, which are protected at the skin interface with a Silastic (R) connector (Peckham *et al.*, 1988). One subject was implanted with a surgically implanted receiver/stimulator (Keith *et al.*, 1988). Utilising a laboratory based system (Thrope *et al.*, 1985) stimulation and control parameters were developed to provide the graded, coordinated movements (Peckham, 1982). These parameters were then programmed into memory of the portable unit (Buckett *et al.*, 1985). This type of system has been utilised in outpatient studies since 1982. Prior to that time, all portable systems were implemented with analog circuitry which provided essentially equivalent function.

RESULTS

Twenty six subjects have participated in the evaluation of this system. Most subjects have evaluated the system as both inpatients and outpatients, but five were outpatients only. Eleven of the subjets began in 1984 or after, which coincides with the expansion of our outpatient evaluation of the FNS system. Half the subjects have completed a high school education, and all but one of the others has some or completed college education. Two subjects are C4/5; eleven C5; five C5/6; eight C6. Six subjects are female; twenty are male. Subjects average age is 25.8 years, and they have been in the programme for 2.4 years beginning 1.4 years after injury.

Most subjects utilised shoulder position to provide the graded command control signal (Buckett, 1985), but some utilised switch inputs to generate a proportional signal through a gated ramp (Peckham, 1988). Lateral prehension/release only was instituted with 8 subjects; palmar prehension/release only with 4; and both grasps with 16. This coincides with the development periods for these grasp modes in our laboratory. All subjects were right handed prior to injury, but the hand system generally was implemented in their stronger arm. The right arm was used with 16 subjects; the left arm with 7; and both arms with 3, although both arms were not simultaneously being used functionally. Only one subject was unable to transfer dominance.

	w/o FNS or WHO	w/WHO	w/FNS
Feeding (finger foods)	1	0	9
Feeding (utensils)	0	9	8
Writing	0	9	9
Drinking	0	0	9
Page Turning	2	9	9

Table 1 Comparison of activities with and without wrist hand orthoses (WHO) and FNS.

Object	w/WHO	w/FNS
Fork	0	9
Pen	0	9
Diskette	0	9
Napkin	9	9
Small Book	0	9
Telephone Receiver	1	9

Table 2 Functional performance in transition tasks involving pickup position and hold for use, and release.

Functional evaluation has included tests of isolated basic tasks, coordinated tasks, and integrated tasks. Aspects of both the Jebsen Hand Test and the Sollerman Test have been incorporated in the evaluation. Basic tasks performed by the subjects included grasp release of utensils, books, writing instruments, telephone, cups, etc. Integrated tasks included pouring, washing, diskette handling, teeth brushing. Some subjects demonstrated the ability to perform advanced tasks such as threading a needle and self-catheterisation. Nine of eleven subjects presently active in the FNS hand evaluation program have completed the clinical evaluation. Subjects were evaluated with no orthosis, with a passive wrist hand orthosis (WHO), and the FNS system. Table 1 shows the results of this test. Note that subjects performed approximately the same tasks with either FNS or the WHO. That is, if a writing instrument were placed in the cuff of the WHO by an assistant, the subject could write. However, with the WHO he required this attendant to place the object. With FNS, he could pick up and hold the object by himself. This is shown in table 2, which shows performance in transition tasks, such as grasping an object, using it for function, and releasing it. This test demonstrates the greatest improvement that was provided by the FNS system. With the exception of grasping a napkin, which all subjects could accomplish, and grasping a telephone receiver, which one subject could accomplish with the WHO, the FNS enabled them to accomplish transition tasks that they were unable to accomplish with the WHO. Since most activities which they wished to accomplish with the hand involve grasp-release and hold for function, this provided a significant improvement in function and independence.

ENHANCEMENT OF PERFORMANCE AND ACCEPTANCE

The clinical acceptance of a neural prosthesis of this type is expected to be related to several factors including reliability, performance, cosmesis, and serviceability. We believe that this can best be accomplished by implanting the greatest number of components and thus freeing the user of tasks of donning and doffing the prosthesis. Furthermore, this will eliminate many of the practical issues which arise in the maintenance of external cables and hardware. However, we also realise that the implantation of a sophisticated implanted system may be a complex surgical procedure that is both time intensive and complex to repair in the event of a subsystem failure. We have thus begun incorporating hand reconstructive procedures which simplify the number of active motors which must be used to provide controlled hand motion. The surgical procedures include tenodeses, tendon transfers of muscles which are paralysed but have peripheral innervation, and joint fusion. Incorporating these procedures enabled the complexity of the hardware/stimulation requirements to be reduced through the use of fewer channels of stimulation or enabling any single channel of stimulation to produce greater function (eg. movement of several digits rather than one digit). The net result of this approach is enhanced function through the more judicious use of the available hardware.

Multicentre trials of the results of functional enhancement in the upper extremity are beginning in four institutions. In 1984, one subject from Edmonton, Alberta, received a system in Cleveland and was supported for her ongoing program by investigators at the University of Alberta. In late 1986 the University of Toronto/Lyndhurst Hospital began the cooperative study with the implantation of one subject. Subsequently in early 1988 transfer of an entire system to Edmonton was completed and a second subject started. The results of these preliminary studies were favourable and indicated the feasibility of a more broad transfer of the FES upper extremity system in these and additional centres. This study is to begin in late 1988.

FUTURE DIRECTIONS

Future directions in the control of the upper extremity provide many opportunities for the hand surgeon, engineer and therapist to collaboratively develop new methods of enhancing function based upon the clinical results which have been demonstrated to date. The most immediate effort must be placed on transferring the present technology and clinical programme to collaborating institutions, so that additional clinical experience can be gained from a breadth of investigators and their evaluative procedures. While the present clinical focus has been on the individual with spinal cord injury, the translation of similar procedures to other patient populations may well provide additional opportunities for persons with stroke, cerebral palsy, and other central nervous system injuries to have improved hand-arm control.

Development must also proceed in the enhancement of these basic systems which provide hand grasp to more proximal joints. For example, the feasibility of controlling elbow extension in a C5 level quadriplegic individual has already been demonstrated. The ability to control fine movement must proceed along the lines of providing better actuation of the muscle, most likely through the use of improved stimulation techniques and internal closed loop feedback, and by providing the user with enhanced sensory information of the grasp force, position of the digits, temperature of the object being grasped, etc. It is uncertain the extent to which we will be able to derive this information from internal sensors, and what must be provided through artificial sensors. In any event, these must be configured and implemented in a form which is clinically acceptable to the user and the surgeon. Finally, it may be possible to derive command control information from more "natural" sources, perhaps through cortical control signals, that will ultimately ease the user's command task. As these research developments are integrated into clinical care, another methodology will be at the disposal of the clinician for improving upper extremity function.

CONCLUSION

The functional performance demonstrated by the subjects provides them with independence of hand function that was not achieved with the wrist hand orthosis. They were able to perform more tasks independently and lost no abilities as a result of implementing the FNS system. We believe these results demonstrate a major advance in the treatment of the high level tetraplegic hand.

ACKNOWLEDGEMENTS

This research was supported in part by the National Institute of Handicapped Research G008300118, the National Institutes of Health N01-NS-3-2344 RO1-NS-1-7955, and the Veterans Administration Rehabilitation Research and Development Service.

REFERENCES

Buckett J R, Braswell S D, Peckham P H, Thrope G B and Keith M W (1985). A Portable Neuromuscular Stimulation System. IEEE/EMBS Proceedings 314-317.

Peckham P H, Keith M W and Freehafer A A (1988). Restoration of Functional Control by Electrical Stimulation in the Upper Extremity of the Quadriplegic Patient. *J. Bone. Jt. Surg.*, **70A** (1), 144-148.

Peckham P H (1982). IFAC Symposium on Control Aspects of Prosthetics and Orthotics, R M Cambell (ed.), Pergamon Press, New York, 29-33.

Thrope G B, Peckham P H and Crago P E (1985). A computer controlled multichannel stimulator for laboratory use in functional electrical stimulation. IEEE Trans on Biomed Engng. **BME-32** (6), 363-370.

Keith M W, Peckham P H, Thrope G B, Stroh K C, Smith B, Buckett J R, Kilgore K L and Jatich J W (1988). Implantable Functional Neuromuscular Stimulation in the Tetraplegic Hand. *J. Hand. Surg.* in press.

46

ELECTRICAL IMPEDANCE IN THE LIMBS OF PATIENTS WITH

MUSCLE DYSTROPHY AND NORMAL SUBJECTS

M Noshiro and T Morimoto

INTRODUCTION

The noninvasive evaluation of the state of muscle dystrophy is indispensable for the care of patients and development of new treatments. The extent of dystrophy has been evaluated subjectively by physicians, from the cross-sectional area of the muscle obtained from X-ray CT and from the activity of creatine phosphokinase in the blood. Although these are useful, we still need a simple and quantitative method using low-cost devices.

Fat and connective tissues which gradually replace muscle fibres in dystrophy may change the electrical impedance in the limbs. There is, thus, a possibility of evaluating the state of muscle dystrophy by the measurement of the impedance. As a first step in examining whether the evaluation is possible or not, we compared the impedance in patients with muscle dystrophy with that in normal subjects.

METHOD

The four-electrode method was used for impedance measurement to reduce the effects of electrode and skin impedances and of current spreading (figure 1). Two current electrodes 100 to 180 mm apart from each other were put on the leg of each subject. Detecting electrodes were placed 40 mm from the current electrodes. All electrodes were made of stainless steel of 5 mm width. A constant current of 150 µA (rms) amplitude and 2 to 200 kHz frequency was sent through the current electrodes and a reference resistor. Preamplifiers with high input impedances amplified the potential differences between two detecting electrodes and across the reference resistor. To minimise the influence of cable capacitances, the cables from the electrodes to the preamplifiers were kept as short as possible and their shields were driven to signal voltages by voltage followers. The outputs from the preamplifiers were fed to a Hewlett-Packard 3575A gain-phase meter which measured the ratio and the phase difference between the detected and reference potentials.

Kanai *et al* (1983) showed the simplified equivalent circuit for the tissues consisted of the intracellular and extracellular resistances (R_i and R_e) and the membrane capacitance (C_m) in the frequency ranging from 1 to 500 kHz (figure 2). The time constant, $T_o = C_m R_i$, is not completely constant but distributed around its mean. Its distribution function can be approximated by the Cole-Cole function. The admittance of the equivalent circuit (Y) is represented by

$$Y = \frac{1}{R_e} + \frac{1}{R_i} - \frac{1}{R_i}\frac{1}{1 + j\omega T_o^{(1-\alpha)}} \tag{1}$$

where α stands for the Cole-Cole factor. Plotting the admittance Y for various frequencies on the complex plane (admittance locus) yields a part of a semicircle.

The parameters in the equivalent circuit were estimated from the measured frequency characteristics of the impedance as follows. In the first place, the admittance per unit volume was calculated for normalisation. Assuming that the leg between the detecting electrodes has the shape of a circular truncated cone, whose shortest and longest radii and height are r, R and L, respectively, then the admittance per unit volume is given by:

$$y = L/\pi RrZ \tag{2}$$

where Z denotes the measured absolute value of the impedance. After thus calculating the admittance per unit volume, we fitted a part of a semicircle to the admittance locus to estimate R_i, R_e and α. The time constant required for C_m estimation was obtained by fitting the frequency characteristics of the conductance computed from R_i, R_e and α to the measured characteristics. The nonlinear least square method was used for fitting.

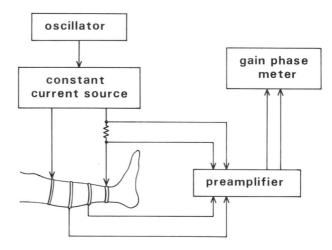

Fig. 1   Diagram of experimental arrangement.

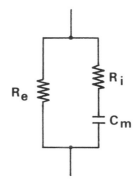

Fig. 2   Equivalent circuit for living tissues in the frequency ranging from 1 - 500kHz.
           R_i : intracellular resistance,
           R_e: extracellular resistance,
           C_m: membrane capacitance.

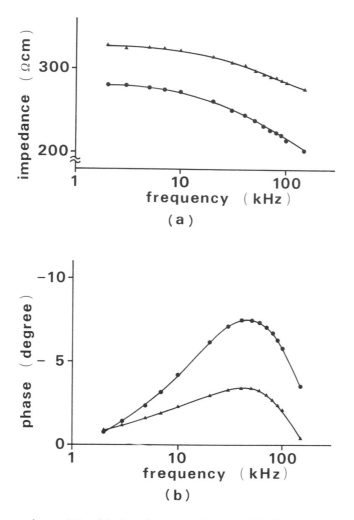

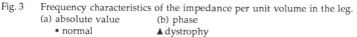

Fig. 3   Frequency characteristics of the impedance per unit volume in the leg.
(a) absolute value   (b) phase
   • normal   ▲ dystrophy

RESULTS

The impedance was measured in 12 patients ranging in age from 3 to 17 years (mean of 10 years) who had Duchenne type dystrophy and in normal subjects ranging in age from 5 to 15 years (mean of 10 years). Figure 3 shows typical frequency characteristics of the absolute value and phase of the impedance in both patients and normal subjects. Figure 4 shows the admittance loci for the same data. The difference in the impedance is clearly seen in these examples. Mean values of the estimated parameters are given in table 1. Statistical testing for the difference between the means indicates that R_i is significantly larger and C_m smaller in the patients (p < 0.01) than in normal subjects.

	$R_i(\Omega cm)^*$	$R_e(\Omega cm)$	$C_m(nF/cm)^*$	α
Normal	765 ± 178	326 ± 56	3.57 ± 1.17	0.158 ± 0.037
Dystrophy	2457 ± 1314	373 ± 142	2.09 ± 0.89	0.169 ± 0.071

Table 1   Values of the estimated parameters in the equivalent circuit; mean ± standard deviation
Asterisks indicate that the means are significantly different ($p < 0.01$).
R_i: intracellar resistance
R_e: extracellar resistance
C_m: membrane capacitance
α: Cole-Cole factor

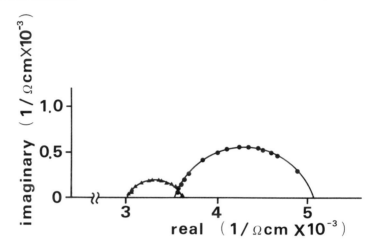

Fig. 4   Admittance loci for the data shown in figure 3.
● normal   △ dystrophy

DISCUSSION

The decrease in the membrane capacitance and the increase in the intracellular resistance produced by dystrophy probably result from destruction of the cell membrane and reduction in the number of cells per unit volume and in the cell size. However, further study is required to specify the cause of the parameter difference.

This study showed that muscle dystrophy altered the parameters in the equivalent circuit for the tissue impedance. We are presently engaged in determining the relationship between the parameters and the state of dystrophy obtained from methods other than impedance measurement.

REFERENCES

Kanai H, Sakamoto K and Haeno M (1983). Electrical measurement of fluid distribution in human legs: estimation of extra- and intra-cellular fluid volume. *J. Microwave Power*, **18**, 234-243.

47

MUSCLE STIMULATION FOR THE PREVENTION OF VENOUS

STASIS DURING SURGERY

W G Kernohan, J G Brown, P E Ward and R A B Mollan

INTRODUCTION

Deep Vein Thrombosis (D.V.T.) is now the commonest complication in the post-operative phase following orthopaedic surgery and is most commonly seen in surgery involving the lower limbs and pelvis. In total hip replacement the D.V.T. rate has been reported between 30% and 50% of all patients, with the further complications of pulmonary embolism at 10% and a mortality of 2% (Salzman and Harris, 1976).

Virchow (1860) was the first to ouline a "triad" of important factors which may lead to D.V.T. formation. Firstly factors causing damage to the vessel wall, secondly changes in the consistency of blood and thirdly venous stasis are indicted. Theurapeutic prophylaxis has been directed at changing the consistency of blood and many regimes have been advocated. Within general surgery Heparin has gained wide acceptance but in joint replacement surgery, increased formation of wound haematoma can lead to infection and ultimate prosthesis failure (Allen *et al.*, 1978). Various physical methods to overcome venous stasis are available. Early mobilisation, leg elevation and graded compression stockings have been advocated in the post-operative phase. To these can be added pneumatic calf compression and calf muscle stimulation in the intra-operative phase. While the patient is anaesthetised venous flow has been shown to be reduced to 50% of normal values (Doran, 1964) and this would appear to be the appropriate time for prophylaxis to commence. In addition, hip surgery has the specific added risk that venous circulation is interrupted by positioning of the leg during operation, further enhancing stasis and thrombus formation (Stamatakis *et al.*, 1977). In any mechanical method of prophylaxis the operative leg must be freely mobile and this immediately ruled out pneumatic compression methods for orthopaedic surgery. As a result we have developed a portable calf stimulator. This has been made into a self-contained unit with the primary aim of overcoming venous stasis at the time of operation.

METHOD

Doran (1964), first introduced the method using a nerve stimulator borrowed from the physiotherapy department. Stimuli were applied using ECG electrodes. To monitor the effect the transit time of radioactive sodium chloride injected at the foot was measured at the groin. He confirmed that stimulation increased venous flow in the legs during anaesthesia. Work that followed confirmed Doran's initial study (Clark and Cotton, 1968; Browse and Negus, 1970) but there were practical problems with the techniques. There was no consistency in the methods proposed, the strength, duration, frequency and site of stimulation being varied. The apparatus providing the stimulus often did not give a reproducible output and in one case shorting of the AC mains stimulator due to diathermy resulted in skin burns. In addition, skin polarisation occurred with repeated stimulation giving decreased effect. The apparatus was obtrusive in theatre and time-consuming to apply. In an attempt to overcome these problems the Powley-Doran stimulator was developed (Powley and Doran, 1973) and became generally available. This was easily applied and self-contained. However the units failed due to water entering the circuitry and short battery life, often resulting in incomplete intra-operative stimulation. Because of the harsh environment and remote electrodes, there were contact problems at the connectors.

Further devices have become available, for example a muscle stimulator unit (Zimmer Inc.) which produce short duration electrical stimulation above and below the knee. Another pneumatic unit (Nova Medic) uses the venous pump of the foot (MacEachern *et al.*, 1986). Both these methods use an external power supply/control unit making them difficult to apply during hip surgery. There is no known published quantitative evaluation carried out on either of these devices in their ability to ameliorate D.V.T.

With the benefit of hindsight it became possible, over the past six years, to develop a portable calf stimulator for more effective use in the operating room.

SYSTEM SPECIFICATION

In the Belfast Calf Stimulator two 45 mm circular stainless steel electrodes, 160 mm apart are attached to a non-conductive plastic case containing the solid state circuitry and a rechargeable battery. From a knowledge of previous work and of physiology it was possible to specify operational parameters. The unit delivers a short pulse of 10 ms duration. This produces a brisk contraction of calf muscle while avoiding a tetanic (cramp-like) prolonged response. The rate of application is 12 stimuli per minute which allows sufficient calf vein filling between pulses. The stimulation voltage is variable from 70-100, depending on the level required to produce plantar flexion at the ankle for a given patient. Polarity reversal occurs to prevent tissue polarisation. Each unit is powered by a small rechargeable battery with a full charge life of 24 hours to prevent intra-operative failure. The case is water proof with the electrodes directly attached. There is a separate mains battery charger which incorporates a unit auto-test facility. In position for use, the upper electrode lies distal to the popliteal fossa with the lower electrode over the initiation of the Achilles tendon. The unit is held in this position encased in an isolating Velcro-fastened holster.

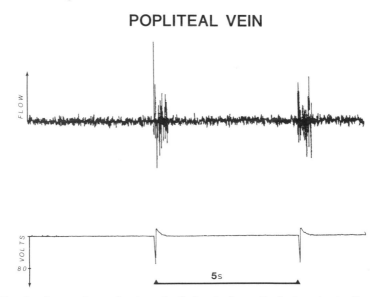

Fig. 1   Doppler ultrasound trace showing pulsatile flow in the popliteal vein and a simultaneous trace of output voltage from the stimulator.

RESULTS

To demonstrate the effectiveness of the unit, flow produced in the popliteal vein has been measured in two cases: a 45 year old arthroscopy patient and a 70 year old undergoing fractured neck of femur fixation. Initially a portable Doppler ultrasound flow meter (Huntleigh

Technology) was used with the patients anaesthetised and the stimulator set to produce brisk plantar flexion at the ankle. Flow measurements were taken in the popliteal vein, figure 1 being an example of the traces produced, with the simultaneous output from the stimulator recorded below. Clearly pulsatile flow is generated in the popliteal vein and, by implication, in the calf vein.

A further study on three cases, (aged 40, 60, 65) admitted for trauma or hip surgery, was performed using a more sophisticated ultrasound technique to measure the actual flow rates produced. Real time ultrasound visualisation was performed, using a Diasonics Duplex machine, and the average velocity was calculated with automatic correction for attitude angle so that valid comparisons could be made between patients. To compare results of calf stimulation with a physiological method the calf was compressed manually. In the anaesthetised patient (control) the baseline velocity was measured. In the stimulated patients and those with manual compression the rates revealed an increase in flow compared to the controls (see table 1). Results also show that there was no diminution of flow rate after one hour of calf stimulation.

Method	Measured velocity (ms^{-1})
Calf stimulation	0.39 - 0.52
Manual compression	0.42 - 0.60
Control	0.10 - 0.16

Table 1 Increase in blood velocity during different prophylaxes

The stimulator has been in successful operation for 18 months. A trial has been carried out with 150 sequential consenting patients undergoing total hip replacement being subjected to calf stimulation and Dextran-70 therapy post-operatively. The D.V.T. rate was confirmed by venography in all cases and found to be 14.2%.

CONCLUSION

Stimulation of the calf muscles causes them to contract. This facilitates the return of blood in the deep veins towards the heart. The Belfast Calf Stimulator is a development of existing techniques which have been enhanced in an attempt to overcome venous stasis in this manner. This limited study has indicated the efficacy of the device. However further research is suggested to prove a consequent reduction in the D.V.T. rate. Physical methods of preventing D.V.T. will be implemented in conjunction with therapeutic regimes such as heparin and its derivatives, dextran and aspirin. Functional electrical muscle stimulation is an important additional adjuct against D.V.T. formation.

REFERENCES

Allen N H, Jenkins J D and Smart C J (1978). Surgical in patients given subcutaneous heparin as prophylaxis against thromboembolism. *Br. med. J.*, **1**, 1326.
Browse N L and Negus D (1970). Prevention of post-operative leg vein thrombosis by electrical muscle stimulation. An evaluation of 125-I labelled fibrinogen. *Br. med. J.*, **3**, 615-618.
Clark C and Cotton L T (1968). Blood-flow in deep veins of the leg. *Br. J. Surg.*, **55**, 211-214.
Doran F S A (1964). A simple way to combat the venous stasis which occurs in the lower limbs during surgical operation. *Br. J. Surg.*, **51**, 436-452.
MacEachern A G, Fox R H, Gardner A M N and Ling R S M (1986). The venous foot pump. *J. Bone. Jt. Surg.*, **68 B**, 667.
Powley J M and Doran F S A (1973). Galvanic stimulation to prevent deep vein thrombosis. *Lancet*, **1**, 406-407.
Salzman E W and Harris W H (1976). Prevention of venous thromboembolism in orthopaedic patients. *J. Bone. Jt. Surg.*, **58A**, 903-913.
Stamatakis J D, Kakkar V V and Sagar S (1977). Femoral vein thrombosis and total hip replacement. *Br. med. J.*, **2**, 223-225.
Virchow R (1860). *Cellular Pathology*: 2nd edition (Trans. F. Chance), Churchill, London.

48

FINITE STATE CONTROL OF HYBRID FES SYSTEM

B J Andrews, C A Kirkwood, R W Barnett, G F Phillips, P. Mowforth and
R H Baxendale

INTRODUCTION

Finite State Control

Our approach to the synthesis of a controller was to decompose the multi-goal and
multivariable control problem into subsystems organised in basically three hierarchical levels
(Andrews, 1988) as illustrated in figure 1. At the top level, the subject directly interacts with
the control system through a command interface comprising manual switches and associated
software. For example, in the present system the patient selects a particular locomotion mode
using a multi-way switch. At the second level each locomotion mode is coordinated with
reference to a finite state model of the process. This model serves to change the control strategy,

Hierarchical Control System

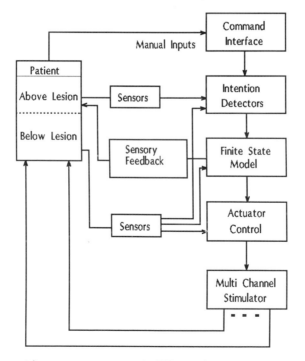

Fig. 1   Block diagram of the main components in the FES control system.

as required, at different stages or phases of the locomotion cycle. Each stage in the gait cycle is represented in the model by a unique state. On entry into a state, an associated predetermined control strategy comprising a set of low level (third level) open or closed loop actuator controllers are enabled. This new state control strategy then supersedes the previously active state control strategy.

It is the lowest level, the actuator control level, that directly interfaces with the multichannel FES stimulator and for some of the actuator control loops sensor feedback may be required. These actuator control loops have been based on artificial reflexes (Tomovic, 1984; Andrews and Baxendale, 1986; Bekey and Tomovic, 1986; Tomovic *et al.*, 1987) and more traditional control systems techniques (Chizeck *et al.*, 1988). Sensory feedback to the patient is provided by a suitable display to a sensate region above the level of the sensory lesion. We are presently exploring the value of a simple electrocutaneous feedback modality to assist the patient to avoid standing postures that may cause the FES system to trip on in response to incipient knee buckling (Andrews *et al.*, 1988). Sensors play a vital role in the control system. They are used in both the above and below lesion parts of the patient-machine system. These sensors are mainly used in two ways as indicated in figure 1. Firstly, sensor signals are used for direct feedback at the actuator control level. Secondly, sensor data are input to rule-based pattern recognition algorithms that detect events in the gait cycle that are then used to update the finite state reference model of the locomotor task. Some of these events predict the patient's locomotor intentions by detecting voluntary preparatory movements that are uniquely associated with the intention.

The finite state model must change state, in real time, on the occurrence of key gait events that signal the need to change control strategy. The next state that the model can assume will be one of a permitted set. A transition will only occur when the first of a set of expected state transition events is detected. Such state transition events are detected by rule-based pattern recognition algorithms operating, in real time, on the sensor signals. The above event and intention detectors embody detailed knowledge of the gait patterns leading up to these events.

THE HYBRID SKAFO ORTHOSIS

The general design objectives for this orthosis were that:
It should enable rest breaks to be taken whilst standing quietly as indicated in figure 2. This rest posture is often referred to as the 'C' posture in which the patient leans slightly forward with the hips and knees fully extended.
The mechanical brace should shunt potentially damaging force actions that may be applied to delicate joint structures and ligaments of the knee, ankle and foot.
The mechanical brace component should be unobtrusive and should not be the limiting factor, when combined with FES, in achieving an acceptable level of static and dynamic cosmesis for level ground walking.
Whilst standing quietly, the paraplegic patient was trained to adopt the 'C' posture. Vertical stabilisation was achieved using preserved muscle actions of the trunk and upper limbs. The hips were stabilised in full hyperextension against the ilio-femoral ligament. In this posture the ground reaction force (GRF) is applied in the metatarsal region and is directed in front of the knee joint and behind the hip joint. In this posture, the leg is mechanically stable without muscle activation. The dorsiflexion stopped ankle joint and rigid footplate also provide a measure of vertical stabilisation. Any tendency to lean forward will shift the region of support ahead of the vertical projection of the centre of gravity, producing a moment tending to restore verticality.

If the GRF should shift behind the axis of rotation of the knee, then the knee would buckle. In response to such an event, the FES control system should immediately activate the knee extensors. In order to respond the control system must sense the buckling action. The supra-patellar and posterior straps are adjusted to prevent potentially harmful hyper-extension of the knee. Tension in the supra-patellar strap indicates the presence of resultant knee extending moment if the knee flexes this tension will be zero. In the present experiments, the lateral upright of the brace, above the knee, was instrumented using strain gauges to sense anterior bending when the brace knee joint was fully extended and ankle joint stopped. This provided a voltage signal proportional to the tension in the supra-patellar strap. If the subject leans slightly forward, the centre of pressure will shift to the metatarsal region and the knee extending moment exerted by the GRF will be at a maximum. A zero signal was registered whenever the brace joint flexed indicating that the resultant knee moment was zero or was in the flexion direction.

Fig. 2   Schematic diagram of the 'C' posture adopted during standing rest breaks. Also shown are the foot   and crutch ground reaction forces.

CONTROL OF STANDING

An appropriate knee stabilising scheme during stance would be to activate the quadriceps only in the absence of a significant resultant knee extending moment. In this way the quadriceps activity would be minimised during stance. The resultant knee extending moment comprises the sum of the moments due to the GRF and that due to the FES activation of quadriceps. In the experiments reported here a preset stimulus was applied to the quadriceps according to the following (simplified) control rules. If knee stabilization is required then:

> If   (quads are switched off and Vs < T)
>   Then (switch on stimulation) ˙   ... rule (a)
>
> If   (quads are switched on and Vs > Q + T)
>   Then (switch off stimulation)   ... rule (b)

where Vs is the strain gauge signal and T is an adjustable threshold value. Q is the value of the maximum strain gauge signal produced by the quadriceps during an initial calibration procedure. Thus FES is switched on and off as the corresponding rule fires. The threshold T was introduced to switch on the quadriceps stimulation just before the knee buckles and to introduce hysteresis.

In our experiments, Q was determined for each leg during an initial calibration procedure, in which the subject was seated with the brace donned. The quadriceps were stimulated with the knee fully extended and the maximum strain gauge signal Vs recorded.

We refer to this on/off GRF activated control loop as an artificial knee extension reflex (KER) (Andrews and Baxendale, 1986). This does not imply that it mimics any natural reflex of the intact central nervous system, rather that it is a low level automatic pre-programmed response to an artificial sensory input that can be either enabled or inhibited from within the control hierarchy.

In addition, a second independent artificial knee extending reflex was implemented as a backup to the primary loop. In this loop the flexible knee goniometer was used to apply stimulation to quadriceps in response to knee buckling according to the following simple rule:

> If   (knee angle > set value of knee flexion)
> Then (switch on stimulation)
> Else (switch off stimulation)   ... Rule (c)

This simple bang-bang loop can oscillate if the GRF is maintained close to the knee joint axis. When the user leans forward slightly, shifting the GRF to the forefoot, the oscillations cease.

Another artificial reflex was used to assist the patient in maintaining the 'C' posture with hyper-extended hips and lumbar spine by switching on the stimulus to the hip extensor and erector spinae musculature whenever the hip joint flexed. For this purpose flexible hip goniometers (type G180 supplied by Penny and Giles Ltd, UK) were used to sense hip flexion. The stimulation was applied to each leg independently when the hip joint flexed, and switched off when the hip was again extended beyond the set threshold according to the rules:

> If   (hip extension > A degrees)
> Then (switch off hip and spine extensors)   ... rule (d)
> If   (hip extension < B degrees)
> Then (switch on hip and spine extensors)   ... rule (e)

where A and B are the preset upper and lower thresholds respectively. This reflex will be referred to below as the hip extending reflex (HER). The reflexes KER and HER were loosely coupled to improve stability by the condition that for each leg, if rule (e) fires then the rule (b) is not applied until after rule (d) fires.

CONTROL OF FORWARD PROGRESSION

The sequential phases of locomotion used in the finite state model for this gait pattern are illustrated in figure 3 together with the corresponding finite states s1 to s5. The gait phase and state name are listed in table 1 together with the control goals within each state and the corresponding state control strategies presently implemented. The gait pattern proceeds as follows. The subject will normally commence from the double support state s1. Suppose that he wishes to make a right step, he first raises the left crutch clear of the ground, moves it forward and then replaces it on the ground. He then shifts his body weight over onto his left leg. Immediately following weight shift, the right leg can be flexed, using FES, thereby entering the initial swing state s2. To complete the swing phase the knee and hip are extended and the terminal swing state s3 entered in preparation for the double support state s1. On entry to the terminal swing state s3 the flexion stimulus is maintained, for a short period (typically 0.2 seconds), overlapping with the extension of the leg. This was found to be helpful in achieving the required stride length. In the present experiments flexion of the leg was achieved by eliciting the preserved flexion withdrawal reflex by stimulating the ipsilateral common peroneal nerves.

The patient has the option to control his gait either manually using handgrip switches or automatically. For the above right leg step manual control proceeds as follows. In state s1 both hand switches are released. State s2 is entered by depressing the right handgrip switch and state s3 is entered upon releasing the switch. The double support state s1 is again re-entered immediately after the overlap period.

States and State Control Strategies

State	Phase	FES Control Goals	Control Algorithms
s1	Double Support	Maintain L&R knee and hip joints extended.	Enable L&R KER and HET reflexes.
s2, [s4]	Initial Swing	Maintain L, [R] hip and knee joints in extension Flex R, [L] leg.	Enable L, [R] KER and HER Apply R, [L] withdrawal stimulus
s3, [s5]	Terminal Swing	Maintain L, [R] hip and knee joints in extension. Extend R, [L] knee. Maintain hip flexion and ankle dorsiflexion	Enable L, [R] KER and HER Apply stim to R, [l] Quads Maintain R, [L] withdrawal stimulus for short period (typically 0.2s)

Table 1   State control goals and strategies.

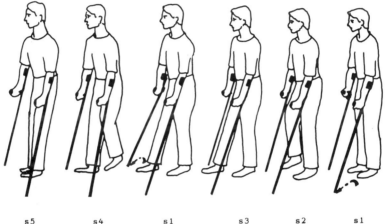

| s 5 | s 4 | s 1 | s 3 | s 2 | s 1 |

Fig. 3   Schematic diagram indicating the typical sequence of stages (states) represented in the finite state model.

In addition to the above brace strain gauges monitoring knee extending force actions, and the flexible hip flexion-extension angle goniometers two other types of sensor were used. A foot contact and load sensor comprising four force sensing resistive (FSR) films (supplied by Interlink Electronics Inc., USA) were incorporated into thin, flexible, low cost printed circuit insoles (Andrews *et al.*, 1987; Kirkwood and Andrews, 1988). The FSR's were positioned in the heel, medial and lateral metatarsal and big toe regions of the foot. To monitor limb load the output of each FRS was summed after A-D conversion. A simple crutch contact and load sensor based on the use of a single FSR was mounted onto the upper surface of the crutch or rollator handgrips (Andrews *et al.*, 1987). This sensor produced an analog signal that was related to the magnitude of the crutch load.

The automatic control of stepping has been described elsewhere (Andrews *et al.*, 1987; Andrews *et al.*, 1988; Andrews, 1988). Briefly, this was implemented as indicated in the state transition diagram of figure 4 in which locomotor task states are shown as circles and state transition events as squares. The cycle for making a step with the right leg proceeds as follows. The intention to step with the right leg i.e. change from state s1 to s2 is detected by repeatedly asserting the following rule whilst in state s1 and concurrent with the state control strategy indicated in table 1;

If

Condition (A) - the left crutch is raised for a period greater than a preset interval t_1 seconds

And if

Condition (B) - the left crutch contacts the ground after (A) is true.

And if

Condition (C) - the left to right leg load ratio exceeds a preset threshold within a preset time window from either (C) or (B).

Then the intention to make a right step has been detected, (e1) is true.

With each assertion the logical value of each condition is tested with the most recent analog to digital conversions of the involved sensors. For example, condition (A) and (B) can be tested based on the transitions in the output signal of the left crutch load sensor; condition (C) was based on the ratio of the limb load insole signals.

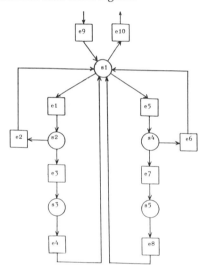

Fig. 4   State transition diagram for 4-point gait. States are represented by circles and transition events by squares.

A similar rule to detect the intention to make a left step is also repeatedly asserted whilst in state s1. The logical conditions of these intention detectors depend only on voluntary preparatory actions that are normally undertaken during 4-point gait. In this way the conscious involvement of the trained user is minimised.

Potential conflicts were resolved by allowing state transition to occur on the detection of the first valid event. Thus the next transition to either s1 or s3 is made on whichever event e2 or e3 occurs first. The event e2 is a simple timeout and is true when a preset time has elapsed since entering s2 before e3 occurs. The event e3 is true when the right hip joint has reached a preset angle of flexion. State s3 is exited on the preset time period e4 set for the above mentioned flexion stimulus overlap.

PRELIMINARY PATIENT TESTS

We report here preliminary laboratory tests performed on two spinal cord damaged volunteers.

Patient A Male aged 22 yr, mass 75 kg, height 1.6 m, traumatic lesion T6/7 motor and sensory complete, 1.5 yr post injury.

Patient B Male aged 24 yr, mass 70 kg, height 1.8 m, traumatic lesion C6 incomplete hemi-transection, 7 yr post injury.

Both patients were previously able to stand using purely FES systems. In the present tests, each patient was requested to remain standing in parallel bars. The test was terminated when either leg showed evidence of uncontrollable buckling, due to fatigue, or after one hour had elapsed. In the test using only FES both patients from time to time increased the stimulus intensity up to a preset maximum (Monophasic, rectangular, current regulated 120 mA, 20 Hz, pulse width 0.3 ms) in order to increase recruitment to accommodate for fatigue and prolong standing time. With the hybrid system the stimulus intensity was maintained at a preset level throughout the tests.

Patient (A) had used calipers regularly for standing. He stated that he liked the SKAFO braces and would use them regularly if they were made available to him. He was fitted bilaterally and with two channels of stimulation applied to quadriceps using surface electrodes. He had previously completed a programme of FES muscle restrengthening exercises for his quadriceps. He was able to stand for periods of up to 15 minutes using the two channel stimulator prior to being fitted with the SKAFO braces. When fitted with the braces he was able to remain standing for the one hour test period. Quadriceps activation was typically less than 5%. This activation was due to occasional gross posture changes (raising up on the bars using the upper limbs) performed to relieve discomfort/boredom. The degree of activation was considerably reduced (typically less than 1%) when the subject was asked to remain standing quietly, for periods of up to 15 minutes, and if possible not to adjust his posture. At the present time this patient has not been trained to use an FES walking system.

Patient (B) had a paralysed left leg having no functional control of the hip knee and ankle. He had preserved some skin sensation and proprioception. In addition, his right leg was partially paralysed with reduced sensation. He also had partially paralysed trunk musculature and some weakening of the triceps. He had preserved sufficient strength and control to enable him to grip forearm crutches and once upright in them to remain so, supported by his right leg, for periods up to 20 minutes. However, he was not able to make a single step. He had previously used a conventional cosmetic KAFO, having a fixed ankle and lockable knee joint, for short range ambulation supplemental to the use of his wheelchair. With the KAFO he could remain standing quietly for periods in excess of one hour. Since 1985 he has used a manually controlled 2 channel FES walking aid for his left leg using quadriceps stimulation for stance support and the flexion withdrawal for stepping (Bajd *et al.*, 1985). The system was controlled by a handgrip mounted, finger operated, switch such that when the switch was pressed the stimulation changed over from quadriceps to flexion reflex and vice versa when the switch was released. This simple FES system provided good dynamic cosmesis and enhanced function e.g. ability to reciprocally negotiate uneven ground and steps. On level ground his preferred 4-point walking speed was 0.41 m/s with good step length symmetry. Although he has used this system on a daily basis FES fatigue of quadriceps has limited its functional usefulness to him compared with his KAFO. He continues to use the system primarily for exercise purposes. In 1986 he was fitted unilaterally with the hybrid orthosis and was pleased with its cosmesis compared with his KAFO. He was able to stand for periods in excess of the test period with quadriceps activation typically less than 2% of the stance time. He has since used the system, in the laboratory, for level ground walking using both the manual and automatic stepping modes described above. The quality of his walking did not appear to be degraded, as judged by visual observation and his comments; his cadence was not significantly affected in manual mode but was slower (typically 0.3 m/s) when using the automatic mode. During the walking tests it was observed that he was more confident with the familiar manual mode than with the automatic mode. He preferred the manual mode because of speed. In both modes, the KER was active during the stance phase of gait and the qudriceps stimulation was switched off shortly after foot flat. The quadriceps were activated for typically less than 15% of the stance phase. The patient also demonstrated an ability to walk slowly (typically 0.1 m/s), whilst taking short steps (typically 0.2 m) with the brace but without FES.

FURTHER DEVELOPMENT OF THE FINITE STATE MODEL

An interesting, and so far unexplored, alternative to handcrafting gait event detectors is to use AI induction methods to automatically learn the rules from previously recorded example sensor data. Research in the field of expert systems has indicated the superiority of computer induction versus handcrafting of rules (Michalski and Chilausky, 1980).

To use induction algorithms such as Quinlan's ID3 (Quinlan, 1979), a collection of representative examples of the events to be detected or classified must be prepared. It is from this collection that the induction algorithm will learn-by-example and produce the required decision rules for the event detectors. We have performed preliminary experiments using inductive learning techniques to determine decision trees that can be used to determine the state of the system from the uncalibrated sensor inputs. A previously reported experiment (Kirkwood and Andrews, 1988) illustrated the use of inductive learning in discriminating various static postures. Here we describe a further experiment that illustrates the use of these techniques in detecting dynamic postures. The experiment was conducted on a normal subject whilst walking at his preferred rate on level ground. The following data were acquired; hip and knee angles using potentiometric goniometers; eight foot pressures using a pair of insole transducers. The data were acquired by IBM PC computer using a 12 bit A/D converter sampling each signal at 50 Hz. Four gait cycles were acquired from each walk test. The induction algorithm developed for these tests is described elsewhere (Kirkwood *et al.*, 1989).

The 12 signals (attributes) referred to in this chapter are defined in table 2.

Attribute			Symbol
		Foot pressures	
(1)		Left Heel	LH
(2)		Left medial metatarsal	LM
(3)		Left lateral metatarsal	LL
(4)		Left big toe	LT
(5)		Right heel	RH
(6)		Right medial metatarsal	RM
(7)		Right lateral metatarsal	RL
(8)		Right big toe	RT
		Joint angles	
(9)		Left hip	ALH
(10)		Right hip	ARH
(11)		Left knee	ALK
(12)		Right knee	ARK
		Definition	
(1)	RHE	Maximum right hip angle until maximum left knee angle.	
(2)	LKE	Maximum left knee angle until maximum left hip angle.	
(3)	LHE	Maximum left hip angle until maximum right knee angle.	
(4)	RKE	Maximum right knee angle until maximum right hip angle.	

Table 2   Abbreviations for attributes and stages of the gait cycles.

The gait cycle was divided into four sequential stages, as illustrated in figure 5, which were used as the classes to be recognised by the inductive learning algorithm. The abbreviations used for these classes are also shown in table 2.

The four gait cycles were manually divided into the four stages using an interactive graphics package on the IBM PC. For each of these gait cycles four different points (examples), across the range of each stage, were selected and the amplitudes corresponding to those points were taken as the attributes. This set of examples formed the training set for the induction algorithm. This set of examples formed the training set for the induction algorithm. Data from four runs were used to provide a training set of 256 examples i.e. 4 runs * 4 gait cycles * 4 stages * 4 points).

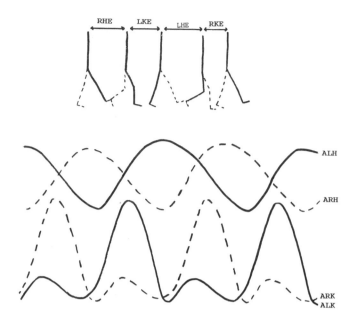

Fig. 5   The four stages of normal gait and typical sensor signals used to demonstrate the rule induction method.

The decision tree produced by the induction algorithm is shown in figure 6. This tree is 100% accurate in classifying the training set and uses the minimum number of nodes possible to distinguish 4 classes i.e. 7 nodes.

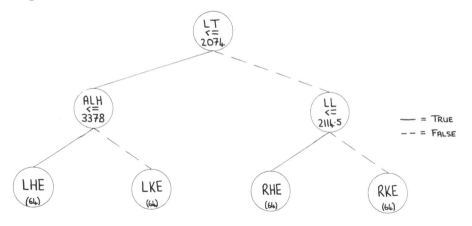

Fig. 6   Decision tree for identification of the four gait stages produced by the inductive learning.

DISCUSSION

Finite State Control

The hierarchical finite state approach was found to provide a flexible structure for integrating the episodic changes in control strategies that were required during locomotion. The simple on/off (bang-bang) control produced movements that were acceptably smoothed by the inertia of the limb segments. Presently finite state models are used mainly to select state control strategies, therefore, the minimum number of states in a locomotion cycle will correspond to the number of changes of state control strategies required. The complexity of the finite state model will be determined on the level of biomechanical knowledge of the gait pattern and how many stages can be reliably discriminated by event detectors and what state control strategies are presently possible or practical for a particular category of patient. In turn, this will depend on practical issues which include the availability of sensors and controllability of muscles.

Handcrafted Rules

The use of handcrafted rules was found to be a natural way for those expert in pathological gait to specify event detector requirements. The state transition graphs were found to be useful visual aids in showing which event detectors will be active during a particular system state and the next possible system states. The rules and facts (knowledge base) can be easily interpreted and modified without requiring extensive computer expertise. This is a promising basis for the development of software tools to interactively aid the manipulation of rules for developing and tuning the control system to the needs of individual patients.

Rules Induction

Machine learning potentially may affect a number of areas in the control of hybrid systems. For example:
> Detection of state transitions that may not be intuitively apparent from the available sensor data.
> To optimise the choice of sensor set, particularly when the choice of sensor or the possible anatomical mounting sites are restricted for practical reasons.
> To develop systems that use built in redundancy to implement fault tolerant strategies.
(As an aside this method may also have application in automated gait analysis systems).

Laboratory Tests

The standing tests suggest that the brace may be used to circumvent quadriceps muscle fatigue. The level ground, laboratory gait patterns previously achieved by two of the patients were found to be significantly affected by the presence of the brace. Both patients stated a preference for the cosmetic appearance of the brace compared with standard cosmetic KAFO's.

Our laboratory trials suggest that by using the brace the present limitations of muscle fatigue and reflex habituation may be avoided by facilitating sufficient rest breaks. However, further work is required to determine the minimum frequency and duration of such rest breaks for different forms of ambulation and the practicality of using the system outside the laboratory.

ACKNOWLEDGEMENTS

We should like to thank the Mr P A Freeman FRCS, Dr M Delargey and staff of the West of Scotland Spinal Injuries Unit at Phillipshill Hospital in Glasgow for their support. The financial support of the Scottish Home and Health Department is acknowledged. One of the authors (C.A.K) is a recipient of a Medical Research Council postgraduate studentship.

REFERENCES

Andrews B J and Baxendale R H (1986). A hybrid orthosis incorporating artificial reflexes for spinal cord damaged patients. *J. Physiol.*, **38**, 19.
Andrews B J, Baxendale R H, Barnett R, Phillips G F, Paul J P and Freeman P A (1987). A hybrid orthosis for paraplegics incorporating feedback control. *Proc. 9th Advances in External Control of Human Extremities*, Dubrovnik, Yugoslavia, 297-311.

Andrews B J, Baxendale R H, Barnett R, Phillips G F, Yamazaki T, Paul J P and Freeman P A (1988). Hybrid FES orthosis incorporating closed loop control and sensory feedback. *J. Biomed. Engng.*, **10** (2), 189-195.

Andrews B J (1989). Rule based control of hybrid FES orthoses, in C Cobelli and L Mariani (eds.), *Proc. 1st IFAC Symp. Modelling and Control in Biomedical Systems*, **1**, 287-294.

Bekey G A and Tomovic R (1986). Robot control by reflex actions. *Proc. IEEE int. Conf. robotics and automation*, **1**, 240-247. IEEE Computer Society Press.

Chizeck H, Kobetic R, Marsolais E B, Abbas J J, Donner I and Simon E (1989). Control of functional neuromuscular stimulation systems for standing and locomotion in paraplegics. *IEEE Trans Biomed. Engng..* in press.

Kirkwood C A and Andrews B J (1988). A Flexible printed circuit board for monitoring patterns of foot loading, *Proc. ICAART 88,,* Montreal, Canada, 488-489.

Kirkwood C A, Andrews B J and Mowforth P (1989). Automatic Detection of Gait Events: a Case Study Using Inductive Learning Techniques. *J. Biomed. Engng.*, in press.

Michalski R S, Chilausky R L (1980). Knowledge acquisition by encoding expert rules versus computer induction from examples. *Int. J. for Man-machine Studies*, **12**, 63-87.

Quinlan J R (1979). Discovering rules by induction from large collections of examples, in D Michie (ed.), *Expert Systems in the Micro-electronic Age.*, Edinburgh University Press, Edinburgh.

Tomovic R (1984). Control of assistive systems by external reflex arcs, *Advances in Control of Human Extremities ETAN*, pp 7-21. Published by Yugoslav Committee for ETAN, Belgrade, Yugoslavia.

Tomovic R, Popovic D, Tepavac D (1987). Adaptive reflex control of assistive systems., *Advance in ECHE* **9**, 207-213, published by Yugoslav Committee for ETAN, Belgrade, Yugoslavia.

49

FUTURE DEVELOPMENT IN FUNCTIONAL ELECTRICAL

STIMULATION

A Kralj and T Bajd

CURRENT FES ACHIEVEMENTS

Past and current achievements in the field of FES may serve as guidelines while discussing future trends and speculating about new developments. The historic development of FES is interesting (Reswick, 1973; McNeal, 1977), but of more importance are the latest developments, which started with the proposal for the foot drop FES device (Liberson et al., 1961). In the early seventies the main interest was focused on the therapeutic and orthotic utilisation of FES in stroke patients and later in the eighties also for head trauma and CP patients (Vodovnik et al., 1981). The interest for FES orthotic use in quadriplegic patients' hand function restoration and locomotion restoration in spinal cord injured (SCI) patients has grown steadily in the last 20 years (Peckham and Mortimer, 1977; Kralj and Grobelnik, 1973). Since 1980 the interest in FES utilisation for SCI patients has rapidly increased and at times unrealistic and false hopes have been given. During that time important progress was achieved, for example prolonged FES enabled standing of SCI patients (Kralj et al., 1986), restoration of a simple reciprocal gait pattern with the utilisation of flexion reflex by afferent FES (Kralj et al., 1980). The number of FES channels has steadily increased and reached in current systems a magic number of 32 channels (Marsolais, 1987). The control aspects were only to a minor degree improved through microprocessor utilisation (Thoma et al., 1983; Petrofsky and Phillips, 1986). Also in 1983 ascending and descending of stairs was demonstrated (Kobetic et al., 1986).

The FES technology and methodology was steadily improved and new possibilities for application in incomplete SCI patients was presented (Bajd et al., 1986b). In some SCI patients combination of FES and passive orthotic bracing is utilised. Here, FES is providing propulsion forces while classical orthotic systems are ensuring the required stability. The hybrid approach was thus realised in practically usable systems (Andrews and Bajd, 1985). In parallel, the stimulation hardware was improved. Self-adhesive skin electrodes were introduced, and percutaneous wire electrodes were improved as well as the implantable FES systems. At the same time surgical techniques, electrode design and fabrication technology, electrode leads, encapsulation methods and general understanding of implants problems increased substantially. Also the FES systems for quadriplegic hand function restoration advanced and made important improvements, in particular regarding patient selection criteria, system application and control. Along with increasing function complexity and the control issues problems were also growing with the advancements in FES applications for manipulation, locomotion or posture restoration. Problems like control concepts, quality of control, dexterity, selectivity of FES, together with questions regarding complexity versus simplicity and functionality, medical and technological problems, with patient population selection were also growing because of missing fundamentals for FES control and selective FES stimulation (McNeal and Bowman, 1985). These unsolved problems dominate the FES field at present. Proprioceptive and exteroceptive sensory feedback and generation of reliable control signals are additional difficulties of current FES applications.

To brighten these cloudy outlooks we can state that FES was established in the last 20 years as a restorative and rehabilitation method and its feasibility has been proved. It is expected that further development will continue advancing the application technology and methodology. Regarding the gait control, it is striking that for the last 20 years the field was unable to improve the control approach and to move away from the simple trigger switch proposed by several authors nearly 30 years ago see Reswick, 1973 and Liberson *et al.*, 1961 indicating that innovative and new concepts are indeed needed together with broader FES patients utilisation. The proposals for closed loop control (Crago *et al.*, 1980) need demanding costly and difficult to develop artificial sensors (Crago *et al.*, 1986; Troyk *et al.*, 1986, Bowman and Mendl, 1986) or need to utilise and detect the information from preserved neural sensors (De Luca, 1975; Hoffer and Sinkjaer, 1986). In the area of electrode-neural tissue interface innovative ideas are also necessary for improving control signal's quality, delivering sensory feedback for sensory restoration by means of FES and for increased dexterity of movements produced. Here, it must be noted that FES has accomplished within the last decade interesting developments in therapeutic applications like muscle modification (Salmous and Henriksson, 1981), restrengthening (Bajd *et al.*, 1986a), tissue regeneration like nerve regeneration and decubitus healing (Vodovnik *et al.*, 1987) and spasticity control (Alfieri, 1982; Bajd *et al.*, 1985). Therefore, our discussion of the expected future developments of FES will include some of these aspects.

Expected Development Trends

As a patient model, when discussing the expected development trends in FES, the SCI patient is the most suitable because they represent a complex and very devastating injury. Thus any conclusions drawn will be to a larger extent valid and applicable if generalised to all other major populations of paralysed patients like stroke, head injury, cerebral palsy and quadriplegic patients.

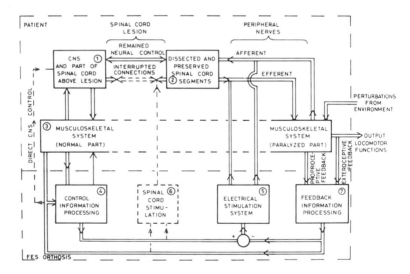

Fig. 1   Logic diagram for FES orthosis control system.

For systematic discussion of possible developments in the FES field figure 1 (Kralj and Bajd, 1989) could be helpful. It must be kept in mind that FES is meant to be used for therapy and restoration of lost functions. Therefore, a logic concept would include the preserved patient structures to the maximal extent. The present FES is utilising the upper motor neuron lesioned muscles, joints and bones, but very little of the preserved neural (pathologically organised) structures and natural neural sensors. Future and current developments are and will be concentrated on control signals detection interfacing (figure 1, blocks 1 and 3) and required information processing (block 4). Observing the FES application locations in figure 1, the interface between blocks 6 and 2 is very interesting in addition to the interfaces of blocks 5 to 2 and 5 to 3. Regarding sensory feedback and sensory restoration the interface and sensors for exteroceptive feedback and proprioceptive feedback information detection are of utmost interest in regard to the closed loop control of FES systems (blocks 4 to 3 interface).

While critically discussing the present surface and implanted gait restoration FES systems in general, several problems are noticeable: the very unsophisticated and clumsy control, the issues regarding complexity versus practicality and cost and all the patient-device related questions that can be expressed as "gadget tolerance". From a FES development point of view the field is facing a dilemma, whether to continue with the development of artificial sensors required in feedback systems and to design an artificial control approach or to rely on natural neural sensors, detected information and the preserved neural system which due to the spinal cord lesion is pathologically organised. The selective nerve stimulation is also probably opening new means for restoration and modification of neural disorders by FES. The presented dilemma: artificial versus natural neural sensors is gaining in importance if we observe the required surgery involvement as displayed in figure 2a and b. Figure 2a is drawn for an implanted system having 4 hip, 4 knee and 3 ankle channels with bipolar muscle electrodes. The length of electrode lead is given in case the stimulator is located in the lower abdomen. In figure 2b the implanted stimulator is located in the axillary region and electrodes are used for shoulder, elbow, wrist and hand stimulation. Totally 20 channels are foreseen, resulting in electrode leads of 7,2 m length. The present number of FES channels that different investigations have proposed for the paraplegic patient gait restoration, is substantial larger than the number of channels used in figure 2a diagram. For instance if 24 channels are used (Rushton, 1988), the electrode leads would increase in length to about 10 m and for 70 electrodes as proposed by Marsolais (1988) the estimated length could be enormous and consequently result in poor reliability and high cost. Obviously, the surgery involved is not simple. Therefore, a sound and realistic alternative is to study better electrodes/nerve interfaces for achieving improved selectivity and probably decreasing the surgical problems because in the latter case more proximal nerves could be utilised. Also to study pathological neural organisation and reflexes is an important task because preserved reflexes can be efficiently included in a restoration scheme, for enabling better function and reducing the hardware requirements.

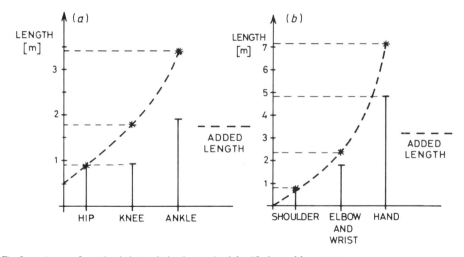

Fig. 2 a) Length of electrode leads required for 19-channel leg system.
        b) Length of electrode leads required for 20-channel upper extremity system.

Regarding functional application, FES is going to be increasingly utilised in the coming years also for posture and sitting assistance. Systems incorporating unidirectional FES (Sweeney and Mortimer, 1986) will be the next step toward more selective FES applications. Nerve conduction blocking can additionally be employed for suppressing or blocking of unwanted neural conductions with the aim to modify the preserved neural functions. FES may also be introduced in the future for closed loop control of movement and of internal organs function like lungs, kidneys, etc., for modification of autonomous systems controlling different body functions like blood pressure, venous return, secretion of different glands etc. In figure 3 the FES control for movement is presented. It is expected that new designs of closed loop control including sensory restoration will be the interest of future research. FES also appears to be promising for muscle properties transformation and shaping for special functions like substituting for lost cardiac muscle portions (First Purdue Conference on Cardiac Assistance with Skeletal Muscle, October 1988).

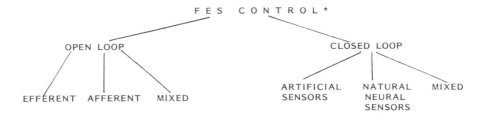

* WITH OR WITHOUT SENSORY RESTORATION

Fig. 3   Types of FES control system.

The selectivity of FES application and selective nerve information detection (Edell, 1986) together with natural sensor utilisation is opening unforeseen new developments not only in the FES field but also in closely related applications where current trends are already indicating the move towards closed loop system utilisation. Let us mention some of these applications: closed loop phrenic nerve stimulation, closed loop control of hypertension by FES of carotid sinus nerve. Also closed loop heart pacing based on PCO_2 and ventilation may be feasible. The whole area of bladder control and many other areas may gain by the developed selective nerve stimulation and information detection advances (Spencer *et al.*, 1986). The selective interaction of nerve conduction and information transmission is also very useful for diagnostic purposes and a very noble means for therapeutic interventions like spasticity modification or moderation of the autonomous system disfunction even in a closed loop sense. Such technological advances also open numerous new interdisciplinary areas of work and when discussed with the advancements and accomplishments of nerve regeneration and wound healing which are also related to the problem of selective application of electric currents with electric and magnetic fields, then medical sciences may gain a means for interacting with the natural processes of cell growth and communication. Such an approach raises hopes that even cell membrane properties (Giaever and Keese, 1986) or cell behaviour might be locally and grossly modified by micro-selective and selective FES across appropriately placed electrode arrays supplied with adequately processed information. One may also speculate that here is one key to the cellular behavioural control, which once reached can satisfy even the most demanding expectations in reconstruction, regeneration and entire rehabilitation of body functions.

ACKNOWLEDGEMENTS

The research in FES was supported by the Slovene Research Community, Slovenia, Ljubljana, Yugoslavia and the National Institute for Disability and Rehabilitation Research, Dept. of Education, Washington D.C., U.S.A.

REFERENCES

Alfieri V (1982). Electrical treatment of spasticity. *Scand J. Rehabil. Med.*, **14**, 177-xxx.
Andrew B J and Bajd T (1985). Paraplegic locomotion: a hybrid FES approach. *XIV ICMBE 7 ICMP*, Espoo, Finland, 404-405.

Bajd T, Gregoric M, Vodovnik L and Benko H (1985). Electrical stimulation in treating spasticity resulting from spinal cord injury. *Arch. Phys. Med. Rehabil.*, **66**, 515-517.

Bajd T, Kralj A, Benko H and Gros N (1986a). Comparison of passive and electrical stimulation training of atrophied spastic muscle. *Proc. 4th Mediterranean Conf. Med Biol. Eng.*, Sevilla, Spain, 357-360.

Bajd T, Kralj A, Turk T and Benko H (1986b). FES rehabilitative approach in incomplete SCI patients. *Proc. 9th RESNA Conf.*, Minneapolis, 316-318.

Bemeut S L, Wise K D, Andrson D J, Njafi K and Drake K L (1986). Solid-State Electrodes for Multichannel Multiplexed Intracortical Neuronal Recording. *IEEE Trans BME*, **33** (2), 230-241.

Bowman L and Meindl J D (1986). The Packaging of Implantable Integrated Sensors. *IEEE Trans. BME*, **33** (2), 248-255.

Crago P E, Chizek J H, Neuman M R and Hambrecht F T (1986). Sensors for use with functional neuromuscular stimulation. *IEEE Trans. Biomed. Eng.*, **33**, 256.

Crago P, Mortimer T and Peckham H (1980). Closed loop control of force during electrical stimulation of muscle. *IEEE Trans. on BME*, **27** (6), 306-312.

De Luca C (1975). Considerations for using the nerve signal as a control source for above-elbow prostheses. *Proc. 5th Int. Symp. on External Control of Human Extremities*, *ETAN*, Yugoslavia, Dubrovnik, pp 101-109.

Edell J D (1986). A peripheral nerve information transducer for amputees: long-term multichanned recordings from rabbit peripheral nerves. *IEEE Trans. on BME*, **33** (2), 203-214.

Giaever J and Keese C R 91986). Use of electric fields to monitor the dynamical aspect of cell behaviour in tissue culture. *IEEE Trans. BME*, **33** (2), 242-247.

Hoffer J A and Sinkjaer T A (1986). A natural 'force sensor' suitable for closed-loop control of functional neuromuscular stimulation. In *Proc. 2nd Vienna Int. Workshop Functional Elektrostimulation*, Vienna, Austria, 47.

Kobetic R, Carroll S G and Marsolais E B (1986). Paraplegic stair climbing assisted by electrical stimulation. in *Proc. 39 ACEMB Conf.*, Baltimore, MD.

Kralj A and Bajd T (1989). Functional electrical stimulation: Standing and walking after spinal cord injury. *CRC Press Inc.*, to be published.

Kralj A, Bajd T and Turk R (1980). Electrical stimulation providing functional use of paraplegic patient muscles. *Medical Progress Through Technology*, Springer Verlag, **7**, pp 3-9.

Kralj A, Bajd T, Turk R and Benko H (1986). Posture switching for prolonging functional electrical stimulation standing in paraplegic patients. *Paraplegia*, **24**, 221-230.

Kralj A and Grobelnik S (1973). Functional electrical stimulation - a new hope for paraplegic patients. *Bull. Prosth. Res.* BPR 10-20, Fall, 75-102.

Lieberson W T, Holmquest H J, Scott D and Dow A (1961). Functional electrotherapy: Stimulation of the peroneal nerve synchronized with the swing phase of the gait of hemiplegic patients. *Arch. Phys. Med. Rehab.*, **42**, 101.

Marsolais E B (1988). According to presentation made at the Engineering Foundation Conference. *Neural Prostheses: Motor Systems*, July, Potosi, Missouri.

Marsolais E B (1987). Establishing and fulfilling criteria for practical FNS systems. *Proc. of Advances in External Control of Human Extremities*, Yugoslav Committee for Electronics and Automation, Belgrade, 105-109.

McNeal D R (1977). 2000 years of electrical stimulation, in J D Reswick and F T Hambrecht (eds.) *Functional Electrical Stimulation, Applications in Neural Prostheses*, Marcel Dekker, New York, pp 3-35.

McNeal D R and Bowman B R (1985). Selective activation of muscles using peripheral nerve electrodes. *Med. Biol. Engng. Computing*, pp 249-253.

Peckham H and Mortimer T (1977). Restoration of hand function in the quadriplegic through electrical stimulation, in J B Reswick and F T Hambrecht (eds.) *Functional Electrical Stimulation, Applications in Neural Prostheses*, Marcel Dekker, New York, pp 83-95.

Petrofsky J and Phillips C (1986). Electrically controlled movement of muscle: A potential aid to muscle paralysis, in D Ghista and H Frankel (eds.) *Spinal Cord Injury Medical Engineering*, Charles C Thomas Publishers, Springfield, pp 393-437.

Proc. of First Conference on Cardiac Assistance with Skeletal Muscle (1988), Purdue University, West Lafayette, Indiana.

Reswick J B (1973). A brief history of functional electrical stimulation, in W S Fields and L A Leavitt (eds.) *Neural Organisation and its Relevance to Prosthetics*, Intern. Medical Book Co., New York, 3.

Rushton D N (1988). MRC neurological prosthesis unit. According to the discussion and presentation delivered at *Engineering Foundation Conference: Neural Prostheses: Motor System*, Potosi, Missouri.

Salmous S and Henriksson J (1981). The adaptive response of skeletal muscle to increased use. *Muscle Nerve*, **4**, 94.

Sweeney J D and Mortimer T J (1986). An asymmetric two electrode cuff for generation of unidirectionally propagated action potential. *IEEE Trans. BME*, **33** (6), 541-549.

Spencer L B, Kensall D W, Anderson D J, Najafi K and Drake K (1986). Solid-State Electrodes for Multichannel Multiplexed Intracortical Neuronal Recording. *IEEE Trans. BME*, **33** (2), 230-241.

Thoma H, Frey M, Gruber H, Holle J, Kern H, Reiner E, Schwanda G and Stöhr H (1983). First implantation of a 16-channel electric stimulation device in human. *Trans. Am. Soc. artif. intern. Organs*, **29**, 301-306.

Troyk P R, Jaeger R J, Haklin M, Payezdala J and Bajzek T (1986). Design and implementation of an implantable goniometer. *IEEE Trans. Biomed. Engng.*, **33**, 215.

Vodovnik L, Bajd T, Kralj A, Gracanin F and Strojnik P (1981). Functional electrical stimulation for control of locomotor systems. *CRC Critical Review in Bioengineering* **6**, 63-131.

Vodovnik L, Stefanovska A, Rebersek S, Gros N, Gregoric M and Malezic M (1987). Effects of electrical stimulation: from transistorised placebo to change in gene expression, in J D Andrade (ed.), *Artificial Organs*, VCH Publishers, New York, pp 543-558.

50

DISCUSSION:

TECHNOLOGICAL ADVANCES

Gorham opened the discussion by asking how much damage or scarring was produced in the area of a tattoo removed by laser. *Evans* replied that a transient reaction occurred with vacuolation in the dermis round the site of the pigment, but in general this caused no noticeable scarring. He commented that there had been one exception associated with a densely pigmented professional tattoo which had required repeated treatment. *Reid* added that an alteration in pigmentation, generally pigment loss, was seen but that this recovered in a few weeks or months. *Paul* asked about the use of lasers for removing or breaking up deposits or tissues in the vasculature *Evans* replied that the concept of selective absorption could be applied by staining target tissues which did not naturally differ optically from the surrounding tissue. He also commented that the more rigid materials and tissues could be removed by short pulses of light which caused fragmentation, rather than the denaturation and phagocytosis which occurred in soft tissues.

Small asked about the specificity of the halothene sensor shown by *Hall*, who replied that it would detect anything which dissolved in silicone oil, but could be used successfully under conditions where there was only one chemical species which would dissolve.

Thoma then asked the general question of who would pay for devices for locomotor electrical stimulation. *Peckham* replied that in America it was necessary first to demonstrate, on a suitable number of patients, that the devices produced useful function. After this, it was useful to run a prototype programme in a number of centres to confirm the effectiveness of the technology, following which the FDA would give premarket approval. In general the devices would then be paid for privately, although the Veterans' Administration would support the systems in their own centres. *Andrews* suggested that a basic system had yet to be produced for standing, after which it had to be made commercially available for clinical trials; walking systems were still at the research level. He suggested that in a few years systems for standing, transfer and exercise would be available within the U.K. at a cost comparable with that of calipers. *Kralj* contrasted patients who generally purchase what medical professionals prescribe, with paraplegic patients who know what they require themselves and would be willing to pay personally. It was essential that the stimulator systems should be cost effective and simple, which is not the case at present. *Weber* suggested that augmentation systems which were available were either extremely simple or highly sophisticated. In France, the provision of cochlear implants provided a useful analogy, where simplicity and reliability were important as was a commercial source of such devices. The cost of cochlear implants was met from social security funds. He further commented that although there were about 3000 cochlear implants he thought that there would only be about 100 for functional electrical stimulation, and hence the cost was small compared to the cost of cochlear implants. The problem of cost and who would provide the money was essentially political.

Malagodi commented that cortical electrodes might be the best way to close the control loop and asked what was the current state of research in this area. *Peckham* replied that there was little current work; *Kralj* agreed but suggested there was exploratory work on stimulation at more central sites, closer to the spinal cord.

Granat asked to what extent the CALIES project used the existing implantable devices and what was the future direction of the project. *Weber* replied that he represented the industrial rather than the research part of the project but commented on the non-academic problem of producing electrodes, leads and connectors. He suggested the future direction of the project was at present unclear. *Kralj* stressed the importance and complexity of software development for functional electrical stimulation.

273

Kirtley asked about the need for reconstructive procedures before functional nerve stimulation was applied to the tetraplegic hand. *Peckham* suggested that the anatomy of the normal hand was rather complex and in order to use the natural anatomy to carry out tasks of daily living it was necessary to understand how the muscles should be activated. In practice in many clinical cases there was neurological injury such that these muscles were no longer available to produce their normal function and it would be necessary to alter the anatomy to reproduce the physiological movement.

THE 1988 ADAM THOMSON LECTURE:

PERSONAL RECOLLECTIONS ON A QUARTER

CENTENNIAL

R M Kenedi

PREAMBLE

Seminar participants have kindly and knowingly assembled to recognise the institutional mile-stone of 25 years in the history of Strathclyde's Bioengineering Unit; unknowingly they are also participating in a personal milestone of my own: the anniversary of my arrival in this country as an "overseas" student to join a fore-runner of this University. Not, I hasten to add, the ancient foundation of Anderson's University as some of my colleagues would have you believe, but 50 years ago in September 1938, the then Royal Technical College.

My remit for this chapter is "to present a learned history of the Bioengineering Unit's 25 years since its inception". My presentation will cover 4 main areas: Historical Highlights; Management and Administration; People and finally Activities.

From the material presented you will I think see that ACCIDENT and COINCIDENCE had a powerful influence on the fortunes of the Unit. Such phenomena are nothing new and have been so very sensitively described in the following:

> An accident befell my brother Jim,
> Somebody threw a tomato at him,
> Tomatoes are soft and don't hurt the skin,
> But this one was specially packed in a tin!

Despite their trauma at the time such happenings did create opportunities which in their turn were converted into successful events by the "quick on the trigger" capability of the Unit's members in exploiting these opportunities both for the Unit's and for their own individual benefit.

HISTORICAL HIGHLIGHTS

Bioengineering at Strathclyde did not in fact commence with the setting up of the Unit - it was "trailblazed" (or more aptly perhaps trailcooled) in the late 1950s by Dr Jimmy Brown of Mechanical Engineering whose speciality was refrigeration.

Major heart operations were then in their pioneering days and were carried out under hypothermia, that is the lowering of body temperature to levels that slowed the vital functions significantly. This was effected by a form of refrigeration and the then Professor of Surgery was concerned that his patients, while successfully weathering the operation tended to end up with frostbite in miscellaneous portions of their anatomy!

Jimmy Brown's refrigerative expertise was then called upon and his collaboration with the medics was the first pioneering venture in which the innovative art of the surgeon was successfully integrated with the scientific back-up of the engineer. This pleased everyone and particularly the patient, whose sojourn in the operating theatre ceased to be an uncontrolled arctic adventure.

Dr Brown's pioneering efforts were carried out at no small cost to himself. He was plunged into the stresses of the operating theatre with little prior preparation and being an individual of vivid imagination he apparently spent a significant time of his indoctrinative period in the operating theatre more in the prone than the upright position, being periodically revived to active capability as the progress of the operation demanded it. One should perhaps also relate that as he became inured to the operative vicissitudes, he got his own back by retailing his experiences to his engineering colleagues so vividly that they in return became overcome just through listening to him. In those days the administration of first aid in the staff room of the Engineering Department became in consequence recurrent routine, attracting appreciative audiences from all over the University.

In 1960 I was a Reader in Experimental Stress Analysis in the then Department of Mechanical, Civil and Chemical Engineering, a genuine "colossus" of an entity presided over by Professor Adam Thomson.

I only hope that he will be as generously forgiving of my transgressions as he has always been in the past!

However to return to my theme: members of the academic staff of the then Royal College of Science and Technology (the University did not come into being until 1964) were pressed into service to give Christmas lectures to school children. The manner of selection was something similar to Russian Roulette and that year I was the one shot with the choice.

To make the lecture more interesting I strain-gauged a human skull, hit it and displayed the resulting effects on an oscilloscope. Schoolkids being then as now pretty sophisticated, were not particularly excited about this and took it in their stride.

However it so happened that the then Professor of Pharmacy, J P Todd came into the lecture, saw and was intrigued. He had a friend, a plastic surgeon, who for some time has been seeking some mechanical technique that would permit him to assess the closability of operation wounds. As a consequence, the surgeon Tom Gibson and I were introduced. Now it sometimes happen that two individuals match each other and the result of that introduction has been a collaboration that still continues today and spawned as one of its early successes the present Bioengineering Unit.

On hindsight it was also the FIRST of these influential coincidences that I mentioned in my Preamble.

Following some two years of collaboration Tom (Gibson) and I put a modest application together for support of biomechanical research on skin in the sum of around £8000 and submitted it to the Medical Research Council (MRC).

Here now obtains coincidence the SECOND.

Wholly unknown to us the MRC was then actively considering investing money in an engineering group to see what kind of contribution, if any, engineering could make to medicine. Our application landing with them at this potentially crucial time provided them with a candidate group and after investigating us (individually AND institutionally) very thoroughly decided to adopt us for extended financing much beyond the application originally submitted.

One of the conditions attached to the MRC's support was a "take-over" of Bioengineering by the University after the initial period of four years MRC financing, provided naturally the effort was successful.

It was at this point that a critically decisive role was played by the then newly arrived founding Principal and Vice-Chancellor of the University Sir Sam Curran. He was willing to chance his arm, possibly because being newly arrived he did not know us all that well and was willing to think the best of us. The University prompted by him underwrote the venture and the Bioengineering Unit was born.

The Unit was financed at an initial strength of 1 Research Professor (myself), 2 academics: Tom Duggan, Rudolf Zalter (for one year) replaced by Jim McGregor; 2 technical support staff Walter Gilmour and Gibby Wingate and lastly but perhaps most importantly an Administrator/Secretary Russell Ritchie.

The present Head Professor Paul joined the Unit full-time in 1969 - from 1960 until then he carried out his now internationally esteemed research work on gait analysis as an Associate of the Unit.

The remit of the Unit was full-time research orientated to "activities of benefit to medicine with due regard to significant clinical problems of the human patient so as to explore and demonstrate the contributions that engineering can provide in this field as a constituent profession of multidisciplinary clinic teams".

The work of the Unit was overseen from its initiation by an Advisory Committee, then by a Steering Committee and eventually by the "Standing Committee on Bioengineering". More of these later.

The "stabilised" size of the Unit in January 1971 stood at 27 academics with 25 technical and 4 secretarial support staff and some 50 postgraduate students. Corresponding with its expansion the Unit's accommodation was beginning to bulge at the seams, which sets the scene for accident/coincidence the THIRD.

To cater for this space need, which by the late 1960s became genuinely acute, it was decided to apply to the Wolfson Foundation for a building grant, somewhat in excess of £0.25 million. In response the then Director of the Foundation kindly indicated that he was going to visit us to discuss our application in detail.

On my phoning his office in London I ask how I would identify him on arrival at Glasgow airport, I was told to look for a Wellingtonian nose locomoting some 6ft 4 inches from the ground. He was thus easily identified but was found to have arrived in a somewhat disturbed frame of mind: - the flight was a rather rough one and he bore traces on his clothing of his seat neighbour's "mal de l'air", an unlooked for and certainly undesirable complication at the very commencement of our association.

The as yet unidentified guardian saint of Strathclyde Bioengineering then took a hand: - the University's chauffeur appreciating the situation at a glance took us to a small chemist's shop run by a pal of his who apparently was a whizz kid at handling just this kind of emergency. After a brief visit the Director emerged, much restored both in physical appearance and equanimity of spirit. This episode incidentally did us no harm at all showing that as potential grant recipients, we were at least capable of action of a pertinent kind in an emergency.

I always had a sneaky feeling that somewhere in the Wolfson Centre that was opened in 1972 and where the Unit holds "Open House" today, there really should be some kind of a commemoration of our unintentional benefactor, the airsick passenger - a kindly gargoyle perhaps?

In designing the building, the architect Mr Jim Morris of Edinburgh was asked to provide maximum flexibility in both space and services. It is very pleasing to report that this remit was very successfully fulfilled and that the building is continuing to provide an excellent environment for the wide range of activities it contains.

Let me now come to coincidence the FOURTH.

One of the major areas of the Unit's research has been, prosthetics and orthotics. This area was initiated through the collaboration and guidance of Professor George Murdoch (recently retired from the Department of Orthopaedic Surgery of Dundee University).

Largely through his initiative the Scottish Office set up a Working Party in the late 1960s, on "The Future of the Artificial Limb Service in Scotland". The Denny Working Party reported in 1970 and among its recommendations was one for the setting up of a Scottish National Centre to provide basic and professional training and education for prosthetists and orthotists.

In 1971 "Action Research for the Crippled Child" sponsored an exhibition of toys for handicapped children in Lewis's (the former Glasgow Department store). This exhibition was attended by George Robertson, the Senior Administrator at the Scottish Home and Health Department (SHHD) then responsible for the general area of rehabilitation of the disabled in Scotland and in this context for the implementation of the Denny report. George knew the Strathclyde Unit since the SHHD financed a significant number of our projects through his office. I also attended this exhibition and we met, quite accidentally in the lift as we were leaving.

The exhibition was staged on the sixth floor of the building and during our descent George indicated that the SHHD decided to establish the National Centre and he wandered if Strathclyde would be willing to "father" the venture. To this I responded with the "enterpreneurial" affirmative - this obtains when one decisively indicates agreement without having any authority whatsoever to do so!

In the event the National Centre for Education and Training in Prosthetics and Orthotics did materialise, it did come to Strathclyde and it did become one of the most successful of the University's departments serving the needs of the disabled at regional, national and international levels.

The Centre 's Director, Professor John Hughes, another local product, researched in the Unit as a postgraduate in association with Professor Murdoch, joined the staff of the Unit in 1965 and took over the Centre in 1972. The Centre commenced its operations in Balmano, the former Victorian workhouse on Rottenrow, (just demolished) and after one intermediate move has been established since 1982 in its own purpose designed palazzio in the Curran Building on the campus.

So what started as a small research group sponsored by the MRC for a limited period a quarter of a century ago has grown into two established and major departments recognised as Centres of Excellence not only the world over but what is far more important and impressive also honoured as such on their own home ground.

MANAGEMENT AND ADMINISTRATION

The Unit's original application to the MRC was supported by a distinguished group of clinicians - when the Unit came into being this group became the Unit's "Advisory Committee". It was chaired by Professor Roland Barnes and its Membership consisted of Professors Gibson, Sir Charles Illingworth, Sir Andrew Kay, McGirr, Wyburn and Dr Weymes from outwith and Sir Samuel Curran, Professors Fletcher, Kenedi, Ross, Scott and Thomson from within the University.

From 1968, when the Unit's "take-over" by the University was completed, the SHHD began to provide significant support to the Unit for a range of projects of importance to Scotland. This led to the setting up of a Steering Committee in addition to the Advisory Committee, charged specifically to oversee the SHHD sponsored activities. This Committee did its work through three working groups: - Artificial Organs, Prosthetics/Orthotics and Obstetrics corresponding to projects in progress.

In 1971 the Western Regional Hospital Board (now the Greater Glasgow Health Board, GGHB) agreed to authorise appropriate members of the Bioengineering Unit to work with patients in National Health Service Hospitals within the Region.

Behind this simple innocuously sounding statement hide protracted negotiations concerning the position of the University's Bioengineers vis a vis the Medical Physicists in the Health Service.

When the Unit came into being its medical sponsors were the profession's "crème de la crème" and in consequence members of the Unit were introduced to clinical practice on the basis of authority delegated to them by the senior consultants with whom they worked.

This produced no "waves" while the Unit and its activities remained small. However as the Unit's clinically orientated work expanded and particularly when SHHD sourced monies became significant, the position began to raise concern in the Department of Medical Physics of the then Western Regional Hospital Board. This department incidentally was the largest and indubitably the best group of its kind in the UK.

Anyhow, what today on hindsight appears as a "nit-picking" demarcation dispute, arose and was enthusiastically even perhaps enjoyably pursued by both sides. The outcome was intriguingly very satisfactory, again for both sides: - the Unit was formally recognised by the Health Board and virtually every Medical Physics Department in the Health Service throughout the UK added "Bioengineering" to its title!

Further as so often happens with initially determined adverseries - enthusiastic antagonisms come to be replaced by mutual respect and trust. Medical Physicists and Bioengineers have both established their rightful place in Health Care and nowadays collaborate within a framework of mutual trust and respect.

Consequent on the formal recognition of the Unit by the NHS in 1971 the Advisory and Steering Committee were amalgamated and reformed as the "Standing Committee on Bioengineering" consisting of representatives of the SHHD (3), the GGHB (3), the University of Strathclyde (6) and a further segment of "expert" members up to a total of 6. The infrastructure of working groups was increased to 6 with the addition of Cardiology, Tissue Mechanics and Trends & Developments.

PEOPLE

"Organisations" are characterised by their "people" content - when I speak of the "Unit" I therefore mean the people who form it.

The basic constituent, particularly in a postgraduate group such as the Strathclyde Unit is an amalgam of students and staff.

The original concept of the Unit visualised a closely integrated group of staff and students, who worked together and who in an interactive way irritated and spurred each other onwards and thus produced results. Such a unified coexistence, which I am glad to say was achieved is a peculiarly persistent thing and becomes like a drug, habit forming.

Speaking personally it is for this reason that, having retired as many times as my colleagues' goodwill permitted (my score todate is four with repeat performances pending), I have returned to a modicum of teaching in the Unit and in the National Centre. Frankly I miss the questioning impact of young minds, who want answers to questions and will not accept prevaricating evasion. This return to teaching, I should point out also has its traumatic side effects - this past year I was assessed for the first time in my life for teaching capability and effectiveness. Frankly the assessment was so sophisticated that I have not yet managed to figure out whether the overall outcome in my case was favourable or otherwise!

The Unit todate graduated some 460 individuals, who are dispersed in 45 countries (in addition to the UK) world-wide. Roughly 63% of our graduates are UK based, 8% in continental Europe, 7% in the Americas, 14% in Asia, 6% in Africa and 2% in Australia. They occupy positions of responsibility and influence in Academia, Industry and Commerce world-wide. Members of the Unit who travel abroad find this of particular convenience since they can rely on being welcomed by former colleagues and/or students, wherever they go.

When my wife and I left Hong Kong in 1984, (having performed one of my recent retirements there) we had a seven weeks tour of Australia and New Zealand and were welcomed in every place we visited by former students and associates. It was an absolute riot of an experience!

In thinking of past students there are my recollections - random examples: - the attractive young lady from the Americas, who two weeks after her arrival in the Unit came to see me and indicated her considered judgement that the Unit required reorganisation from top to bottom. She then stayed on for some three years to do this and also did an excellent Ph.D as sideline entertainment (she is with us today); the very bright ginger haired youngster who tragically developed multiple sclerosis while doing a Masters - this he finished and then passed on a few years later, retaining however the capability of enjoying life in his own slightly sardonic way to its very end; the wild ones from "down under" who turned in excellent PhDs but how they made everyone of us suffer! - I met one of them in 1984 in Australia and to my surprised regret he has become a pillar of respectability, disapproving of the lack of discipline in Australian life and remembering nostalgically the orderliness and rationality of his behaviour and life in Scotland (!) and many others.

With colleagues I had similar experiences of which I shall say nothing - primarily since they have taken me back with tolerance and forbearance and this I want to preserve. It is also pleasant to record that 25 years on they seem not to have changed at all - they are still precisely as I remember them?!

One of the great regrets is that during my period with the Unit we lost Tom Grassie in 1980, one of our up and coming young staff members whose contributions to the Unit augured so auspiciously for his future which tragically however was not to be.

ACTIVITIES

"What has the Unit actually done, has it for example produced effective solutions to practical problems of significance?"

The answer to such a question is an emphatic yes both in the general and in the particular. As an aside the very continuation of the Unit's existence is a direct demonstration that engineering indeed can and does contribute in improving medical diagnosis and treatment.

On the general level the most important reason for the Unit's wide ranging success has been the basic work concept pioneered from inception that all activity must be undertaken in a framework of fully integrated collaboration between the clinicians and the bioengineers as members of the one multidisciplinary clinic team, which naturally must also include the patient!

This philosophy of approach, hammered into members of the Unit by its clinical collaborators has obtained in all the activities undertaken and has always led to success. Separation of the clinical and bioengineering activities and the patient's bedside is on the other hand an unfailing guarantee of failure.

On the particular level, you will see the present most commendable range of work undertaken by the Unit displayed ready for viewing in its "Open House" period this afternoon.

Two examples picked at random, representing widely differing problems: - the Souter/Strathclyde artificial elbow, some 500 of these have been implanted with an overall symptom free success rate of 89%, its development took 4 years from inception (1973) to prototype implantation (1977), it is now a money earner for its clinical and bioengineering innovators and for the University; - the technique of tattoo removal by pulsed laser, a potentially most important contribution to social rehabilitation of the young, with the resources made available within the Health Service in Glasgow permitting the treatment of some 250 "clients" per year in contrast with a waiting list of 2500.

The philosophy of multidisciplinary integration has also been implemented in education and training at both undergraduate and postgraduate levels.

The popularity and impact of the National Centre's education and training programmes, which incidentally are at the present time serving as models being followed in England (Professor John Hughes and his colleagues have truly become prophets in their own country) are manifest world-wide.

Intriguingly it now also seems that England in their developments may steal a march on Scotland!

Two weeks ago I happened to be at the University of Salford where the Biological Engineering Society held its 1988 Annual Scientific Conference.

In an address to the conference Professor Ashworth, Salford's Vice-Chancellor outlined his University's plans for honours degree programmes in Prosthetics/Orthotics (P/O) together with others covering a wide spectrum of the para-medical therapies - physio, occupational etc. He appreciatively acknowledged Strathclyde's pioneer programmes in the P/O area which they intended to follow.

Personally I think it is a great pity that the honours level degree programmes in the therapies mentioned have been stalled in Scotland and that such pass degree programmes that do exist have been permitted to fragment in various Colleges in Glasgow and Edinburgh by being harnessed to London based CNAA validated qualifications.

Having at various times been concerned with the development of honours degree proposals for Physiotherapy and Occupational Therapy I am very concerned indeed at an apparent short-sightedness that exists in the powers that be. They seem to refuse to recognise the extraordinary potential for forward looking developments that could be accomplished through properly planned degree validation links between the SCOTTISH Colleges and the SCOTTISH Universities. I suppose in this regard one can only wait, hope and now and again foam a bit at the mouth, as I am now doing!

In conclusion, I hope that I have managed to convey to you that work in Bioengineering at Strathclyde during the last 25 years has not only produced achievements of value to patients but that at all times it has also been fun to do!

Seminar Participants and
Author Index

(*a*: participating author, *a**: non-attending author, *c*: chairman
and chairman associate)

Allen, Mr J A D, Bioengineering Unit,
University of Strathclyde, Glasgow, G4 0NW, UK
a (page 256) Andrews, Dr B J, Bioengineering Unit,
University of Strathclyde, Glasgow, G4 0NW, UK
Armstrong, Dr C J, Dept of Mechanical Engineering,
Queen's University of Belfast, Belfast, UK

*a** (page 267) Bajd, Dr T, Faculty of Electrical Engineering,
Edvard Kardelj University, Trzaska 25, 61000 Ljubljana, Yugoslavia
Bakken, Ms N G, Thorax Dept, Kie A,
Rikshospitalet, 0027 Oslo 1, Norway
c, a (pages 48, 62) Barbenel, Prof. J C, Bioengineering Unit,
University of Strathclyde, Glasgow, G4 0NW, UK
a (page 256) Barnett, Mr R W, Bioengineering Unit,
University of Strathclyde, Glasgow, G4 0NW, UK
*a** (page 256) Baxendale, Dr R H, Institute of Physiology,
University of Glasgow, G12 8QQ, UK
Beith, Mr M A, Medical Production Ltd,
9/11 Albion Way, East Kilbride, G75 0YN, UK
*a** (page 48) Belch, Dr J J F, Ninewells Hospital and Medical School,
University of Dundee, Dundee, DD1 9SY, UK
Bell, Dr F, Dept of Physiotherapy,
Queen Margaret College, Clerwood Terrace, Edinburgh,
EH12 8TS, UK
Berthinussen, Mr A P, Senior Exec. Prosthetics,
Hugh Steeper Ltd, 237 Roehampton Lane, London, SW15 4LB, UK
Biggs, Dr M S, Hydro Polymers Ltd,
Vinyls Division, Newton Aycliffe, County Durham, DL5 6EA, UK
a (page 120) Bignardi, Dr C, Dip. Meccanica,
Politecnico di Torino, Cso Duca degli Abruzzi 24,
10129 Torino, Italy
*a** (page 128) Binnington, Dr A G, Ontario Veterinary College,
University of Guelph, Guelph, Ontario, N1G 2W1, Canada
Blackhurst, Mrs M, Bioengineering Unit,
University of Strathclyde, Glasgow, G4 0NW, UK

Blass, Mr C R, Hydro Polymers Ltd,
Vinyls Division, Newton Aycliffe, County Durham, DL5 6EA, UK
Bowry, Dr S, Akzo/Enka AG,
Membrana, Öhder Strasse 28, D5600 Wuppertal, FRG
Bransby-Zachary, Mr M A P, Orthopaedic Department,
Royal Infirmary, Glasgow, G4, UK
Brown, Ms G S, Klinik für Innere Medizin,
University of Rostock, Rostock, GDR
Brown, Dr I A, Zimmer Ltd,
Dunbeath Road, Elgin Industrial Estate, Swindon, Wilts, SN2 6EA, UK

a* (page 253)	Brown, Mr J G, Queen's University of Belfast,
Dept of Orthopaedic Surgery, Musgrave Park Hospital,
Belfast, BT9 7JB, UK

a* (page 56)	Brubaker, Dr C E, Rehabilitation Engng Center,
University of Virginia, P.O. Box 3368, Charlottesville, VA 22903, USA

a* (page 28)	Buck, Dr R, Gambro AB,
P.O. Box 10101, S-220 10 Lund, Sweden

a* (page 245)	Buckett, Dr J R, Dept Biomed. Eng. & Orthopaedics,
Case Western Reserve University, Cleveland, Ohio 44106, USA
Burnie, Dr J, Pilkington CRS,
23–24 Colomendy Ind. Estate, Denbigh, Clwyd, LL16 5TA, UK
Butterfield, Ms M,
29 Kenley Road, St Margaret's, Twickenham, TW1 1JR, UK

a (pages 120, 179)	Calderale, Prof. P M, Dip. Meccanica,
Politechnico di Torino, Cso Duca degli Abruzzi 24, 10129 Torino, Italy
Campbell, Miss E, Building Div. CSA,
Clifton House, Clifton Place, Glasgow, UK

a* (page 104)	Campbell, Dr P, Implant Retrieval Lab.,
Division of Orthopaedics, UCLA Medical School,
Los Angeles, CA 90024, USA

a* (page 68)	Cardi, Mr M, Center for Rehab. Technology,
Helen Hayes Hospital, Route 9W, West Haverstraw, NY 10993, USA

a (page 210)	Childress, Prof. D S, Northwestern University,
RIC Rm 1441, 345 E Superior St, Chicago, Illinois 60611, USA
Clark, Mr C, Room 110C Block 1,
GVMT Buildings, Warbreck Hill Road, Blackpool, FY8 3AD, UK
Clark, Dr M, Nursing Practice Research Unit,
University of Surrey, Guildford, Surrey, GU2 5XH, UK

a (page 104)	Clarke, Dr I C, Kinamed Inc, 10780 Santa Monica Blvd,
Suite 100, Los Angeles, California 90025, USA
Cliquet Jr, Mr A, Bioengineering Unit,
University of Strathclyde, Wolfson Centre, Glasgow, G4 0NW, UK

a (page 74)	Condie, Mr D N, Tayside Rehab. Eng. Services,
Dundee Limb Fitting Centre, 33 Queen Street, Broughty Ferry,
Dundee, DD5 1AG, UK

Miller, Mr J H, 80 Henderland Dr.,
Westerton, Bearsden, Glasgow, G61 1JG, UK
Miller, Mrs T M, Bioengineering Unit,
University of Strathclyde, Glasgow, G4 0NW, UK
Millington, Dr P F, Bioengineering Unit,
University of Strathclyde, Glasgow, G4 0NW, UK
Mitchell, Dr D C, 12 Compton Close,
Southcrest Meadows, Redditch, Worcestershire, UK

a* (page 151) Mitsui, Mr T, Dept of Orthopaedic Surgery,
Aichi Medical University, Nagakute-cho, Aichi-gun, Aichi-ken, 80-11, Japan

a* (page 253) Mollan, Mr R A B, Queen's University of Belfast,
Dept of Orthopaedic Surgery, Musgrave Park Hospital, Belfast, BT9 7JB, UK

a (page 216) Morecki, Prof. A, Technical University of Warsaw,
Al. Niepodlegosci 222, 00-663 Warsaw, Poland

a* (page 249) Morimoto, Dr T, Department of Paediatrics,
Faculty of Medicine, Ehime University, Shizugawa, Shigenobu, Onsen-gun, Ehime, Japan

a (page 13) Morrice, Dr L M A, University Dept of Medicine,
Ninewells Hospital & Medical School, Dundee, DD1 9SY, UK

a* (page 96) Morrison, Dr J B, School of Kinesiology,
Simon Fraser University, Burnaby, British Columbia, Canada

a* (page 256) Mowforth, Dr P, Turing Institute,
University of Strathclyde, 36 North Hanover Street, Glasgow, G1, UK

c, a (page 188) Murdoch, Prof. G, 2 Muirfield Crescent, Dundee, DD3 8PT, UK
Murphy, Mr M, Bioengineering Unit,
University of Strathclyde, Glasgow, G4 0NW, UK

Newstead, Dr J H, Bioengineering Unit,
University of Strathclyde, Glasgow, G4 0NW, UK

a (pages 166, 174) Nicol, Dr A C, Bioengineering Unit,
University of Strathclyde, Glasgow, G4 0NW, UK
Nicol, Mrs S M M, Bioengineering Unit,
University of Strathclyde, Glasgow, G4 0NW, UK

a (page 170) Nieuwenhuis, Ir F J M, Delft Univ. of Technology,
Lab. for Measurement & Control, Mekelweg 2, 2628 CD Delft, The Netherlands

a (page 151) Niwa, Prof. S, Dept of Orthopaedic Surgery,
Aichi Medical University, Nagakute-cho, Aichi-gun, Aichi-ken, 480-11, Japan

a (page 249) Noshiro, Dr M, Instit. Med. & Dental Engng,
Tokyo Med. & Dental University, 2-3-10 Kanda-Surugadai, Chiyoda-ku, Tokyo 171, Japan

Robertson, Mr A, Bioengineering Unit,
University of Strathclyde, Glasgow, G4 0NW, UK
Robertson, Mr D W, Research & Development Services,
University of Strathclyde, Glasgow, G4 0NW, UK
a (page 21)   Robertson, Dr L M, Bioengineering Unit,
University of Strathclyde, Glasgow, G4 0NW, UK
Ross, Miss A, Bioengineering Unit,
University of Strathclyde, Glasgow, G4 0NW, UK
a (page 179)   Rossetto, Dr M, Dip. Meccanica,
Politechnico di Torino, Cso Duca degli Abruzzi 24, 10129 Torino,
Italy
Rowe, Mr P J, Bioengineering Unit,
University of Strathclyde, Glasgow, G4 0NW, UK
Rumley, Mrs A, Department of Medicine,
Royal Infirmary, Glasgow, G4, UK
Ruston, Dr S A, Dept of Physiotherapy,
Queen's College, Glasgow, UK
a (page 21)   Ryan, Dr C J, Dept of Surgery,
Royal Postgraduate Med. School, Hammersmith Hospital,
Ducane Road, London, W12 0HS, UK

Sacchetti, Mr R, Prosthetic Research Center,
INAIL, 40054 Budrio, Italy
a* (page 13)   Saniabadi, Dr A R, Ninewells Hospital and Medical School,
University of Dundee, Dundee, DD1 9SY, UK
Scales, Prof. J T, 17 Brockley Avenue, Stanmore, Middlesex,
HA7 4LX, UK
a (page 131)   Schreiber, Prof. A, Orth. Universitatsklinik,
Balgrist, Forchstrasse 340, 8008 Zurich, Switzerland
a* (page   Sessa, Dr V, 20 Wynford Drive, Suite 105,
   128)   Don Mills, Ontario, M3C 1J4, Canada
a (page 174)   Shah, Dr K M, University Dept of Orthopaedics,
Western Infirmary, Glasgow, G11 6NT, UK
Shaw, Mr C G, Rehab. Engineering Centre,
University of Tennessee, 682 Court Avenue, Memphis, TN 38163, USA
Shepherd, Dr M T, Coopers Animal Health Ltd,
9 Church Hill, Cheddington, Nr Leighton Buzzard, LU7 05X, UK
Shivas, Dr A, Ethicon Ltd,
P.O. Box 408, Bankhead Avenue, Edinburgh, EH11 4HE, UK
c   Simpson, Prof. D C, 2 Dalrymple Crescent, Edinburgh,
EH9 2NU, UK
Small, Dr C F, Dept of Mechanical Engineering,
Queen's University, Kingston, Ontario, K7L 3N6, Canada
Smart, Mrs S, Bioengineering Unit,
University of Strathclyde, Glasgow, G4 0NW, UK

a (page 238)　Vance, Dr C A, Laser Products Division,
　　　　　　　Spectra-Physics Ltd, Boundary Way, Hemel Hempstead,
　　　　　　　Herts, HP2 7SH, UK
　　　　　　　Vasselet, Dr R, Lab. Biologie Cutanee,
　　　　　　　Faculty de Medecine et Pharmacie, 2 place St Jacques,
　　　　　　　25030 Besançon, France
　　　　　　　Veres, Mr G, National College of Prosthetics,
　　　　　　　Trondheimsveien 132, 0570 Oslo 5, Norway
　　　　　　　Vithal, Dr, Institute of Physiology,
　　　　　　　University of Glasgow, Glasgow, UK

*a** (page　　Waddell, Mrs F, Princess Margaret Rose Hospital,
　166)　　　　41–43 Frogston Rd West, Edinburgh,
　　　　　　　EH10 7ED, UK
　　　　　　　Walker, Mr A C, Bioengineering Unit,
　　　　　　　University of Strathclyde, Glasgow, G4 0NW, UK
　　　　　　　Walker, Dr J, City University,
　　　　　　　Northampton Square, London, EC1V 0HB, UK
　　　　　　　Walsh, Dr M E, Accident & Orthopaedic Division,
　　　　　　　Royal Infirmary, Glasgow, G4, UK
*a** (page　　Ward, Dr P E, Queen's University of Belfast,
　253)　　　　Dept of Orthopaedic Surgery, Musgrave Park Hospital,
　　　　　　　Belfast, BT9 7JB, UK
　　　　　　　Watt, Dr A H, Room 5, St Andrew's House, Regent Road,
　　　　　　　Edinburgh, EH1 3DE, UK
　　　　　　　Webb, Mrs C, Dept of Rehabilitation Sciences,
　　　　　　　Hong Kong Polytechnic, Hung Hom, Kowloon,
　　　　　　　Hong Kong
　　　　　　　Weber, Dr J L, Bertin et Compagnie,
　　　　　　　B. P. 22, Rue Louis Armand, Z. I. d'Aix-en-Provence,
　　　　　　　13762 Les Milles Cedex, France
　　　　　　　Whateley, Dr T L, Department of Pharmacy,
　　　　　　　University of Strathclyde, Glasgow, G4 0NW, UK
*a** (pages　Wheatley, Prof. D J, Dept of Cardiac Surgery,
　41, 238)　　Royal Infirmary, Glasgow, UK
　　　　　　　Wilkinson, Mr J J, Bioengineering Centre,
　　　　　　　University College London, Roehampton, London, UK
a (page 160)　Wilkinson, Dr R, Bioengineering Unit,
　　　　　　　University of Strathclyde, Glasgow, G4 0NW, UK
　　　　　　　Wilson, Mrs J D, Bioengineering Unit,
　　　　　　　University of Strathclyde, Glasgow, G4 0NW, UK
　　　　　　　Wing, Mr D, Technical Dept,
　　　　　　　Downs Surgical plc, Church Path, Mitcham, Surrey,
　　　　　　　CR4, UK

Yacoob, Mr P, Medical Equipment Centre,
Ministry of Health, P.O. Box 12, Bahrain, Arabian Gulf

Zioupos, Dr P, Bioengineering Unit,
University of Strathclyde, Glasgow, G4 0NW, UK

Subject Index

**Scottish
Development
Agency**

*T*he Scottish Development Agency was set up in 1975 to help build
a strong economy and a better quality of life.

*A major area of its activity is in the development and transfer of
technology from research in academia into industry.*

*This is a challenging job and the Agency has built up teams of
scientists, engineers and business people to tackle the tasks together.
These teams are experienced in advanced engineering, healthcare,
biotechnology, electronics, and technology transfer.*

*The SDA depends on the energy and abilities of everyone in Scotland,
including those working in universities, colleges and research institutes,
and was delighted to support the conference.*

*Scotland has great strengths and an international reputation in many
areas of bioengineering and the Agency's aim is to assist in the
development of this expertise.*

Scottish Development Agency, 120 Bothwell Street, Glasgow G2 7JP.
Tel: 041 248 2700. Fax: 041 221 3217.

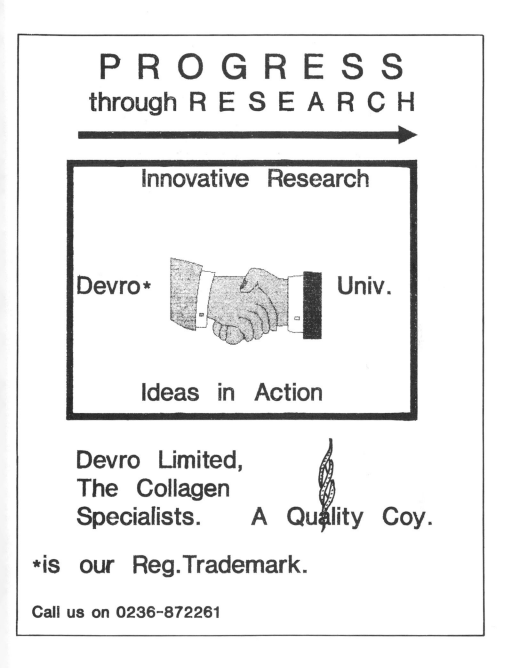

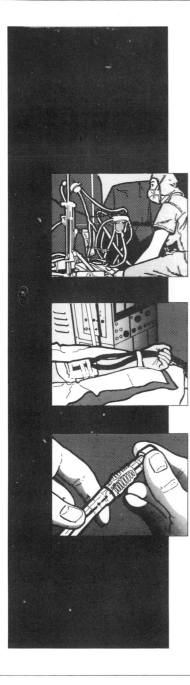

hy·vin®

PVC COMPOUNDS

DEVELOPED WITH
MEDICAL APPLICATIONS IN MIND

Hydro Polymers; one of Europe's leading producers of PVC resins and compounds, has always pursued areas of high technology. As a consequence this policy has led to the development of a comprehensive range of PVC medical compounds.

Such has been the effectiveness of their Hy-vin PVC Medical Compounds that they have gained worldwide acceptance; and Hydro regularly support users with an unbeatable and ever increasing application expertise.

Medical grades are available for both extruding and moulding life saving products such as blood transfusion/dialysis kit components and heart/lung bypass kits. Special properties are featured such as low plasticiser extraction during prolonged contact with human blood and resistance to radiation sterilisation.

Natural, clear and tinted effects can be achieved.

WORLDWIDE ACCEPTANCE

FORMULATED FOR PRE-STERILISED DISPOSABLE DEVICES

UK DEPARTMENT OF HEALTH AND US F & DA REGISTRATION

CERTIFIED QUALITY

 **HYDRO POLYMERS**

Vinyls Division

For more details contact:-

HYDRO POLYMERS LTD.,
Newton Aycliffe, Co. Durham, DL5 6EA, England. Telephone:(0325) 300555. Fax:(0325) 300215. Telex: 58322.

BP/A13737

Science and Skill
in Surgical wound closure

Ethicon Limited, the leading U.K. manufacturer of wound closure
and tissue support products are pleased to be associated with
Progress in Bioengineering at the University of Strathclyde.

Gambro® Polyamide
- a most versatile membrane.

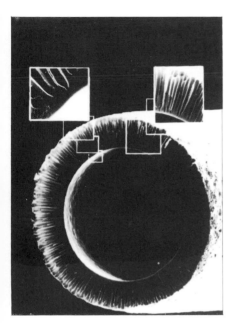

Developed and manu-
factured exclusively by
Gambro, our polyamide
membranes have been
found to provide out-
standing performance for
a wide range of appli-
cations:

- **hemofiltration**
- **autotransfusion**
- **depyrogenization**
- **blood concentration**
- **ultrafiltration**

And with the introduction of our new high - performance
polyamide membrane designed for:

- **high-flux dialysis**
- **hemodiafiltration**

Gambro has once again demonstrated its resourcefulness in
dialysis technology.

*Congratulations on
the occasion of the
25th Anniversary of
Strathclyde Bio-engineering Unit
from*

*Manufacturers of
vascular prostheses*

Newmains Avenue Inchinnan Renfrewshire PA4 9RR Scotland
Telephone: 041-812 5555 Telex: Vastek 776553
Fax: 041 812 7650

∩euroTech

Neuromuscular Electrical Stimulation System

The NeuroTech Neuromuscular Stimulation System is the most advanced in the world today. It provides the medical professional with optimum versatility and flexibility of patient treatment. NeuroTech's product innovations and continuing commitment to service sets new standards for Neuromuscular Stimulation.

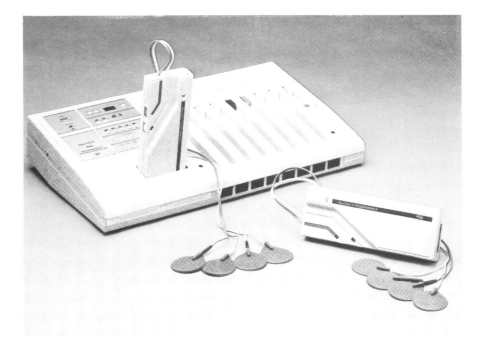

System Features:

NEUROTECH LIMITED
P.O. Box 379
HARROW
Middlesex HA1 1AA
Tele: 01-427 7031
Fax: 01-427 1337
Telex: 9419837 NTECH G

Digitally Programmable Parameters:	Allows accuracy and repeatability of treatment.
Eight Channels:	Treatment set-up no longer limited by equipment. Can treat in functional patterns or more than one limb at a time.
Pre-set Programs:	Provide ease of use and quick set-up time for five common protocols.
Alternating Outputs:	Agonist/antagonist muscle groups can be treated in more functional alternating patterns.
Programmable Portable Unit:	NT-4 is as powerful and accurate as the NT-16 with an interrogation feature that allows monitoring of unsupervised use.
Unique Voltage Limiting Circuits:	Electrodes can be lifted off or moved about without shocks or stinging.

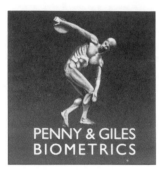

PENNY & GILES
BIOMETRICS

A unique system incorporating solid state electronics enabling cost effective, accurate and rapid measurements of limb movements which can be displayed on a digital indicator, or recorded on a compact data recorder

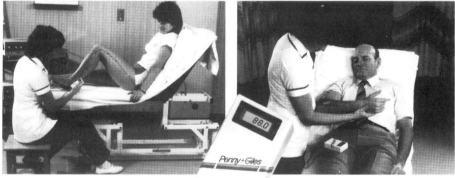

Penny+Giles

One unicondylar stands alone

When it comes to deciding which unicondylar knee to use, one unicondylar stands out from the rest - The Robert Brigham Uni-Condylar Knee. Since the introduction of its original design in 1974, clinical evidence has accumulated to demonstrate its success.

- "Recovery and rehabilitation following operation is less prolonged and less complicated than with tibial osteotomy."[1]

- At three to five year follow-up of 50 cases, it was found that no revisions had been required.[2]

- At an average review of five and a half years, "...92 per cent of the knees...had a good or excellent result."[1]

- The Uni-Condylar Knee "...is an attractive alternative for the treatment of unicompartmental osteoarthritis."[3]

- Patients can achieve an average flexion of 120°.[4]

This success is based on the design of its components *and* instrumentation. In addition to being designed to allow considerable range of motion the prosthesis has a tab in the femoral component for enhanced rotational stability and a titanium alloy tibial tray to provide optimal support for the UHMWPE component. The straightforward, easy-to-use instrumentation is designed to give accurate, reproducible bone cuts. Used with suitable patient selection, it is an excellent first choice which will also retain your freedom of action for the future.

As an alternative to tibial osteotomy, in the elderly, unicompartmental knee arthroplasty offers much faster post-surgical rehabilitation without the complications of non-unions.

As an alternative to other unicondylar knees, The Robert Brigham Uni-Condylar Knee stands alone.

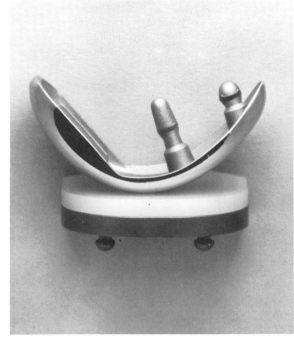

References

1. Marx, C.L. and Scott, R.D. Unicompartmental Total Knee Arthroplasty (A Four to Six Year Follow Up Utilizing Metal-Backed Tibial Components). Paper presented at the British Orthopaedic Association Congress, Plymouth, 1988.

2. Clinical data - Brigham and Women's Hospital, Boston, MA.

3. Scott, R.D. and Santore, R.F. Unicondylar Uncompartmental Replacement for Osteoarthritis of the Knee. J. Bone Joint Surg. [Am] 1981; 63-A; 4; 536-544

4. Westin, C.D. and Scott, R.D. Metal-Backed Unicondylar Replacement. Paper presented at American Academy of Orthopaedic Surgeons, 1983

The Robert Brigham Uni-Condylar Knee

Further information may be obtained from the Customer Services Department at

ORTHOPAEDICS LIMITED

a Johnson&Johnson company

STEM LANE, NEW MILTON, HANTS. BH25 5NN.
TEL. (0425) 620888   © J&J 1989

For many patients having a leg ulcer has meant years of immobility, pain, endless treatment and a general feeling of ill health.

But now Granuflex can help get them back on their feet.[1]

Once in place, the dressing gives rapid pain relief.[2] And Granuflex's special formulation also promotes the healing process and protects against bacterial contamination.[3,4]

What's more, Granuflex is waterproof, easy to apply, painless

GRANUFLEX. FOR THOSE WHO PREFER TO KEEP THEIR FEET ON THE GROUND.

to remove and needs no secondary dressing to hold it in place.

And, because it doesn't need changing as often as conventional dressings – on average every 4 days[5] – it's not only your patients' quality of life that can improve with Granuflex.

Granuflex*
HYDROCOLLOID DRESSING
— IMPERMEABLE —

A real advance in the management of leg ulcers and pressure sores.
Available on prescription (10cm x 10cm S141)

ConvaTec SQUIBB
... pioneers in hydrocolloid technology

References 1. Davies, J. (1988) Community Outlook: August; 11-12. 2. Pottle, B. (1987) Nursing Times: March 25; 54-58. 3. Mertz, P.M. et al. (1985) J Am Ac Derm: **12** (4): 662-668. 4. Meredith, K., Gray, E. (1988) J District Nursing: September; 8-10. 5. Johnson, A. (1989) Brit J Pharm Pract (in press).
*Trademarks of the Squibb Group (E.R. Squibb & Sons, Inc. and its subsidiary companies). Authorised user of the Trademarks.
For further information contact:
ConvaTec Limited, Squibb House, 141-149 Staines Road, Hounslow, Middlesex TW3 3JA, England.

TAKING THE TRAUMA OUT OF FRACTURE

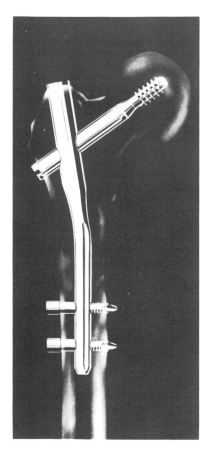